D1755579

# DISTRIBUTIONS, ULTRADISTRIBUTIONS AND OTHER GENERALIZED FUNCTIONS

# MATHEMATICS AND ITS APPLICATIONS

*Editor*: B.W. CONOLLY

Emeritus Professor of Mathematics (Operational Research), Queen Mary College, University of London

| | |
|---|---|
| Anderson, I | COMBINATORIAL DESIGNS Construction Methods |
| Arczewski, K. & Pietrucha, J. | MATHEMATICAL MODELLING OF COMPLEX MECHANICAL SYSTEMS: Volume 1: Discrete Models |
| Arczewski, K. & Pietrucha, J. | MATHEMATICAL MODELLING OF COMPLEX MECHANICAL SYSTEMS: Volume 2: Continuous Models |
| Baker, A.C. & Porteous, H. L | LINEAR ALGEBRA AND DIFFERENTIAL EQUATIONS |
| Balcerzyk, S. & Josefiak, T. | COMMUTATIVE RINGS |
| Barnett, S. | SOME MODERN APPLICATIONS OF MATHEMATICS |
| Beaumont, G.P. & Knowles, J.D. | STATISTICAL TESTS: An Introduction with Minitab Commentary |
| Bejancu, A. | FINSLER GEOMETRY AND APPLICATIONS |
| Bell, G. M. & Lavis, D.A. | STATISTICAL MECHANICS OF LATTICE MODELS, Vols. 1 & 2 |
| Breiteig, T., Huntley, I. & Kaiser-Messmer, G. | TEACHING AND LEARNING MATHEMATICS IN CONTEXT |
| Bunday, B.D. | STATISTICAL METHODS IN RELIABILITY THEORY AND PRACTICE |
| Berry, J.S., Burghes, D.N., Huntley, I.D., James, D.J.G. & Moscardini, A.O. Bunday, B.D. | STATISTICAL METHODS IN RELIABILITY THEORY AND PRACTICE MATHEMATICAL MODELLING COURSES |
| Berry, J.S., Burghes, D.N., Huntley, I.D., James, D.J.G. & Moscardini, A.O. | MATHEMATICAL MODELLING METHODOLOGY, MODELS AND MICROS |
| Berry, J.S., Burghes, D.N., Huntley, I.D., James, D.J.G. & Moscardini, A.O | TEACHING AND APPLYING MATHEMATICAL MODELLING |
| Berry, J.S., Graham, E. & Watkins, A.J.P. | LEARNING MATHEMATICS THROUGH DERIVE |
| Burghes, D.N. & Borrie, M. | MODELLING WITH DIFFERENTIAL EQUATIONS |
| Beaumont, G.P. & Knowles, J.D. | STATISTICAL TESTS: An Introduction with Minitab Commentary |
| Cartwright, M. | FOURIER METHODS: for Mathematicians, Scientists and Engineers |
| Cerny, I. | COMPLEX DOMAIN ANALYSIS |
| Chorlton, F. | VECTOR AND TENSOR METHODS |
| Cohen, D.E. | COMPUTABILITY AND LOGIC |
| Crapper, G.D. | INTRODUCTION TO WATER WAVES |
| Cullen, M.R. | LINEAR MODELS IN BIOLOGY |
| Doucet, P.G. & Sloep, P.B. | MATHEMATICAL MODELING IN THE LIFE SCIENCES |
| Dunning-Davies, J. | MATHEMATICAL METHODS FOR MATHEMATICIANS, PHYSICAL SCIENTISTS AND ENGINEERS |
| Eason, G. Coles, C.W. & Gettinby, G. | MATHEMATICS AND STATISTICS FOR THE BIOSCIENCES |
| Farrashkhalvat, M & Miles, J.P. | TENSOR METHODS FOR ENGINEERS AND SCIENTISTS |
| Faux, I.D. & Pratt, M.J. | COMPUTATIONAL GEOMETRY FOR DESIGN AND MANUFACTURE |
| French, S. | SEQUENCING AND SCHEDULING: Mathematics of the Job Shop |
| French, S. | DECISION THEORY; An Introduction to the Mathematics of Rationality |
| Firby, P.A. & Gardiner, C.F. | SURFACE TOPOLOGY: Second Edition |
| Gardiner, C.F. | ALGEBRAIC STRUCTURES |
| Gilbert, R.P. & Howard, H.C. | ORDINARY AND PARTIAL DIFFERENTIAL EQUATIONS WITH APPLICATIONS |
| Goodbody, A. M. | CARTESIAN TENSORS |
| Griffel, D.H | APPLIED FUNCTIONAL ANALYSIS |
| Griffel, D. | LINEAR ALGEBRA AND ITS APPLICATIONS: Vol. 1, A First Course; Vol. 2, More Advanced |
| Griffiths, P. & Hill, I.D. | APPLIED STATISTICS ALGORITHMS |
| Griffiths, H.B. & Oldknow, A. | MATHEMATICAL MODELS OF CONTINUOUS AND DISCRETE DYNAMIC SYSTEMS |
| Guest, P.B. | LAPLACE TRANSFORMS AND AN INTRODUCTION TO DISTRIBUTIONS |
| Hart, D. & Croft, A. | MODELLING WITH PROJECTILES |
| Hartley, R. | LINEAR AND NON-LINEAR PROGRAMMING |
| Hoskins, R.F. | STANDARD AND NONSTANDARD ANALYSIS |
| Hoskins, R.F. & Sousa Pinto, J.J.M. | DISTRIBUTIONS, ULTRADISTRIBUTIONS AND OTHER GENERALIZED FUNCTIONS |
| Irons, B.M. & Shrive, N.G. | NUMERICAL METHODS IN ENGINEERING AND APPLIED SCIENCE |
| Janacek, G. & Swift, | TIME SERIES: Forecasting, Simulation, Applications |
| Johnson, R.M. | THEORY AND APPLICATIONS OF LINEAR DIFFERENTIAL AND DIFFERENCE EQUATIONS |
| Johnson, R | CALCULUS: Theory and Applications in Technology and the Physical and Life Sciences |

*series continued at the back of the book*

# DISTRIBUTIONS, ULTRADISTRIBUTIONS AND OTHER GENERALIZED FUNCTIONS

**R. F. Hoskins**
Formerly
Department of Applied Computing and Mathematics
Cranfield Institute of Technology, UK
and
**J. Sousa Pinto**
Associate Professor, Department of Mathematics
University of Aveiro, Portugal

**ELLIS HORWOOD**
NEW YORK   LONDON   TORONTO   SYDNEY   TOKYO   SINGAPORE

First published 1994 by
Ellis Horwood Limited
Market Cross House, Cooper Street
Chichester
West Sussex, PO19 1EB
A division of
Simon & Schuster International Group

© Ellis Horwood Limited 1994

All rights reserved. No part of this publication may be reproduced, stored in a retrieval system, or transmitted, in any form, or by any means, electronic, mechanical, photocopying, recording or otherwise, without prior permission, in writing, from the publisher

Printed and bound in Great Britain by
Bookcraft, Midsomer Norton

Library of Congress Cataloging-in-Publication Data

Available from the publisher

British Library Cataloguing in Publication Data

A catalogue record for this book is available from the British Library

ISBN    0-13-217936-9    (hbk)

1 2 3 4 5    98 97 96 95 94

To Ana and João
and
David and Anthony

# Contents

Preface ..................................................... ix

List of Symbols ............................................. xiii

1 **Introduction to Distributions** ........................... 1
   1.1 FUNCTIONS AND GENERALISED FUNCTIONS ......... 1
   1.2 THE SCHWARTZ THEORY OF DISTRIBUTIONS ........ 20
   1.3 CALCULUS OF DISTRIBUTIONS ..................... 35
   1.4 THE J.S.SILVA THEORY OF DISTRIBUTIONS .......... 43
   1.5 GENERALISED POWERS AND PSEUDO-FUNCTIONS ..... 48

2 **Further Properties of Distributions** ..................... 63
   2.1 CONVOLUTION AND DIRECT PRODUCTS ............ 63
   2.2 CONVERGENCE OF DISTRIBUTIONS ................. 70
   2.3 SEQUENTIAL THEORIES OF DISTRIBUTIONS ......... 80
   2.4 THE MULTIPLICATION PROBLEM .................... 85

3 **Generalised Functions and Fourier Analysis** ............. 97
   3.1 REVIEW OF THE CLASSICAL FOURIER TRANSFORM .... 97
   3.2 GENERALISED FOURIER TRANSFORMS .............. 108
   3.3 FOURIER TRANSFORMS IN $\mathcal{E}'$ ....................... 113
   3.4 FOURIER TRANSFORMS AND ULTRADISTRIBUTIONS ... 134
   3.5 PERIODIC ULTRADISTRIBUTIONS .................... 147
   3.6 ULTRADISTRIBUTIONS OF EXPONENTIAL TYPE ....... 150

## 4 Analytic Representation of Generalised Functions — 159
- 4.1 THE CAUCHY TRANSFORM .................... 159
- 4.2 ANALYTIC REPRESENTATION OF DISTRIBUTIONS ...... 175
- 4.3 SATO HYPERFUNCTIONS ON THE LINE ............ 184
- 4.4 ULTRADISTRIBUTIONS ...................... 190
- 4.5 THE FOURIER-CARLEMAN TRANSFORM ........... 203

## 5 Irregular Operations and Colombeau Generalised Functions — 209
- 5.1 INTRINSIC PRODUCTS OF DISTRIBUTIONS .......... 209
- 5.2 GENERALISED FUNCTIONS AND OPERATORS ........ 225
- 5.3 CALCULUS OF GENERALISED FUNCTIONS .......... 239
- 5.4 THE SCHWARTZ PRODUCT AND MULTIPLICATION IN $\mathcal{G}(\mathbb{R})$ — 249

## 6 Nonstandard Treatments of Generalised Functions on $\mathbb{R}$ — 253
- 6.1 INTRODUCTION TO NONSTANDARD ANALYSIS ........ 253
- 6.2 NONSTANDARD REPRESENTATIONS OF DISTRIBUTIONS .. 265
- 6.3 PRODUCTS OF $^\Xi$DISTRIBUTIONS ................ 278
- 6.4 THE DIFFERENTIAL ALGEBRA $^\omega\mathcal{G}(\mathbb{R})$ ............. 279
- 6.5 CANONICAL EMBEDDING OF $^\Xi\mathcal{C}_\pi(\mathbb{R})$ INTO AN ALGEBRA .. 286

## Bibliography — 293

## Index — 299

# Preface

The publication in the 1950's of the classic *Théorie des Distributions* of Laurent Schwartz marks a convenient starting point for the theory of generalised functions as a subject in its own right. In this a systematic treatment of a wide and important class of generalised functions was developed, unifying much of the earlier work by Hadamard, Böchner, Sobolev and others. Since then an enormous literature dealing with both theory and applications has grown up, and the subject has undergone extensive further development. The original Schwartz treatment defined a distribution as a linear continuous functional on a space of test functions. That is, a distribution in the sense of Schwartz is essentially a member of the dual of a certain linear topological space, and since the appearance of the Schwartz text there has been a tendency to treat the term "generalised function" as synonymous with "distribution" in this sense. On the one hand this has tended to make the theory rather difficult to master, particularly for physicists and engineers. As a result, a number of alternative approaches have been developed over the years, giving rise to some differences in notation and terminology. On the other hand it was recognised at an early stage that distributions were not enough and that there remained a need for other classes of generalised functions. In particular the extended definition of the Fourier transform made necessary the introduction of a somewhat different family of functionals, the so-called *ultradistributions*; similarly, generalisation of the analytic representation of ordinary functions gave rise to the *hyperfunctions* of Sato. More recently, within the last decade, nonlinear theories of generalised functions have appeared in response to the problems posed by the attempts to define products of distributions (and the results of other, similarly irregular, operations).

The present text is intended as a broadly based introduction to the major theories of generalised functions which currently exist. The classical Schwartz theory of distributions as linear continuous functionals on spaces of test functions is described in the first two chapters. A brief review of basic material on topological vector spaces is included in the hope of making the treatment reasonably self-contained while remaining simple and straightforward. In addition there are parallel discussions of the altenative sequential treatments of the theory which have been developed by Mikusinski, and by Temple, Lighthill and Jones. An account of the much less well known axiomatic formulation proposed by Sebastião e Silva is also included.

Chapter 3 deals with the Fourier transform and its extensions. After a brief review of the classical Fourier transform the concept of the tempered distribution and of the corresponding distributional transform is introduced. Applications to periodic distributions and generalised Fourier series are supplemented by some material on sampling theorems which are not generally discussed in this context. The chapter continues with an account of the generalised transform for arbitrary distributions and the definition of ultradistributions. Material on ultradistributions is not usually given much space in expository texts on distributions and the treatment here is accordingly more extended.

In Chapter 4 we consider the Cauchy transform and its generalisations, and the analytic representation of distributions. Once again, although there is an extensive research literature on such material, it is not often treated in any detail in text books on generalised functions as such; the books by Bremermann, and by Beltrami and Wohlers are almost the only readily available sources for the general reader. The account given in the present text leads on to the definition of Sato hyperfunctions, the analytic representation of ultradistributions and the discussion of the significance of "support" in the context of ultradistributions.

The theory of distributions is essentially a *linear* theory, and the problem of devising satisfactory and comprehensive definitions for such operations as multiplication has been a major research pre-occupation over the years. Chapter 5 begins with a discussion of the definition of intrinsic distributional products and of the limitations of such a definition. There is a very extensive literature devoted to possible extensions of the elementary Schwartz product and it is not possible in the space available to give more than a cursory account of some of the principal theories which have been advanced. More attention has been paid to sequential theories of distributional products and to non-classical extensions of them since, apart from their intrinsic interest, they bear an obvious relation to the recent non-standard treatments of the theory of distributions as discussed in Chapter 6. But the major part of Chapter 5 is taken up with a presentation of the nonlinear theories of generalised functions recently put forward by Colombeau, and developed and extended by Rosinger, Biagioni and others. The *New Generalised Functions* of Colombeau include the distributions of Schwartz as a special case, and represent one of the most important new developments of the subject area. However, the ideas involved are novel and unfamiliar, and the Colombeau theory has therefore seemed difficult to understand, at least for the potential user trained in classical techniques of analysis. The approach adopted in Chapter 5 attempts to remedy this by basing the theory on the relatively familiar concept of operator.

The theory and techniques of Nonstandard Analysis (NSA) have been in circulation for some three decades or so and, despite their application to a wide variety of disciplines, still remain unknown, and untried, to many mathematicians. It is not easy to predict with confidence how large a part NSA is likely to play in the future development of mathematics in general but its value and importance in certain specific areas is by now well established. In particular there is a strong case for claiming that a nonstandard approach to generalised functions is the most natural and likely to prove the most productive. In Chapter 6 we give a self-contained account of an elementary form of NSA which is adequate for a nonstandard theory of distributions, and for a subsequent treatment of Colombeau's new generalised functions, at least in the simplified form presented by Biagioni. A sketch of some of the more advanced aspects of NSA and of its applications to generalised functions is included.

The book is set at final year or graduate level and presupposes only general mathematical background with little prior acquaintance of the specialised subject matter. To keep the presentation simple we work throughout in terms of a single real variable, and avoid the complications posed by generalised functions of several variables. Limitations of space have also prevented any discussion of applications. The object of the book has been to equip the reader with an adequate grounding in the major modern theories of generalised functions, without unduly favouring any one approach at the expense of others. In addition to those sources quoted explicitly in the main text the bibliography contains a selection of related material which should be helpful for further study.

Thanks are due to the Departamento de Matemática da Universidade de Aveiro for providing the necessary facilities for preparation of the manuscript in printed form. The authors would particularly like to acknowledge the advice and assistance of Jorge Sá Esteves and António Batel Anjo.

# List of Symbols

$\mathcal{B}^0_{\pi\Omega}$, $\mathcal{B}^0(\mathbb{C})$, 117
$\mathcal{B}_{\pi\Omega}(\mathbb{C})$, $\mathcal{B}(\mathbb{C})$, 118, 119
$\mathcal{C}^p(\mathbb{R})$, $\mathcal{C}^p_0(\mathbb{R})$, $\mathcal{C}^p_K(\mathbb{R})$, $\mathcal{C}^\infty(\mathbb{R})$, 5
$\mathcal{C}^\infty_0(\mathbb{R})$, $\mathcal{C}^\infty_K(\mathbb{R})$, 6
$\mathcal{C}_\infty(\mathbb{R})$, 46
$^\star\mathcal{C}^\infty(\mathbb{R})$, 263
$^s\mathcal{C}(I)$, $^s\mathcal{C}^\infty(I)$, 271, 272
$^\omega\mathcal{C}^k(\mathbb{R})$, $^\omega\mathcal{C}^\infty(\mathbb{R})$, 282
$^\Xi\mathcal{C}_\infty(I)$, $^\Xi\mathcal{C}_\pi(I)$, 274, 276
$\overline{\overline{\mathbb{C}}}$, $\overline{\overline{\mathbb{C}}}_0$, 240, 241
$^\star\mathbb{C}_M$, 289
$\mathcal{D}(\mathbb{R})$, $\mathcal{D}_K(\mathbb{R})$, 6, 7
$\mathcal{D}'(\mathbb{R})$, $\mathcal{D}'_{\text{fin}}(\mathbb{R})$, 12, 13
$^s\mathbf{D}(I)$, $^\pi\mathbf{D}(I)$, 272, 275
$\mathcal{E}(\mathbb{R})$, $\mathcal{E}'(\mathbb{R})$, 113
$\mathcal{E}'_{\pi\Omega}(\mathbb{R})$, 130
$\mathcal{E}[\mathcal{D},\mathcal{C}^\infty]$, $\mathcal{E}[\Phi,\mathcal{C}^\infty]$ 229, 232
$\mathcal{E}_M[\Phi,\mathcal{C}^\infty]$, $\mathcal{E}_s[\Phi,\mathcal{C}^\infty]$, $\mathcal{E}_{s,M}[\Phi,\mathcal{C}^\infty]$, 233, 281
$\mathcal{E}_0$, $\mathcal{E}_M$, 239
$^\Xi\mathcal{E}(\mathbb{R})$, 278
$^\omega\text{fin}(^\star\mathbb{R})$, 282
$\mathcal{G}(\mathbb{R})$, $\mathcal{G}_s(\mathbb{R})$, $^\omega\mathcal{G}(\mathbb{R})$, 234, 281, 283
$\mathcal{H}(\mathbb{C})$, $\mathcal{H}_e(\mathbb{C})$, 117
$\mathcal{H}_\mathcal{S}$, $\mathcal{H}_\mathcal{S}[K]$, 184, 187
$\mathcal{H}_{e,p}$, $\mathcal{H}^0_{e,p}$, $\mathcal{H}_{S,\exp}$, 203
$\mathcal{H}_{\mathcal{S},\Lambda}$, 186

## xiv List of Symbols

$^\omega\inf(^\star\mathbb{R})$, 282
$\mathcal{K}$, 150
$L^{i,1}(\mathbb{R})$, 166
$\mathcal{N}[\Phi, \mathcal{C}^\infty]$, 234
$^\star\mathbb{N}$, $^\star\mathbb{N}_\infty$, 256
$^\omega\mathcal{N}_0(\mathbb{R})$, $^\omega\mathcal{N}(\mathbb{R})$, 283
$\mathcal{O}_{-1}(\mathbb{R})$, $\mathcal{O}'_{-1}(\mathbb{R})$, 173
$\Pi_m$, 46
$\mathcal{P}_{2\pi}(\mathbb{R})$, 122
$\Phi$, 231
$\mathcal{Q}_{2\pi}(\mathbb{R})$, 147
$\overline{\mathbb{R}}$, 15
$^\star\mathbb{R}$, $^\star\mathbb{R}_{\mathrm{bd}}$, $^\star\mathbb{R}_\infty$, 255, 256
$^\omega\mathbb{R}$, 283
$\mathcal{S}(\mathbb{R})$, $\mathcal{S}'(\mathbb{R})$, 102, 108
$\mathcal{T}_{2\pi}(\mathbb{R})$, 122
$\mathcal{U}_p$, $\mathcal{U}$, 153
$\mathcal{V}_{2\pi}(\mathbb{R})$, 147
$\mathcal{Z}$, $\mathcal{Z}'$, $\mathcal{Z}'_{\mathrm{fin}}$, 136, 138, 144
$\mathcal{Z}_{\exp,j}$, $\mathcal{Z}_{\exp}$, $\mathcal{Z}'_{\exp}$, 154, 157

# 1

# Introduction to Distributions

## 1.1 FUNCTIONS AND GENERALISED FUNCTIONS

The term "generalised function" is most familiar in connection with the Heaviside unit step function $H$ and the so-called delta function $\delta$. Heaviside himself introduced both the function which bears his name and the delta function as its derivative and referred to the latter as the **unit impulse**. The properties attributed to the delta function, and the familiar notation $\delta(x)$, appear explicitly for the first time in the classic text on quantum mechanics by Paul Dirac [15]; for this reason it is often called the **Dirac delta function**. Nevertheless generalised functions have a history which antedates both Dirac and Heaviside. In his comprehensive and fascinating study of the historical development of the theory of distributions Lützen [69] suggests that the delta function appeared in implicit form as early as 1822, in Fourier's *Théorie Analytique de la Chaleur*, although the first mathematical definition seems to be due to Kirchhoff. This is in agreement with George Temple's judgment that Kirchhoff's researches in the equation of wave propagation (ca.1882) reveal the first explicit use of what may legitimately be described as a generalised function (cf. Temple [107]). The work of Heaviside, and subsequently of Dirac, in the systematic exploitation of the step and delta functions have made these particular examples of generalised functions familiar to applied mathematicians, physicists and engineers, although on an informal and largely non-rigorous basis. The first comprehensive and thoroughly coherent treatment of delta functions and other generalised functions appeared in the 1950s with the publication of Laurent Schwartz's *Théorie des Distributions* [97], [98] (although the earlier work of Sobolev anticipated some of the key ideas). The object of this chapter, and its successor, is to present a simplified account of the Schwartz theory of distributions, together with some of the alternative treatments which have subsequently appeared.

### 1.1.1 Step functions and delta functions

We begin with a brief review of the basic facts concerning the Heaviside step function $H(x)$ and the delta function $\delta(x)$. Their definition as *generalised* functions depends on their action on certain types of ordinary functions which, in this context, are usually referred to as **test functions**. The **support** of a function $\varphi : \mathbb{R} \to \mathbb{C}$ is defined to be the closure of the set $\{x \in \mathbb{R} : \varphi(x) \neq 0\}$; hence $\varphi$ has compact support if and only if it vanishes identically outside some finite interval $[a, b]$. Functions which are continuous on $\mathbb{R}$ and which have compact support can be used as test functions for the Heaviside step function $H(x)$.

For any given $c \in \mathbb{R}$ let $H_c$ denote the function defined by

$$H_c(x) = \begin{cases} 1 & \text{for } x > 0 \\ c & \text{for } x = 0 \\ 0 & \text{for } x < 0 \,. \end{cases} \tag{1.1}$$

Then for any continuous function $\varphi$ with compact support we have the result

$$\int_{-\infty}^{+\infty} \varphi(x) H_c(x) dx = \int_0^{+\infty} \varphi(x) dx. \tag{1.2}$$

That is to say, the family $\{H_c\}_{c \in \mathbb{R}}$ defines a mapping which carries each test function $\varphi$ into the number $\int_0^{+\infty} \varphi(x) dx$. This mapping is independent of the number $c$ and therefore of the particular function $H_c$ which we may choose to put in the integrand on the left-hand side of (1.2). We use the symbol $H$ (with no subscript $c$) to represent the mapping

$$H : \varphi \leadsto \int_0^{+\infty} \varphi(x) dx \tag{1.3}$$

which is well defined by (1.2) for all continuous test functions of compact support. This means that $H$ represents something other than an ordinary function. It is associated with an entire family or equivalence class of ordinary functions since, for example, any one of the ordinary functions $H_c(x)$ can be taken as a representative of $H$ in (1.2).

Any locally integrable function $f : \mathbb{R} \to \mathbb{C}$ can be said to generate a generalised function $F$ in the same sense, namely as a mapping:

$$F : \varphi \rightsquigarrow \int_{-\infty}^{+\infty} \varphi(x) f(x) dx \qquad (1.4)$$

where $\varphi$ is any continuous function of compact support. Once again $F$ is a symbol associated with an equivalence class of functions, since we may replace $f$ in the right-hand side of (1.4) by any function $f_1$ such that $f(x) = f_1(x)$ almost everywhere. It is a common and relatively harmless abuse of notation to identify the ordinary function $f$ with this generalised function $F$. Suppose now that both $F$ and $\varphi$ are continuously differentiable; if $\varphi$ vanishes outside $[a, b]$ then integration by parts gives

$$\begin{aligned}
\int_{-\infty}^{+\infty} f'(x)\varphi(x) dx &= \int_a^b f'(x)\varphi(x) dx \\
&= [f(x)\varphi(x)]_a^b - \int_a^b f(x)\varphi'(x) dx \\
&= \int_{-\infty}^{+\infty} f(x)[-\varphi'(x)] dx .
\end{aligned} \qquad (1.5)$$

This suggests that we might define the derivative $DF$ of the generalised function $F$ as the mapping

$$DF : \varphi \rightsquigarrow \int_{-\infty}^{+\infty} [-\varphi'(x)] f(x) dx. \qquad (1.6)$$

$DF$ may then be identified with the classical derivative $f'$ whenever the latter exists. Equation (1.5) remains valid if we relax the conditions on $f$ and assume only that it is absolutely continuous. In this case $f$ may not have a classical derivative $f'$ which exists everywhere. For example, the function

$$x_+^{\frac{1}{2}} = \begin{cases} +\sqrt{x} & \text{for } x \geq 0 \\ 0 & \text{for } x < 0 \end{cases}$$

is an absolutely continuous function which is differentiable everywhere except at the origin. It has the classical derivative

$$\frac{1}{2} x_+^{-\frac{1}{2}} = \begin{cases} +1/2\sqrt{x} & \text{for } x > 0 \\ 0 & \text{for } x < 0 \end{cases}$$

which is unbounded in any neighbourhood of the origin. Nevertheless, equation (1.5) holds for any continuously differentiable function $\varphi$ of compact support. Thus $\frac{1}{2} x_+^{-\frac{1}{2}}$

is a representative of the derivative $DF$ of the generalised function $F$ generated by the ordinary function $x_+^{\frac{1}{2}}$.

On the other hand equation (1.5) will not hold in general if we take for $f$ any representative $H_c$ of $H$. Any such function $H_c(x)$ will have a classical derivative $H_c'(x)$ which is equal to zero for all $x \neq 0$ but is undefined at the origin. Hence

$$\int_{-\infty}^{+\infty} \varphi(x) H_c'(x) dx = \lim_{\varepsilon \downarrow 0} \int_{-\infty}^{-\varepsilon} \varphi(x) H_c'(x) dx + \lim_{\varepsilon \downarrow 0} \int_{\varepsilon}^{+\infty} \varphi(x) H_c'(x) dx = 0$$

for every continuous function $\varphi$ of compact support, and therefore certainly for every continuously differentiable function $\varphi$ of compact support. On the right-hand side of (1.5) however, we get

$$\int_{-\infty}^{+\infty} [-\varphi'(x)] H_c(x) dx = \int_0^{+\infty} [-\varphi'(x)] dx = [-\varphi(x)]_0^{+\infty} = \varphi(0)$$

a result which is again independent of the particular value $c$ assigned to $H_c(0)$. This suggests that just as we use the symbol $H$ to stand for the mapping (1.3) we can allow $DH$, or in more familiar terms $\delta$, to represent the mapping

$$\delta : \varphi \rightsquigarrow \varphi(0) \tag{1.7}$$

for all continuously differentiable functions of compact support. Once again the symbol $\delta \equiv DH$ represents a generalised function, and one which is the derivative, in some generalised sense, of $H$. It is often convenient to retain the familiar notation of the integral calculus and to write this so-called **sampling operation** as a formal application of integration by parts:

$$\int_{-\infty}^{+\infty} \varphi(x) DH(x) dx \equiv \int_{-\infty}^{+\infty} \varphi(x) \delta(x) dx$$
$$= \int_0^{+\infty} [-\varphi'(x)] dx = \varphi(0). \tag{1.8}$$

However it is easily seen that there can be no ordinary function $\delta(x)$ such that equation (1.8) remains valid for all continuously differentiable test functions $\varphi$. Instead the significance of $\delta$ must be sought for in terms of the sampling operation itself, that is in terms of the mapping (1.7).

The generalised functions $H$ and $\delta$ may therefore be given a proper description as **functionals**, that is as numerically valued mappings, defined on suitably chosen linear spaces of test functions. This is the basis of the Schwartz theory of distributions. In 1944 Laurent Schwartz [97] developed a comprehensive theory of generalised functions using just this approach. In what follows we give a simplified overview of the resulting Schwartz Theory of Distributions, together with some of the alternative treatments which have subsequently appeared.

**Exercise 1.1 :**

The sampling operation associated with the delta function can be legitimately expressed in terms of a true integration process if we generalise the concept of integral. The Riemann-Stieltjes integral of a continuous function $f$ on an interval $[a,b]$ with respect to a monotone function $\psi$ (called the integrator) can be expressed as the limit

$$\int_a^b f(x)d\psi(x) = \lim_{\Delta \to 0} \sum_{k=1}^{n(\Delta)} f(\tau_k)[\psi(x_k) - \psi(x_{k-1})]$$

where $[a,b]$ is subdivided by points

$$a \equiv x_0 < x_1 < \ldots < x_{n-1} < x_n \equiv b,$$

$\tau_k$ is an arbitrarily chosen point in $[x_{k-1}, x_k]$ and $\Delta \equiv \max_{1 \leq k \leq n}(x_k - x_{k-1})$.

**1.** If $\psi$ is monotone increasing and continuously differentiable on $[a,b]$ show that

$$\int_a^b f(x)d\psi(x) = \int_a^b f(x)\psi'(x)dx.$$

**2.** If $\psi(x) = H_c(x)$, where $0 \leq c \leq 1$, and $a < 0 < b$, show that

$$\int_a^b f(x)dH_c(x) = f(0).$$

### 1.1.2 Fundamental spaces of test functions

For $p = 0, 1, 2, \ldots$, and compact $K \sqsubset \mathbb{R}$, we use the following standard notation:

$\mathcal{C}^p \equiv \mathcal{C}^p(\mathbb{R})$ : linear space of all complex-valued functions on $\mathbb{R}$ with continuous derivatives at least up to order $p$,

$\mathcal{C}_0^p \equiv \mathcal{C}_0^p(\mathbb{R})$ : linear subspace of $\mathcal{C}^p$ comprising all functions with compact support,

$\mathcal{C}_K^p \equiv \mathcal{C}_K^p(\mathbb{R})$ : linear subspace of $\mathcal{C}_0^p$ comprising all functions with support contained in the same fixed compact $K$.

Similarly, for $p = \infty$, we define

$\mathcal{C}^\infty \equiv \mathcal{C}^\infty(\mathbb{R})$ : linear space of all complex-valued functions on $\mathbb{R}$ which have continuous derivatives of all orders,

We wish to extend the definitions of $\mathcal{C}_0^p(\mathbb{R})$ and $\mathcal{C}_K^p(\mathbb{R})$ in the same way to include the case $p = \infty$ but it is not obvious that infinitely differentiable functions of compact support exist. However a useful family of such functions, which vanish outside arbitrarily small intervals, can be defined as follows: let

$$\psi(x) = \begin{cases} \exp[(x^2-1)^{-1}] & \text{for } |x| < 1 \\ 0 & \text{for } |x| \geq 1 \end{cases} \quad (1.9)$$

and for $n = 1, 2, \ldots$, set

$$\psi_n(x) \equiv \frac{n}{A} \psi(nx) = \begin{cases} \frac{n}{A} \exp[(n^2 x^2 - 1)^{-1}] & \text{for } |x| < \frac{1}{n} \\ 0 & \text{for } |x| \geq \frac{1}{n} \end{cases} \quad (1.10)$$

where $A \equiv \int_{-1}^{+1} \exp[(x^2-1)^{-1}]\,dx$. Then $\psi_n$ has support contained in the finite interval $[-1/n, +1/n]$, and it is easily confirmed that each $\psi_n$ has continuous derivatives of all orders on $\mathbb{R}$. Note that our definition also ensures that

$$\int_{-\infty}^{+\infty} \psi_n(x)dx = 1 \;, \quad \forall n \in \mathbb{N}.$$

Hence we may define

$\mathcal{C}_0^\infty \equiv \mathcal{C}_0^\infty(\mathbb{R})$ : linear subspace of $\mathcal{C}^\infty$ comprising all infinitely differentiable functions with compact support,

$\mathcal{C}_K^\infty \equiv \mathcal{C}_K^\infty(\mathbb{R})$ : linear subspace of $\mathcal{C}_0^\infty$ comprising all infinitely differentiable functions with compact support contained in the same fixed compact $K$.

The fundamental space of test functions in the Schwartz theory of distributions is the linear space $\mathcal{C}_0^\infty(\mathbb{R})$ equipped with a certain locally convex topology and as such is usually denoted by $\mathcal{D}(\mathbb{R})$, or simply by $\mathcal{D}$. It is enough at this stage to limit the discussion to the mode of convergence which the topology in question generates. This is straightforward in the relatively familiar case of a topology which is generated by a **norm**. A norm on a linear space $E$ is a real-valued, non-negative function $\| \cdot \|: E \to \mathbb{R}$ which satisfies the conditions

**N1** $\| \varphi \| = 0$ if and only if $\varphi$ is the null element of $E$,
**N2** $\| a\varphi \| = |a| \cdot \| \varphi \|$ for every $\varphi \in E$ and every scalar $a$,
**N3** $\| \varphi + \psi \| \leq \| \varphi \| + \| \psi \|$, for all $\varphi, \psi \in E$.

A linear space $E$ equipped with a norm $\| \cdot \|$ is said to be a **normed linear space**. Convergence of a sequence $(\varphi_n)_{n \in \mathbb{N}} \subset E$ to a limit $\varphi \in E$ in the sense of the norm is understood to mean that

$$\lim_{n \to \infty} \| \varphi - \varphi_n \| = 0 \;.$$

Each linear space $C_K^p$ is a normed linear space with respect to the norm

$$\|\varphi\|_{[p,K]} \equiv \max_{0 \leq r \leq p} \left\{ \sup_{x \in K} |\varphi^{(r)}(x)| \right\}. \tag{1.11}$$

A sequence $(\varphi_n)_{n \in \mathbb{N}} \subset C_K^p$ converges to $\varphi$ in the sense of the norm $\|\cdot\|_{[p,K]}$ if and only if each of the sequences $(\varphi_n^{(r)})_{n \in \mathbb{N}}, r = 0, 1, \ldots, p$, of the derivatives of $\varphi_n$ converges *uniformly* to the correspondingly derivative $\varphi^{(r)}$ of the limit function $\varphi$. (For brevity we often say that the sequence $(\varphi_n)_{n \in \mathbb{N}}$ converges $p$-**uniformly** to $\varphi$).

For a given $K$ the space $C_K^\infty$ is a linear subspace of each space $C_K^p$. If we let $p$ tend to infinity then we cannot define a norm $\|\cdot\|_{[\infty,K]}$ as the limit of the norms $\|\cdot\|_{[p,K]}$. But the limiting mode of convergence is well defined. We say that a sequence $(\varphi_n)_{n \in \mathbb{N}}$ of functions in $C_K^\infty$ converges $\omega$-**uniformly** to zero in case each of the sequences $\left(\varphi_n^{(r)}\right)_{n \in \mathbb{N}}$, for $r = 0, 1, 2, \ldots$, of the $r$-th derivatives of the $(\varphi_n)_{n \in \mathbb{N}}$ converges uniformly to zero. Then we can define the space $\mathcal{D}_K \equiv \mathcal{D}_K(\mathbb{R})$ to be the linear space $C_K^\infty$ equipped with $\omega$-uniform convergence. That is to say a sequence $(\varphi_n)_{n \in \mathbb{N}}$ converges to zero in the space $\mathcal{D}_K$ if and only if

(a) all the $\varphi_n$ belong to the space $C_K^\infty$, and

(b) for each $p = 0, 1, 2 \ldots$, we have $\lim_{n \to \infty} \|\varphi_n\|_{[p,K]} = 0$.

Then the mode of convergence appropriate to the Schwartz space $\mathcal{D}$ itself is as follows: a sequence $(\varphi_n)_{n \in \mathbb{N}} \subset C_0^\infty$ is said to converge to zero in $\mathcal{D}$ if and only if

(a) all the $\varphi_n$ have support contained in the same fixed compact $K$ (that is, all the $\varphi_n$ belong to the same subspace $C_K^\infty$), and

(b) the sequence $(\varphi_n)_{n \in \mathbb{N}}$, converges $\omega$-uniformly to zero.

Thus we can say that each $\mathcal{D}_K$ is a subspace of $\mathcal{D}$, and that $\mathcal{D}$ is the union of all possible spaces $\mathcal{D}_K$ as $K$ runs over all compacts of $\mathbb{R}$. We also have the result:

**Theorem 1.1** *If $\varphi$ is any function belonging to $C_0^p$ for some $p$ such that $0 \leq p \leq \infty$ then there exists a sequence $(\varphi_n)_{n \in \mathbb{N}}$ of functions in $C_0^\infty$ which converges $p$-uniformly to $\varphi$. Further, all the $\varphi_n$ have supports contained in the same finite interval.*

**Proof:** Suppose first that $\varphi$ is a function which is continuous everywhere and which vanishes identically outside some finite interval $[a, b]$; then $\varphi$ is a function belonging to $C_0^0$. For $n = 1, 2, 3, \ldots$, define functions $\varphi_n$ by

$$\varphi_n(x) = \int_{-\infty}^{+\infty} \varphi(t) \psi_n(x-t) dt = \int_{-\infty}^{+\infty} \varphi(x-t) \psi_n(t) dt$$

where the $\psi_n$ are the infinitely differentiable functions of compact support defined in (1.10). By differentiation under the integral sign we can show that each $\varphi_n$ must have derivatives of all orders; also $\varphi_n$ vanishes identically outside the interval $[a-1/n, b+1/n]$; therefore $\varphi_n \in C_0^\infty$ for $n = 1, 2, 3, \ldots$. Now for any point $x$ we have

$$|\varphi_n(x) - \varphi(x)| \leq \int_a^b |\varphi(t) - \varphi(x)|\psi_n(x-t)dt.$$

Since $\psi_n(x-t)$ is non-zero only if $|x-t| < 1/n$, and since $\varphi$ is uniformly continuous it follows that $\varphi$ is the uniform limit of the sequence $(\varphi_n)_{n \in \mathbb{N}}$. Moreover all the functions $\varphi_n$ have supports contained in the same fixed interval $[a-1, b+1]$.

In the same way, if $\varphi$ happens to be continuously differentiable on $\mathbb{R}$, we can show that

$$|\varphi_n'(x) - \varphi'(x)| \leq \int_a^b |\varphi'(t) - \varphi'(x)|\psi_n(x-t)dt$$

and it follows that $\varphi'$ is the uniform limit of the $\varphi_n'$, and so that $\varphi$ itself is the 1-uniform limit of the sequence $(\varphi_n)_{n \in \mathbb{N}}$.

In general, if $\varphi \in C_0^p$ for some $p = 0, 1, 2, \ldots$, then we can always treat $\varphi$ as the $p$-uniform limit of a sequence $(\varphi_n)_{n \in \mathbb{N}}$ of functions belonging to $C_0^\infty$; further, all the $\varphi_n$ vanish outside the same fixed finite interval. □

**Exercise 1.2** :

1. Confirm that $\| \varphi \|_{[p,K]}$, as defined in equation (1.11), is a norm on the linear space $C_K^p(\mathbb{R})$, and that convergence in the sense of this norm is equivalent to $p$-uniform convergence as defined in the text.

2. Let $C_{\text{inf}}(\mathbb{R})$ denote the linear space of all bounded continuous functions on $\mathbb{R}$ which vanish at infinity (that is, of all continuous functions $\varphi$ such that $\lim_{|x| \to \infty} \varphi(x) = 0$). Prove that $C_{\text{inf}}(\mathbb{R})$ is closed under convergence with respect to the norm

$$\| \varphi \| = \sup_{x \in \mathbb{R}} |\varphi(x)|.$$

(Note that we cannot replace $C_{\text{inf}}(\mathbb{R})$ by $C_0^0(\mathbb{R})$ here).

### 1.1.3 Distributions as continuous linear functionals

A **functional** on a linear space $E$ is a mapping $\mu : E \to \mathbb{C}$ which assigns to each member $\varphi$ of $E$ a certain complex number; the image of $\varphi \in E$ under $\mu$ is usually written as $\mu(\varphi)$ or $<\mu, \varphi>$. We are particularly interested in functionals which are (a) linear and (b) continuous in some sense:

(a) **Linearity:** $\mu$ is said to be a linear functional on $E$ if and only if given any two members $\varphi_1, \varphi_2$ of $E$ and any scalars $a_1, a_2$, we have

$$<\mu, a_1\varphi_1 + a_2\varphi_2> = a_1 <\mu, \varphi_1> + a_2 <\mu, \varphi_2>;$$

(b) **Continuity:** This can be defined provided $E$ is equipped with a specific topology, or at least a specific mode of convergence (generated by some underlying topology). Then $\mu$ is said to be a continuous functional on $E$ if and only if whenever a sequence $(\varphi_n)_{n \in \mathbb{N}}$ converges to zero in $E$ (in the agreed sense) the corresponding sequence of complex numbers $(<\mu, \varphi_n>)_{n \in \mathbb{N}}$ converges to zero in the usual sense.

The set of all continuous linear functionals on a linear space $E$ is denoted by $E'$. It forms a linear space in its own right under the natural componentwise definitions of vector addition and multiplication by scalars:

$$<\mu + \nu, \varphi> \ = \ <\mu, \varphi> + <\nu, \varphi>$$
$$<a\mu, \varphi> \ = \ a <\mu, \varphi>$$

where $\mu, \nu \in E', \varphi \in E$ and $a$ is any scalar. $E'$ is called the **dual space** of $E$.

Suppose in particular that $E$ is a linear space of complex-valued functions on $\mathbb{R}$ (more generally on $\mathbb{R}^m$), satisfying some appropriate conditions of smoothness and of growth. Vector addition and multiplication by scalars are supposed defined in the usual pointwise sense. Then the functions belonging to $E$ will generally be called test functions, while the functionals belonging to $E'$ will be called **generalised functions**. The motivation for this terminology should already be clear from the earlier informal discussion of the step and delta functions in terms of mappings defined on certain classes of functions.

The simplest definition of a **distribution** as a functional (as used, for example, in Gel'fand and Shilov [26], or Zemanian [117]) is that it is a mapping $\mu : \mathcal{D} \to \mathbb{C}$ which is linear and sequentially continuous with respect to convergence in the sense of $\mathcal{D}$:

**Definition 1.1** *A distribution is a mapping $\mu : \mathcal{D} \to \mathbb{C}$ such that*

(1) $<\mu, \varphi>$ *is a well defined complex number for every $\varphi \in \mathcal{D}$,*

(2) *for any $\varphi_1, \varphi_2 \in \mathcal{D}$ and any scalars (real or complex numbers) $a_1, a_2$,*

$$<\mu, a_1\varphi_1 + a_2\varphi_2> = a_1 <\mu, \varphi_1> + a_2 <\mu, \varphi_2>,$$

(3) let $(\varphi_n)_{n \in \mathbb{N}}$ be any sequence of functions in $\mathcal{D}$ which converges to a limit $\varphi$ in the sense of $\mathcal{D}$ (so that each of the $\varphi_n$ and the limit-function $\varphi$ belong to the same subspace $\mathcal{D}_K$, and the convergence is $\omega$-uniform); then $<\mu, \varphi> = \lim_{n \to \infty} <\mu, \varphi_n>$.

An alternative definition (used, for example, by Friedlander [25]) replaces continuity with respect to the convergence in $\mathcal{D}$ by **local boundedness** with respect to the family of norms $(\|\cdot\|_{[p,K]})_{p \in \mathbb{N}_0}$:

A **distribution** on $\mathbb{R}$ is a linear functional $\mu$ on the space $\mathcal{D}$ which is locally bounded in the sense that for each compact $K \sqsubset \mathbb{R}$ there exists a number $M \equiv M(K) > 0$ and an integer $p \equiv p(K) \geq 0$ such that, for every $\varphi \in \mathcal{C}_0^\infty$,

$$|\mu(\varphi)| \leq M \cdot \max_{0 \leq k \leq p} \left\{ \sup_{x \in K} |\varphi^{(k)}(x)| \right\} = M \cdot \| \varphi \|_{[p,K]}. \qquad (1.12)$$

**Theorem 1.2** *A linear functional $\mu$ on $\mathcal{C}_K^\infty$ is continuous with respect to $\omega$-uniform convergence if and only if it is bounded with respect to some norm $\|\varphi\|_{[p,K]}$.*

**Proof:** (i) If $\mu$ is bounded with respect to some norm $\| \varphi \|_{[p,K]}$ on $\mathcal{C}_K^\infty$ then it is certainly continuous with respect to $p$-uniform convergence and therefore also with respect to $\omega$-uniform convergence.

(ii) Suppose that $\mu$ is continuous with respect to $\omega$-uniform convergence on $\mathcal{C}_K^\infty$ but that (if possible) it is not bounded with respect to any norm $\| \varphi \|_{[p,K]}$ on that space. Then for each $n \in \mathbb{N}$ we can find $\varphi_n \in \mathcal{C}_K^\infty$ such that

$$|<\mu, \varphi_n>| > n \cdot \|\varphi_n\|_{[n,K]}.$$

For each $n \in \mathbb{N}$ define a function $\gamma_n \in \mathcal{C}_K^\infty$ by setting $\gamma_n = \varphi_n/(n \| \varphi_n \|_{[n,K]})$. Then for each $p \in \mathbb{N}$ and for all $n \geq p$ we have

$$\|\gamma_n\|_{[p,K]} \leq \|\gamma_n\|_{[n,K]} = \frac{\| \varphi_n \|_{[n,K]}}{n \| \varphi_n \|_{[n,K]}} = \frac{1}{n}.$$

Hence, for each $p \in \mathbb{N}$, $\lim_{n \to \infty} \| \gamma_n \|_{[p,K]} = 0$ and so $(\gamma_n^{(p)})_{n \in \mathbb{N}}$ converges uniformly to zero for $p = 0, 1, 2, \ldots$. By the continuity of $\mu$ it follows that $\lim_{n \to \infty} <\mu, \gamma_n> = 0$. However this gives a contradiction since,

$$|<\mu, \gamma_n>| = \frac{|<\mu, \varphi_n>|}{n\|\varphi_n\|_{[n,K]}} > \frac{n\|\varphi_n\|_{[n,K]}}{n\|\varphi_n\|_{[n,K]}} = 1.$$

□

**Distributions of finite order** Theorem 1.2 shows that the two definitions of distribution given above are equivalent. In particular it shows that for each compact $K \sqsubset \mathbb{R}$ the restriction of a distribution $\mu$ to the subspace $\mathcal{D}_K$ of $\mathcal{D}$ will be a linear functional on the linear space $\mathcal{C}_K^\infty$ which is bounded with respect to some norm $\|\varphi\|_{[p,K]}$. The integer $p$ in $\|\varphi\|_{[p,K]}$ will generally depend on the particular compact $K$ in question, so that $p = p(K)$. It may happen that we can take the same value $p$ for all compact subsets of $\mathbb{R}$, and in this case the distribution $\mu$ is said to be of **finite order**; the smallest value of $p$ for which this is true is then said to be the **order** of the distribution $\mu$. As we shall see later on there is another sense in which the order of a distribution may be defined; to avoid confusion we shall therefore refer to order in the present sense as the Schwartz order of a distribution.

**Definition 1.2** *A distribution $\mu \in \mathcal{D}'$ is said to be of finite Schwartz order if and only if there exists an integer $p \in \mathbb{N}_0$ such that for every compact $K \sqsubset \mathbb{R}$ we have*

$$|<\mu, \varphi>| \leq M \cdot \|\varphi\|_{[p,K]} \qquad (1.13)$$

*for all $\varphi \in \mathcal{D}_K$, where the constant $M \equiv M(K)$ generally depends on $K$ but the integer $p$ is independent of $K$.*

*If $\mu$ is of finite Schwartz order and if $p$ is the smallest integer for which the inequality (1.13) is true then $\mu$ is said to be a distribution of (Schwartz) order $p$.*

### 1.1.4 Examples of distributions

The following examples of distributions are all of finite order.

**Example 1.1.1** Any locally integrable function $h$ will define a distribution $\mu_{(h)}$ as the functional

$$\varphi \rightsquigarrow <\mu_{(h)}, \varphi> \equiv \int_{-\infty}^{+\infty} \varphi(x) h(x) dx \qquad (1.14)$$

for any $\varphi \in \mathcal{C}_0^\infty$. Then $\mu_{(h)}$ is actually well defined as a linear functional on $\mathcal{C}_0^0$ and if $\varphi$ has support contained in the finite interval $K = [a,b]$ we will have

$$|\mu_{(h)}(\varphi)| \leq \left\{ \sup_{a \leq x \leq b} |\varphi(x)| \right\} \cdot \int_a^b |h(x)| dx = M \cdot \|\varphi\|_{[0,K]} \qquad (1.15)$$

where $M = M([a,b])$ is a finite positive constant which depends only on $[a,b]$ and on the fixed function $h$ and is independent of $\varphi$. Hence the function $h$ defines a distribution $\mu_{(h)}$ which is of order 0. Such a distribution is said to be **regular**. Note that we could replace $h$ in the definition of $\mu_{(h)}$ by any other function $h_1$ such that $h_1(x) = h(x)$ almost everywhere. The regular distribution $\mu_{(h)}$ corresponds not to a single locally integrable function but rather to an equivalence class of such functions. Nevertheless it is common practice to identify $\mu_{(h)}$ with a single representative function $h$ and refer to it as *the generalised function* $h(x)$. As remarked in §1.1.1 this is a relative harmless convention, although it is worth noting that the concept of the value of such a generalised function at a point has already lost its usual significance. In particular we denote by $H$ the regular distribution which carries each function $\varphi$ in $\mathcal{C}_0^0$ into the number $\int_0^{+\infty} \varphi(x) dx$. In some texts a different symbol, $\theta$, is used for the (ordinary) function which is equal to 1 for positive values of the argument and equal to 0 for negative values (and is undefined at the origin). This is useful in certain contexts (when, for example, we consider products of ordinary functions and of generalised functions) but it is not really necessary. We shall however be careful to distinguish the distribution $H$ from any typical representative function $H_c(x)$ whenever the need arises.

Finally note that to each number $a \in \mathbb{R}$ (more generally, $a \in \mathbb{C}$) there corresponds a constant function $f_a(x) = a$ for all $x \in \mathbb{R}$. This function is certainly locally integrable and therefore defines a regular distribution which we call a **constant distribution**. Identifying the number $a$ with the corresponding constant distribution $f_a$ shows that $\mathbb{R}$ (respectively $\mathbb{C}$) can be embedded in the space $\mathcal{D}'$ of all distributions.

**Example 1.1.2** Any distribution which cannot be defined in terms of a locally integrable function $h$, as in equation (1.14) above, is said to be **singular**. Thus, the mapping

$$\varphi \rightsquigarrow <\delta, \varphi> = \varphi(0)$$

defines a singular distribution, $\delta$, which ought properly to be referred to as the **Dirac delta distribution**. Nevertheless the term *"delta function"* is very well established and we shall continue to use it whenever convenient throughout the text. For all $\varphi \in \mathcal{C}_0^0$ we have

$$|<\delta, \varphi>| = |\varphi(0)| \leq \sup_{x \in K} |\varphi(x)| \equiv \|\varphi\|_{[0,K]} .$$

where $\text{supp}(\varphi) \subset K \sqsubset \mathbb{R}$. Hence, $\delta$ is a singular distribution of order 0.

**Example 1.1.3** The function $x^{-1}$ is not locally integrable and so does not define a regular distribution; for an arbitrary test function $\varphi$ in $\mathcal{D}$ the integral

$$\int_{-\infty}^{+\infty} \frac{\varphi(x)}{x}\, dx$$

does not generally converge. Nevertheless we can use it to generate a singular distribution by taking the *Cauchy Principal Value* of the divergent integrals which arise. Writing this distribution as $\mathrm{Pv}\{x^{-1}\}$, to distinguish it from the ordinary function $x^{-1}$, we have

$$\begin{aligned}
<\mathrm{Pv}\{x^{-1}\},\varphi> &= \mathrm{Pv}\int_{-\infty}^{+\infty}\frac{\varphi(x)}{x}\,dx \\
&= \lim_{\varepsilon\downarrow 0}\left\{\int_{+\varepsilon}^{+\infty}\frac{\varphi(x)}{x}\,dx + \int_{-\infty}^{-\varepsilon}\frac{\varphi(x)}{x}\,dx\right\} \\
&= \lim_{\varepsilon\downarrow 0}\int_{+\varepsilon}^{+\infty}\frac{\varphi(x)-\varphi(-x)}{x}\,dx.
\end{aligned}$$

Now by Taylor's Theorem, for $x > 0$ we can write

$$\begin{aligned}
\varphi(+x) &= \varphi(0) + x\varphi'(\theta_1 x)\ ,\\
\varphi(-x) &= \varphi(0) - x\varphi'(\theta_2 x)\ ,
\end{aligned}$$

where $0 < \theta_1 < 1$ and $-1 < \theta_2 < 0$. Hence, for some $b > 0$ (depending on the support of $\varphi$),

$$<\mathrm{Pv}\{x^{-1}\},\varphi(x)> = \lim_{\varepsilon\downarrow 0}\int_{+\varepsilon}^{+b}[\varphi'(\theta_1 x) + \varphi'(\theta_2 x)]\,dx$$

and so if $K$ is a compact that contains $\mathrm{supp}(\varphi)$,

$$|<\mathrm{Pv}\{x^{-1}\},\varphi(x)>|\ \leq\ 2b\cdot\|\varphi\|_{[1,K]}$$

which means that $\mathrm{Pv}\{x^{-1}\}$ is a distribution of order 1.

In practice it is the distributions of finite order which most frequently occur and which are the most generally useful in applications. It is convenient to denote by $\mathcal{D}'_{\mathrm{fin}} \equiv \mathcal{D}'_{\mathrm{fin}}(\mathbb{R})$ the linear subspace of $\mathcal{D}'(\mathbb{R})$ comprising all distributions of finite order. Quite often it is sufficient (and certainly more convenient) to develop aspects of the theory of distributions by working wholly in terms of $\mathcal{D}'_{\mathrm{fin}}$, and then extending the results, as required, to the whole of $\mathcal{D}'$. Examples of infinite order distributions can be given quite easily, though they may seem somewhat contrived and artificial. The functional $\sigma$ given by the mapping

$$\varphi \rightsquigarrow <\sigma,\varphi> = \sum_{n=0}^{\infty} n(-1)^n \varphi^{(n)}(n)$$

is well defined for all $\varphi \in \mathcal{D}$ since only finitely many non-zero terms will actually appear in the sum. It represents a distribution of infinite order which, as will appear later when we consider infinite series of distributions, can be written in the form

$$<\sigma,\varphi> = <\sum_{n=0}^{\infty} n\delta^{(n)}(x-n),\varphi> .$$

**Extensions of finite order distributions** A modification of the proof of theorem 1.1 can be made to show that if $\varphi$ is a function in $\mathcal{C}_0^p$ whose support is contained in some specific compact $K$ then we can always choose a sequence $(\varphi_n)_{n\in\mathbb{N}}$ of infinitely differentiable functions converging $p$-uniformly to $\varphi$ such that all the $\varphi_n$ have supports contained in $K$. This means that the space $\mathcal{D}_K$ is dense in the space $\mathcal{C}_K^p$ with respect to the norm $\|\cdot\|_{[p,K]}$. (See, for example, the discussion given in §3.4 of the text by Zemanian [117]). It follows that any distribution of order $p$ can be extended as a linear functional on the whole space $\mathcal{C}_0^p$ which is continuous with respect to the norm $\|\cdot\|_{[p,K]}$. On the other hand let $\mu$ be any linear functional which is defined on some space $\mathcal{C}_0^p$, and which is continuous on each subspace $\mathcal{C}_K^p$ with respect to $p$-uniform convergence. Then the restriction of $\mu$ to $\mathcal{C}_0^\infty$ is linear and continuous with respect to $p$-uniform convergence, and therefore with respect to $\omega$-uniform convergence, on each subspace $\mathcal{C}_K^\infty$. It follows that $\mu$ is a distribution of finite order (specifically, of order $\leq p$).

**Exercise 1.3 :**

1. A sequence $(\varphi_n)_{n\in\mathbb{N}}$ contained in a normed linear space $E$ is said to converge **weakly** to a limit $\varphi \in E$ whenever

$$<\mu,\varphi> = \lim_{n\to\infty} <\mu,\varphi_n>$$

   for all $\mu \in E'$. Prove that if $(\varphi_n)_{n\in\mathbb{N}}$ converges strongly to $\varphi$ (that is, converges in the sense of the norm in $E$) then it necessarily converges weakly to $\varphi$.

2. Let $\mathcal{C}[0,1]$ denote the linear space of all functions $f : \mathbb{R} \to \mathbb{R}$ which are bounded and continuous on $[0,1]$, equipped with the supremum norm $\|f\| = \sup_{0\leq x\leq 1} |f(x)|$. Find conditions which ensure that a sequence $(f_n)_{n\in\mathbb{N}}$ will be weakly convergent to $f$ in $\mathcal{C}[0,1]$.

3. Each of the following expressions defines a functional on the space $\mathcal{D}$. Determine in each case whether the functional is a distribution and, if so, find if it is of finite order or of infinite order,

(a) $<\mu,\varphi> = \int_{-\infty}^{+\infty} |\varphi(x)|dx$;   (b) $<\mu,\varphi> = \varphi(0)\int_{-\infty}^{+\infty} \varphi(x)dx$;
(c) $<\mu,\varphi> = \sum_{n=-\infty}^{+\infty} \varphi(n^2)$;   (d) $<\mu,\varphi> = \sum_{n=-\infty}^{+\infty} \varphi^2(n)$;
(e) $<\mu,\varphi> = \sum_{n=1}^{\infty} \varphi(1/n)$;   (f) $<\mu,\varphi> = \sum_{n=1}^{\infty} n^{-2}\varphi(1/n)$;
(g) $<\mu,\varphi> = \sum_{n=0}^{\infty} \varphi^{(n)}(0)$;   (h) $<\mu,\varphi> = \sum_{n=0}^{\infty} \varphi^{(n)}(n^2)$.

### 1.1.5 Distributions of order 0 and measures

Distributions of order 0 are of particular importance in connection with the standard theory of measure, which we recall briefly here. A collection $\mathcal{T}$ of subsets of $\mathbb{R}$ (respectively of $\mathbb{R}^m$) is called a $\sigma$-**ring** if it is closed under complementation and under countable unions (and therefore under countable intersections also). In particular the collection $B$ of all sets obtainable by carrying out countably many operations of unions, intersections and complements, starting from open sets, is a $\sigma$-ring. The members of this collection $B$ are called **Borel sets**, and $B$ is the smallest $\sigma$-ring which contains all open sets.

It is convenient at this point to introduce the conventional extension of the real number system $\mathbb{R}$ to the system $\overline{\mathbb{R}} = \mathbb{R} \cup \{-\infty, +\infty\}$ equipped with the following conventions for arithmetic operations and order:

$$(+\infty) \cdot x = +\infty, \ (0 < x \leq +\infty) \ ; \ (+\infty) \cdot x = -\infty, \ (-\infty \leq x < 0)$$
$$(-\infty) \cdot x = -\infty, \ (0 < x \leq +\infty) \ ; \ (-\infty) \cdot x = +\infty, \ (-\infty \leq x < 0)$$
$$x/(+\infty) = x/(-\infty) = 0 \ , \ (-\infty < x < +\infty)$$
$$(+\infty) + x = +\infty, \ (x > -\infty) \ ; \ (-\infty) + x = -\infty, \ (x < +\infty).$$

Combinations like $(\pm\infty)+(\mp\infty), (\pm\infty)/(\pm\infty)$ and $(\pm\infty)/(\mp\infty)$ are left undefined.

A set function $\mu$ defined on a $\sigma$-ring $\mathcal{T}$ and taking values in the extended real number system is said to be **completely additive** if for each sequence $(A_n)_{n\in\mathbb{N}}$ of pairwise disjoint sets belonging to $\mathcal{T}$ we have

$$\mu\left(\bigcup_{n=1}^{\infty} A_n\right) = \sum_{n=1}^{\infty} \mu(A_n). \tag{1.16}$$

If $\mathcal{T}$ contains the collection $B$ of all Borel sets then $\mu$ is called a **measure** (strictly, a Borel measure) on $\mathbb{R}$ and the members of $\mathcal{T}$ are said to be **measurable** for $\mu$. If, in addition, we have

$$0 \leq \mu(A) \leq +\infty$$

for every $A \in \mathcal{T}$ then $\mu$ is said to be a **positive measure** on $\mathbb{R}$. Every measure $\mu$ can be expressed as the difference of two positive measures on $\mathbb{R}$.

Now let $f$ be any function defined on $\mathbb{R}$ and taking values in $\overline{\mathbb{R}}$. We say that $f$ is **measurable** with respect to the positive measure $\mu$ if every set of the form

$$\{x \in \mathbb{R} : a \le f(x) < b\},$$

where $a, b \in \overline{\mathbb{R}}$ and $a < b$, belongs to $\mathcal{T}$ (i.e. is measurable for $\mu$). For a bounded function $f$ which is measurable for a positive measure $\mu$ we can define an integral as follows. Given any finite, strictly increasing sequence of real numbers, $\alpha \equiv (\alpha_k)_{1 \le k \le p}$, such that $\alpha_1 \le f(x) < \alpha_p$, we write

$$A_{\alpha,k} \equiv \{x \in \mathbb{R} : \alpha_k \le f(x) < \alpha_{k+1}\}.$$

Then the **Lebesgue-Stieltjes integral** $\int_{\mathbb{R}} f(x) d\mu(x)$ of $f$ with respect to the positive measure $\mu$ can be defined as the least upper bound of the values

$$\sum_{k=1}^{p} \alpha_k \cdot \mu(A_{\alpha,k})$$

as $\alpha$ runs over all possible sequences of the type described.

In particular, the Lebesgue measure on $\mathbb{R}$ is the (positive) measure $m$ which assigns to each open interval $I \equiv (a,b) \subset \mathbb{R}$ the measure $m(I) = b - a$, and the corresponding integral is the Lebesgue integral,

$$m(f) = \int_{-\infty}^{+\infty} f(x) dx.$$

For a continuous function $f$ with compact support contained in an interval $[a, b]$ (and therefore certainly for any test function) this coincides with the ordinary Riemann integral of $f$ from $a$ to $b$,

$$\int_a^b f(x) dx.$$

By contrast, let $a \in \mathbb{R}$ and define a measure $\delta_a$ on the $\sigma$-ring of all subsets of $\mathbb{R}$ by writing,

$$\delta_a(A) = \begin{cases} 1 & \text{if } a \in A \\ 0 & \text{if } a \notin A \end{cases} \qquad (1.17)$$

for any $A \subset \mathbb{R}$. This is often referred to as the **Dirac measure** located at the point $a \in \mathbb{R}$. For a continuous function $\varphi$ with support contained in $[a, b]$ we get

$$\delta_a(\varphi) = \int_{-\infty}^{+\infty} \varphi(x) dH_c(x-a) = \int_a^b \varphi(x) dH_c(x-a) \qquad (1.18)$$

where $H_c$ denotes, as before, an arbitrary function of the form (1.1) and the integral is understood in the elementary Riemann-Stieltjes sense. A direct evaluation of this integral shows that $\delta_a(\varphi) = \varphi(a)$. In particular for $a = 0$ the Dirac measure $\delta_0$ admits a representation as a functional which is identical with the distribution $\delta$. In the 1940's N. Bourbaki and H. Cartan were responsible for a change of emphasis in measure theory which allowed arbitrary Borel measures to be characterised in terms of functionals. Thus, let $\mu$ be a positive Borel measure on $\mathbb{R}$. Then every continuous function of compact support is measurable for $\mu$ and the map

$$\varphi \rightsquigarrow \int_{-\infty}^{+\infty} \varphi(x)d\mu(x) \equiv \mu(\varphi)$$

defines a linear functional on $C_0^0$ which is positive in the sense that $\mu(\varphi) \geq 0$ for every real-valued function $\varphi \in C_0^0$ which is non-negative on $\mathbb{R}$. Such a functional is called a (positive) **Radon measure** on $\mathbb{R}$. It follows at once that for any positive Radon measure $\mu$ we will always have $\mu(\varphi) \geq \mu(\psi)$ whenever $\varphi$ and $\psi$ are real-valued on $\mathbb{R}$ and $\varphi \geq \psi$. What is more we can establish the following result:

**Theorem 1.3** *Every positive linear functional on $C_0^0$ is locally bounded with respect to the uniform norm $\| \varphi \|_{[0,K]}$.*

**Proof:** We have to show that for each compact $K \sqsubset \mathbb{R}$ there exists a number $M = M(K) > 0$ such that

$$|\mu(\varphi)| \leq M \cdot \sup_{x \in K} |\varphi(x)|.$$

Given any compact $K \sqsubset \mathbb{R}$ we can always find a finite closed interval $[a,b]$ such that $K \subset [a,b]$. For any $\varepsilon > 0$ let $I_\varepsilon$ denote the open interval $(a-\varepsilon, b+\varepsilon)$. Define a continuous function $\theta$ such that $\theta(x) = 1$ for $x \in [a,b]$, $\theta(x) = 0$ for $x \notin I_\varepsilon$ and $0 \leq \theta(x) \leq 1$ everywhere. Then, if $\mu$ is any positive linear functional on $C_0^0$ and $\varphi \in C_K^0$, we have for all $x \in \mathbb{R}$,

$$-\| \varphi \|_{[0,K]}\theta(x) \leq \varphi(x) \leq \|\varphi\|_{[0,K]}\theta(x)$$

and so

$$-\|\varphi\|_{[0,K]}\mu(\theta) \leq \mu(\varphi) \leq \|\varphi\|_{[0,K]}\mu(\theta).$$

Therefore we have

$$|\mu(\varphi)| \leq |\mu(\theta)| \cdot \| \varphi \|_{[0,K]}$$

for all $\varphi \in C_0^0$, and we have only to take $M = M(K) \equiv |\mu(\theta)|$. $\square$

This means that every positive Borel measure on $\mathbb{R}$ defines (and may be identified with) a certain distribution of order 0. Conversely it can be shown that every positive linear functional on $C_0^0$ admits an extension to a wider space of functions which includes the characteristic functions of all Borel sets and thereby defines a positive Borel measure on $\mathbb{R}$. This is one form of the classical theorem due to Riesz (see [93], for example):

**Theorem 1.4 (Riesz)** *Every Radon measure $\mu$ generates a uniquely defined measure on $\mathbb{R}$ which is such that*

$$\mu(\varphi) = \int_{\mathbb{R}} \varphi(x) d\mu(x)$$

*for every function $\varphi$ in the space $C_0^0$.*

### 1.1.6  Localisation

We have worked until now with distributions defined as functionals acting on test functions defined on the whole space $\mathbb{R}$; that is, as members of the dual space $\mathcal{D}'(\mathbb{R})$. The whole theory clearly applies to test functions defined on an open set $\Omega \subset \mathbb{R}$, and with an obvious change of notation we can speak of a distribution $\mu$ on $\Omega$, ($\mu \in \mathcal{D}'(\Omega)$), as well as of a distribution $\mu$ on $\mathbb{R}$ ($\mu \in \mathcal{D}'$). It is also clear that each distribution $\mu \in \mathcal{D}'$ admits a **restriction** to a distribution belonging to $\mathcal{D}'(\Omega)$.

A distribution $\mu \in \mathcal{D}'(\mathbb{R})$ is said to be zero on an open set $\Omega$ if $<\mu, \varphi> = 0$ for every $\varphi \in \mathcal{D}$ with support contained in $\Omega$; two distributions $\mu_1, \mu_2$ are then equal on $\Omega$ if and only if $\mu_1 - \mu_2 = 0$ in $\Omega$. It follows that two distributions in $\mathcal{D}'(\mathbb{R})$ are equal in $\Omega$ if and only if their restrictions to distributions on $\Omega$ are equal. In particular a distribution is zero in the neighbourhood of a point $x \in \mathbb{R}$ if it is zero in some open set containing that point. The delta function, for example, is zero in the neighbourhood of every point $x \in \mathbb{R}\setminus\{0\}$; this gives a precise meaning to the informal statement that *$\delta(x)$ has the value zero* for all $x \neq 0$.

The restriction of a distribution $\mu \in \mathcal{D}'(\mathbb{R})$ to an open subset of $\mathbb{R}$ is usually called a **localisation** of $\mu$. We can show that any distribution in $\mathcal{D}'(\mathbb{R})$ can be recovered from its localisations. This result depends upon the fact that there exists a so-called **partition of unity** for $\mathbb{R}$, that is a separation of the constant function 1 into the sum of functions in $\mathcal{D}$ whose supports can be chosen arbitrarily small:

**Theorem 1.5** *Given an open set $\Omega$ of the real line (respectively of $\mathbb{R}^m$), let $\{\Omega_\alpha\}_{\alpha \in J}$ be a (finite or infinite) family of open sets which covers $\Omega$. Then there exists a corresponding family $\{\gamma_\alpha\}_{\alpha \in J}$ of infinitely differentiable functions on $\Omega$ which are such that,*

(a) *the $\gamma_\alpha$ are non-negative on $\Omega$,*

(b) *for each $\alpha \in J$ the support of $\gamma_\alpha$ is contained in $\Omega_\alpha$,*

(c) *on each compact subset $K$ of $\Omega$ only finitely many of the $\gamma_\alpha$ will not be identically zero,*

**(d)** *the equality*

$$\sum_{\alpha \in J} \gamma_\alpha(x) \equiv 1$$

*holds on* $\Omega$.

**Proof:** (See, for example, Schwartz [97]; a detailed discussion is also given by Zemanian [117]. A simpler result which is adequate for the localisation theorem is proved in Friedlander [25].) □

**Theorem 1.6 (Localisation Principle)** *Let $\{\mu_\alpha\}_{\alpha \in J}$ be a family of distributions defined respectively on the members of a corresponding family $\{\Omega_\alpha\}_{\alpha \in J}$ of open sets, and let $\Omega$ be the union of all $\Omega_\alpha$. If for every $\alpha, \beta \in J$ we have that $\mu_\alpha = \mu_\beta$ on $\Omega_\alpha \cap \Omega_\beta$ then there exists a unique distribution $\mu$ on $\Omega$ whose restriction to each $\Omega_\alpha$ coincides with the given distribution $\mu_\alpha$ on $\Omega_\alpha$.*

**Proof:** Let $\{\gamma_\alpha\}_{\alpha \in J}$ be a partition of unity for $\Omega$, and let the support of $\gamma_\alpha$ be contained in $\Omega_\alpha$. For any $\varphi \in \mathcal{D}(\Omega)$ we can write

$$\varphi = \sum_{\alpha \in J} (\gamma_\alpha \varphi) \qquad (1.19)$$

where the sum on the right-hand side of (1.19) has only finitely many terms which do not vanish identically, since only finitely many of the $\gamma_\alpha$ are not identically zero on the support of $\varphi$. The formula

$$<\mu, \varphi> = \sum_{\alpha \in J} <\mu_\alpha, \gamma_\alpha \varphi>$$

defines $\mu$ as a continuous linear functional on $\mathcal{D}(\Omega)$. Further, if $\varphi \in \mathcal{D}(\Omega)$ has support contained in $\Omega_\beta$ then $\gamma_\alpha \varphi$ has support contained in $\Omega_\alpha \cap \Omega_\beta$; $\mu_\alpha$ and $\mu_\beta$ coincide in this intersection so that we have

$$<\mu_\alpha, \gamma_\alpha \varphi> = <\mu_\beta, \gamma_\alpha \varphi>$$

and therefore

$$\begin{aligned}<\mu, \varphi> &= \sum_{\alpha \in J} <\mu_\alpha, \gamma_\alpha \varphi> \\ &= \sum_{\alpha \in J} <\mu_\beta, \gamma_\alpha \varphi> = <\mu_\beta, \sum_{\alpha \in J} \gamma_\alpha \varphi> = <\mu_\beta, \varphi>\end{aligned}$$

which proves the assertion made. □

**Exercise 1.4 :**

1. Show that if $\Omega = (0, +\infty)$ then the functional on $\mathcal{D}(\Omega)$ defined by

$$<\mu, \varphi> = \sum_{k=1}^{\infty} \varphi^{(k)}(1/k)$$

is a distribution in $\mathcal{D}'(\Omega)$, but that there is no $\nu \in \mathcal{D}'(\mathbb{R})$ whose restriction to $\Omega$ is equal to $\mu$.

## 1.2 THE SCHWARTZ THEORY OF DISTRIBUTIONS

The definition of distributions given in section 1.1 has been simplified by confining attention to the one-dimensional case on the one hand and by avoiding topological vector space theory on the other. The extension to the case of distributions defined on $\mathbb{R}^m$ is straightforward and is facilitated by use of the multi-index convention described in §1.2.1 below. The theory of topological vector spaces is an indispensable background for the proper presentation of the Schwartz theory of distributions although for many applications the simplified approach is adequate. We give a brief account of the basic material from functional analysis which is required for the understanding of the issues involved and refer the reader to specialist texts such as that by Tréves [115], or the second volume of Gel'fand & Shilov [26], for further information.

### 1.2.1 Distributions on $\mathbb{R}^m$

The $m$-tuplet $r \equiv (r_1, r_2, \ldots, r_m)$, where the $r_k$ are non-negative integers, is said to be a multi-index of dimension $m$. We define $|r|$ to be

$$|r| = r_1 + r_2 + \cdots + r_m \tag{1.20}$$

and write

$$\begin{aligned} D^r &\equiv \partial^{|r|}/\partial x_1^{r_1} \partial x_2^{r_2} \ldots \partial x_m^{r_m} \\ &= \partial^{r_1 + r_2 + \cdots + r_m}/\partial x_1^{r_1} \partial x_2^{r_2} \ldots \partial x_m^{r_m} \end{aligned} \tag{1.21}$$

and

$$x^r \equiv x_1^{r_1} x_2^{r_2} \ldots x_m^{r_m}. \tag{1.22}$$

Then we have

$$x^r y^q = x_1^{r_1} y_1^{q_1} \ldots x_m^{r_m} y_m^{q_m}$$

and
$$D^r D^q f(x) = D^{r+q} f(x).$$

The general definition for distributions in $\mathbb{R}^m$ can be framed in the following way: the space $\mathcal{D}(\mathbb{R}^m)$ is the linear space of all infinitely differentiable functions $\varphi(x)$, defined on $\mathbb{R}^m$ and taking real or complex values, which have compact support. Convergence in this space (which is equivalent to what we have called $\omega$-uniform convergence over compact subsets in the case $m = 1$) is defined thus: the sequence $(\varphi_n)_{n \in \mathbb{N}} \subset \mathcal{D}(\mathbb{R}^m)$ is a **null-sequence** if and only if the following conditions hold:

(1) there exists a ball $B(0, \varepsilon) \equiv \{x \in \mathbb{R}^m : (\sum_{k=1}^m x_k^2)^{\frac{1}{2}} < \varepsilon\}$ such that each of the functions $\varphi_n(x)$ vanishes identically outside $B(0, \varepsilon)$,

(2) $\lim_{n \to \infty} \{\sup_{x \in B(0,\varepsilon)} |D^r \varphi_n(x)|\} = 0$ for every multi-index $r$ of dimension $m$.

Condition (2) ensures that each sequence $(D^r \varphi_n)_{n \in \mathbb{N}}$ converges uniformly to zero. A sequence $(\varphi_n)_{n \in \mathbb{N}}$ converges to the limit $\varphi$ in the sense of the space $\mathcal{D}(\mathbb{R}^m)$ if and only if $(\varphi - \varphi_n)_{n \in \mathbb{N}}$ is a null-sequence.

A distribution on $\mathbb{R}^m$ is a linear functional $\mu$ on $\mathcal{D}(\mathbb{R}^m)$ which is continuous in the sense that
$$\lim_{n \to \infty} <\mu, \varphi_n> \equiv \lim_{n \to \infty} \mu(\varphi_n) = 0$$
for every null-sequence $(\varphi_n)_{n \in \mathbb{N}}$ in $\mathcal{D}(\mathbb{R}^m)$.

### 1.2.2 Topological vector spaces

The theory of topological vector spaces is a generalisation of the relatively familiar material associated with normed linear space theory and we begin by reviewing such basic topological concepts as are necessary.

In ordinary Euclidean space $\mathbb{R}^m$ the distance between two points $x$ and $y$ is given by
$$d(x, y) = \left[\sum_{k=1}^m (x_k - y_k)^2\right]^{\frac{1}{2}}. \tag{1.23}$$

For any given point $x \in \mathbb{R}^m$ we can then define the open ball $B(x, \varepsilon)$ of radius $\varepsilon > 0$ as the set of all points $y$ whose distance from $x$ is less than $\varepsilon$:
$$B(x, \varepsilon) = \{y \in \mathbb{R}^m : d(x, y) < \varepsilon\}. \tag{1.24}$$

A subset $G$ of $\mathbb{R}^m$ is said to be open if and only if for each point $x \in G$ there exists an open ball $B(x,\varepsilon)$ which is contained in $G$. A subset $F$ of $\mathbb{R}^m$ is said to be closed if and only if its complement $F^c \equiv \mathbb{R}^m \backslash F$ is open.

The family of all possible open sets (or, equivalently, the family of all closed sets) constitutes a structure called the **usual** or the **standard topology** of $\mathbb{R}^m$. Other topologies for $\mathbb{R}^m$ are possible. For example we could define a (rather trivial) alternative measure of the distance between two points $x$ and $y$ by setting

$$d_0(x,y) = \begin{cases} 1 & \text{if } x \neq y \\ 0 & \text{if } x = y \end{cases} \qquad (1.25)$$

This would mean that for every $\varepsilon$ such that $0 \leq \varepsilon \leq 1$, the open ball $B(x,\varepsilon)$ would be the set $\{x\}$ whose only member is the single point $x$. The resulting topology is called the *discrete* topology of $\mathbb{R}^m$, and every subset of $\mathbb{R}^m$ is both open and closed in this topology.

In general a **topology** on a set $E$ is a family $\mathcal{T}$ of subsets of $E$, hereinafter called the open sets of the topology, which have the same basic properties as the open sets in the standard topology of $\mathbb{R}^m$. That is to say,

**T1** $\emptyset \in \mathcal{T}$ and $E \in \mathcal{T}$,

**T2** the union of any collection (finite or infinite) of sets of $\mathcal{T}$ is in $\mathcal{T}$,

**T3** the intersection of any finite collection of sets of $\mathcal{T}$ belongs to $\mathcal{T}$.

A subset $F$ of $E$ is said to be closed if and only if $F^c \equiv E\backslash F$ is a member of $\mathcal{T}$ (i.e. if and only if $F^c$ is an open subset of $E$).

The set $E$, equipped with a specific topology $\mathcal{T}$, is said to be a topological space. The same set $E$ may be equipped with different topologies $\mathcal{T}_1$ and $\mathcal{T}_2$, and therefore give rise to different topological spaces $E_1 \equiv (E, \mathcal{T}_1)$ and $E_2 \equiv (E, \mathcal{T}_2)$. (For example, $\mathbb{R}^m$ equipped with the discrete topology is a different topological space from the Euclidean space $\mathbb{R}^m$ equipped with the standard topology.) If $\mathcal{T}_1$ and $\mathcal{T}_2$ are topologies on $E$ such that $\mathcal{T}_1 \subset \mathcal{T}_2$ then $\mathcal{T}_1$ is said to be **weaker** (coarser) than $\mathcal{T}_2$ and $\mathcal{T}_2$ is said to be **stronger** (finer) than $\mathcal{T}_1$. The two topologies are **equivalent** if $\mathcal{T}_1$ is both stronger and weaker than $\mathcal{T}_2$.

**Neighbourhoods** The role played by the open ball $B(x,\varepsilon)$ in the standard topology of $\mathbb{R}^m$ is generalised in the concept of neighbourhood. If $x$ is a point of the topological space $E = (E, \mathcal{T})$ then any set $U \subset E$ which contains an open set containing $x$ is called a **neighbourhood** of $x$. Similarly if $A$ is any subset of $E$ then a neighbourhood of $A$ is defined as any set $V \subset E$ which contains an open set containing $A$.

A family $\mathcal{B}$ of subsets of $E$ (necessarily open sets) is said to be a basis for a topology $\mathcal{T}$ on $E$ if every open set of $\mathcal{T}$ is the union of sets of $\mathcal{B}$. A family $\mathcal{N}(x)$ of neighbourhoods of $x \in E$ is said to be a **basis of neighbourhoods** at $x$ if every neighbourhood $U$ contains some set belonging to $\mathcal{N}(x)$. The topological space $(E, \mathcal{T})$ satisfies the **First Countability Axiom** if there exists a countable basis of neighbourhoods at each point of $E$; it satisfies the **Second Countability Axiom** if there exists a countable basis for $E$ itself.

**Convergence** Once the concept of neighbourhood has been introduced we can say what is meant by convergence in a topological space. A sequence $(x_n)_{n \in \mathbb{N}}$ of points of $E$ is said to *converge* to a limit $x \in E$ if and only if every neighbourhood of $x$ contains all the points $x_n$ save for at most finitely many of them; that is to say, $(x_n)_{n \in \mathbb{N}}$ converges to $x$ if and only if for each neighbourhood $U$ of $x$ there exists an integer $n_0 = n_0(U) \in \mathbb{N}$ such that $x_n \in U$ for all $n$ greater than or equal to $n_0$.

**Hausdorff space, normal space and regular space** A topological space $E$ is called a **Hausdorff** space if for each pair $x, y$ of distinct points of $E$ there exist neighbourhoods $U, V$, of $x$ and $y$ respectively such that $U \cap V = \emptyset$. In a Hausdorff space every convergent sequence $(x_n)_{n \in \mathbb{N}}$ has a *unique* limit $x$. $E$ is a **regular** space if, for each $x \in E$ and every closed set $F \subset E$ such that $x \notin F$, there are neighbourhoods of $x$ and $F$ which are disjoint. $E$ is **normal** if for every pair $A, B$ of disjoint closed sets there exist disjoint neighbourhoods of $A$ and $B$ respectively.

**Closure; limit points** The closure of a set $A \subset E$ is the smallest closed set which contains $A$, and is denoted by $\bar{A}$. The interior of $A$ is the largest open set which is contained in $A$ and is denoted by $\mathring{A}$. The boundary of $A$ is the set $\bar{A} \setminus \mathring{A}$. A point $x$ is a **limit point** of a set $A \subset E$ if $A$ meets every punctured neighbourhood $U \setminus \{x\}$ of $x$. The set of all limit points of $A$ is called the **derived set** of $A$ and is written $A'$. We always have

$$\bar{A} = A \cup A' \tag{1.26}$$

and it follows that $A$ is closed if and only if $A' \subset A$. A set $A \subset E$ is said to be **dense** in $E$ if $\bar{A} = E$; a topological space $E$ is **separable** if it contains a countable dense subset.

**Induced topology** A subset $F$ of a topological space $E$ can itself be equipped with a topology which derives from the existing topology on $E$. This is called the **induced topology** and is obtained when we define a subset $B \subset F$ to be open in $F$ if and only if $B = A \cap F$ where $A$ is a set open in $E$. The topological space $F$ is then said to be a **topological subspace** of $E$.

**Compactness** A topological space $E$ is a **compact space** if every family $\{G_\alpha\}_{\alpha \in J}$ of open sets which covers $E$ contains a finite subfamily $\{G_{\alpha_i}\}_{1 \le i \le n}$ which also covers $E$. $E$ is said to be **sequentially compact** if every sequence $(x_n)_{n \in \mathbb{N}} \subset E$ contains a subsequence which converges to some point of $E$. A subset $F$ of $E$ is said to be compact (respectively, sequentially compact) if $F$ is compact (respectively, sequentially compact) as a topological subspace of $E$.

**Continuous mappings** If $E$ and $F$ are topological spaces and $f$ is a mapping of $E$ into $F$ then $f$ is said to be **continuous** on $E$ if the inverse image of every set open in $F$ is open in $E$; if $f$ is a one-one mapping of $E$ onto $F$ and if both $f$ and $f^{-1}$ are continuous then $f$ is said to be a **homeomorphism** of $E$ onto $F$. If $f : E \to F$ is continuous and if $x \in E$ is the limit of a sequence $(x_n)_{n \in \mathbb{N}} \subset E$ then $f(x)$ is the limit of the sequence $(y_n)_{n \in \mathbb{N}} \subset F$, where $y_n = f(x_n)$.

**Topological vector space** Now suppose that $E$ is a (real or complex) linear space. If $E$ is equipped with a topology such that the operations of vector addition and multiplication by scalars are both continuous, then $E$ is said to be a **topological vector space** (TVS). Such a topology is completely specified once we are given a basis of neighbourhoods of the null vector $0 \in E$, that is to say, a family $\mathcal{N}$ of neighbourhoods of 0 such that each neighbourhood of 0 contains some member of $\mathcal{N}$. The topology is Hausdorff if and only if the intersection of all $U \in \mathcal{N}$ is the set $\{0\}$. A necessary and sufficient condition for a set $A \subset E$ to be open is that for each point $x \in A$ there exists some set $U \in \mathcal{N}$ such that we have $x + U \subset A$ (where $x + U = \{y \in E : y - x \in U\}$).

**Convex space** Given points $x$ and $y$ of a linear space $E$, the **line segment** joining $x$ and $y$ is defined to be the set of all points of the form $\alpha x + \beta y$, where $\alpha$ and $\beta$ are real, non-negative numbers such that $\alpha + \beta = 1$. A subset $A$ of $E$ is **convex** if it contains the line segment joining $x$ and $y$ whenever it contains $x$ and $y$. The subset $A$ of $E$ is **balanced** if $x \in A$ implies $\alpha x \in A$ for all $\alpha$ such that

$|\alpha| \leq 1$, and **absorbing** if to each $x \in E$ there corresponds some $\beta > 0$ such that $\beta x \in A$.

A topological vector space $(E, \mathcal{T})$ is said to be **locally convex** if there exists a basis of neighbourhoods of 0 which consists of sets which are convex, balanced and absorbing.

### 1.2.3 Norms and normed linear spaces

**Normed space; metric space**  The simplest and most straightforward example of a TVS arises when the topology on a linear space $E$ is defined by a norm (*for the definition of norm see page 6*). In particular the Euclidean norm

$$\|x\| = \left[\sum_{k=1}^{m} x_k^2\right]^{\frac{1}{2}} \tag{1.27}$$

generates the usual topology on $\mathbb{R}^m$.

A linear space $E$ which has been equipped with a norm $\|\cdot\|_E$ is called a **normed linear space**. In such a space we can always define the distance $d(x, y)$ between any two members $x$ and $y$ of $E$ as

$$d(x, y) \equiv \|x - y\|_E. \tag{1.28}$$

The real-valued function $d(\cdot, \cdot)$ defined on $E \times E$ is called a **distance function** or a **metric** on $E$. A metric can actually be defined on a set $E$, whether that set is equipped with a linear space structure or not, as a real-valued function satisfying the conditions

**D1** $d(x, y) \geq 0$, and $d(x, y) = 0$ if and only if $x = y$,

**D2** $d(x, y) = d(y, x)$,

**D3** $d(x, z) \leq d(x, y) + d(y, z)$,

for all $x, y, z \in E$. Any set of objects on which such a function can be defined is said to be a **metric space**. For example the linear space $\mathcal{C}_K^0$ of all continuous functions with support contained in the same compact subset $K$ of $\mathbb{R}$ can be equipped with a metric $d_{[0]}(\varphi, \psi)$ defined by

$$d_{[0]}(\varphi, \psi) = \sup_{x \in K} |\varphi(x) - \psi(x)|.$$

**Metric space topology** Any linear space on which a norm has been defined is automatically a metric space with metric $d(x,y) \equiv \|x-y\|$. Thus the metric $d_{[0]}(\varphi, \psi)$ defined above on $\mathcal{C}_K^0$ arises naturally from the norm $\|\varphi\|_{[0,K]} = \sup_{x \in K} |\varphi(x)|$. As in the case of $\mathbb{R}^m$ the metric generates a natural topology on the space: if $a$ is any point in a metric space $E$ and $\varepsilon$ any positive real number then the set $B(a, \varepsilon) = \{x \in E : d(a,x) < \varepsilon\}$ is called the **open ball** of centre $a$ and radius $\varepsilon$. Define a subset $A$ of $E$ to be open if and only if for each point $a \in A$ there exists an open ball $B(a, \varepsilon) \subset A$. Then the set of all open balls in $E$ forms the basis for a topology on $E$. This topology is Hausdorff and certainly satisfies the first countability axiom; the topological space $E$ will be separable if and only if it satisfies the second countability axiom.

If $E$ is a linear space and the metric is generated by a norm $\|\cdot\|$ then the resulting topology is locally convex. It may happen that the same linear space $E$ can be equipped with more than one norm, each of which generates a topology on $E$. If $\|\cdot\|_1$ and $\|\cdot\|_2$ are norms on $E$ such that for every $x \in E$ we have

$$\|x\|_1 \leq C\|x\|_2,$$

where $C$ is a positive constant, then we say that $\|\cdot\|_2$ is stronger than $\|\cdot\|_1$ or equivalently that $\|\cdot\|_1$ is weaker than $\|\cdot\|_2$. For example the space $\mathcal{C}_K^1$ of all continuously differentiable functions with support contained in $K$ can be equipped with either of the norms

$$\|\varphi\|_{[0,K]} = \sup_{x \in K} |\varphi(x)|$$
$$\|\varphi\|_{[1,K]} = \max\{\sup_{x \in K} |\varphi(x)|, \sup_{x \in K} |\varphi'(x)|\}.$$

Clearly we have $\|\varphi\|_{[0,K]} \leq \|\varphi\|_{[1,K]} \equiv \max\{\|\varphi\|_{[0,K]}, \|\varphi'\|_{[0,K]}\}$ so that $\|\cdot\|_{[0,K]}$ is weaker than $\|\varphi\|_{[1,K]}$. More generally, given any $p$ in $\mathbb{N}$, we can equip the linear space $\mathcal{C}_K^p$ with any one of the norms $\|\varphi\|_{[q,K]} = \max_{0 \leq r \leq q}\{\sup_{x \in K} |\varphi^{(r)}(x)|\}$, where $q = 0, 1, \ldots, p$. If $q_1 < q_2$ then $\|\cdot\|_{[q_1,K]}$ is weaker than $\|\cdot\|_{[q_2,K]}$.

**Cauchy sequences: completeness** A sequence $(x_n)_{n \in \mathbb{N}}$ in a metric space $E$ is said to be a **Cauchy sequence** if $d(x_m, x_n)$ tends to zero as $m$ and $n$ tend to infinity independently. Every convergent sequence is necessarily Cauchy. If it happens that every Cauchy sequence in $E$ converges to a point of $E$ then the metric space is said to be **complete**. A normed linear space which is complete with respect to the metric induced by the norm is called a **Banach space**. For $p \in \mathbb{N}$ and $K \sqsubset \mathbb{R}$ the linear space $\mathcal{C}_K^p$ is a Banach space with respect to the norm $\|\cdot\|_{[p,K]}$.

**Theorem 1.7** *Every normed space $E$ is either complete or can be embedded in a Banach space $\hat{E}$, called the* **completion** *of $E$, without changing the norm of elements in $E$. $E$ is dense in $\hat{E}$.*

The concept of Cauchy sequence (and, similarly, that of completeness) readily generalises to an arbitrary topological vector space: we say that $(x_n)_{n \in \mathbb{N}}$ is a Cauchy sequence in a topological vector space $E$ if for every neighbourhood $U$ of the origin of $E$ we have $x_n - x_m \in U$ for all sufficiently large values of $n$ and $m$.

**Compatible norms** Now let $E$ be any linear space which admits two distinct norms $\|\cdot\|_1$ and $\|\cdot\|_2$, where $\|\cdot\|_1$ is weaker than $\|\cdot\|_2$. Then every sequence in $E$ which is Cauchy with respect to $\|\cdot\|_2$ is also Cauchy with respect to $\|\cdot\|_1$. If $E_1$ and $E_2$ denote the completions of $E$ with respect to $\|\cdot\|_1$ and $\|\cdot\|_2$ then it follows that $E_2 \subset E_1$. Each point $x^{(2)} \in E_2$ is the limit of a Cauchy sequence $(x_n)_{n \in \mathbb{N}}$ of elements of $E$ which, in turn, must have a limit $x^{(1)}$ in $E_1$. The map $x^{(2)} \rightsquigarrow x^{(1)}$ is a linear map of $E_2$ onto $E_1$, but will not in general be injective; distinct elements of $E_2$ may correspond to the same element of $E_1$. This defect is avoided when the norms are compatible:

> Two norms $\|\cdot\|_1$ and $\|\cdot\|_2$ on a linear space $E$ are said to be **compatible** if every sequence which is Cauchy with respect to both norms and which converges to zero with respect to one norm also converges to zero with respect to the other.

Note that if $\|\cdot\|_1$ and $\|\cdot\|_2$ are compatible norms on $E$ and we define

$$\|\cdot\|_3 = \max\{\|\cdot\|_1, \|\cdot\|_2\} \tag{1.29}$$

then $\|\cdot\|_3$ is a norm on $E$ which is compatible with both $\|\cdot\|_1$ and $\|\cdot\|_2$. This fact is of importance when we come to consider topologies defined on a linear space $E$ by *families* of norms on $E$ rather than by individual norms. As an example, take $E = \mathcal{C}^1_K$ and consider again the norms $\|\cdot\|_{[0,K]}$ and $\|\cdot\|_{[1,K]}$ on $\mathcal{C}^1_K$. A sequence $(\varphi_n)_{n \in \mathbb{N}}$ converges to zero in $\mathcal{C}^1_K$ with respect to the norm $\|\cdot\|_{[0,K]}$ if and only if the convergence is uniform; it converges to zero in $\mathcal{C}^1_K$ with respect to the norm $\|\cdot\|_{[1,K]}$ if and only if both $(\varphi_n)_{n \in \mathbb{N}}$ and $(\varphi'_n)_{n \in \mathbb{N}}$ converge uniformly to zero. Hence convergence with respect to the stronger norm $\|\cdot\|_{[1,K]}$ clearly implies convergence with respect to the weaker norm $\|\cdot\|_{[0,K]}$, but not conversely. Although $\mathcal{C}^1_K$ is a Banach space with respect to $\|\cdot\|_{[1,K]}$ it is not complete with respect to $\|\cdot\|_{[0,K]}$ and, indeed, is not even closed under convergence with respect to this latter norm. Nevertheless these two norms are compatible.

**Exercise 1.5 :**

1. If $d(x,y)$ is a real, finite-valued function, defined for all points $x,y$ of a given set $E$, which is such that

   (a) $d(x,y) \leq d(x,z) + d(z,y)$ for every $x,y,z \in E$;
   
   (b) $d(x,y) = 0$ if and only if $x = y$;
   
   prove that $d(\cdot,\cdot)$ defines a metric on $E$.

2. If $E$ is a normed linear space with norm $\|\cdot\|$ and if we define

$$\varphi(\|x\|) = \frac{\|x\|}{1+\|x\|}$$

   show that $\varphi(\|x-y\|)$ defines a metric on $E$ but that $\varphi(\|x\|)$ itself is not a norm on $E$.

3. If $K$ is any compact subset of $\mathbb{R}$ and $p$ a non-negative integer prove that the linear space $\mathcal{C}_K^p(\mathbb{R})$, equipped with the norm

$$\|\varphi\|_{[p,K]} = \max_{0 \leq q \leq p} \left\{ \sup_{x \in K} |\varphi^{(q)}(x)| \right\}$$

   is a Banach space.

### 1.2.4 Countably normed spaces

Let $E$ be a linear space and let $(\|\cdot\|_n)_{n \in \mathbb{N}}$ be a sequence of norms defined on $E$. We can obtain a topology on $E$ by defining a neighbourhood basis at the null vector $0 \in E$. In particular we can get such a basis by choosing sets of the form

$$\mathcal{N}_{p,\varepsilon}(0) = \left\{ x \in E : \|x\|_1 < \varepsilon, \|x\|_2 < \varepsilon, \ldots, \|x\|_p < \varepsilon \right\} \tag{1.30}$$

where $p$ is any integer $\geq 1$ and $\varepsilon > 0$. Then $E$ becomes a locally convex topological space (and one which satisfies the first countability axiom). Clearly, a sequence $(x_n)_{n \in \mathbb{N}} \subset E$ converges to zero if and only if $\lim_{n \to \infty} \|x_n\|_p = 0$ for any $p$; similarly it will be a Cauchy sequence if and only if, for any $p$, $\|x_n - x_m\|_p$ tends to zero as $n$ and $m$ tend to infinity independently.

If each pair of norms in the sequence $(\|\cdot\|_n)_{n \in \mathbb{N}}$ are compatible then the resulting topological space $E$ is called a **countably normed space**. It is easy to show that then the same topology is defined on $E$ by the monotone increasing sequence $(\|\cdot\|'_n)_{n \in \mathbb{N}}$ of norms defined by

$$\|x\|'_p \equiv \max \left\{ \|x\|_1, \ldots, \|x\|_p \right\}.$$

Hence, without any loss of generality, we may assume in future that a countably normed space is a linear space $E$ equipped with the topology induced by a sequence $(\|\cdot\|_n)_{n \in \mathbb{N}}$ of (mutually compatible) norms on $E$ such that

$$\|x\|_1 \leq \|x\|_2 \leq \ldots \leq \|x\|_p \leq \ldots \tag{1.31}$$

This means that we can say that a neighbourhood basis at the origin of $E$ is given by sets of the form

$$\mathcal{N}_{p,\varepsilon}(0) = \{x \in E : \|x\|_p < \varepsilon\}$$

for $p = 1, 2, 3, \ldots$, and all $\varepsilon > 0$. For example, if $K$ is any fixed compact subset of $\mathbb{R}$ then the linear space $\mathcal{C}_K^\infty$ of all infinitely differentiable functions with support contained in $K$ can be equipped with each of the norms

$$\|\varphi\|_{[p,K]} = \max_{0 \leq r \leq p} \left\{ \sup_{x \in K} |\varphi^{(r)}(x)| \right\}$$

for $p = 0, 1, \ldots$. By definition we have that, for each $p$,

$$\|\varphi\|_{[p,K]} = \max_{0 \leq r \leq p} \left\{ \|\varphi\|_{[r,K]} \right\}$$

so that the norms form a monotone increasing sequence and define a topology on $\mathcal{C}_K^\infty$ which makes it a countably normed space. A sequence $(\varphi_n)_{n \in \mathbb{N}}$ in $\mathcal{C}_K^\infty$ converges to zero in the sense of this topology if and only if it converges to zero with respect to each of the norms $\|\cdot\|_{[p,K]}$. This means that $(\varphi_n)_{n \in \mathbb{N}}$ converges to zero if and only if each of the sequences $(\varphi_n^{(k)})_{n \in \mathbb{N}}$ of the $k^{th}$ derivatives of the $\varphi_n$ converges uniformly to zero for $k \in \mathbb{N}_0$. Convergence in the sense of the countably normed space $\mathcal{C}_K^\infty$ is therefore $\omega$-uniform convergence (on $K$).

**Theorem 1.8** *Let $E$ be a countably normed space with respect to the monotone increasing sequence of norms $(\|\cdot\|_n)_{n \in \mathbb{N}}$, and let $E_n$ denote the completion of $E$ with respect to the norm $\|\cdot\|_n$. Then,*

**(1)** $E_1 \supset E_2 \supset \ldots \supset E_n \supset \ldots \supset E$, *and*

**(2)** *the countably normed space $E$ is complete if and only if*

$$E = \bigcap_{k=1}^{\infty} E_k.$$

**Proof:** Since (1) is immediate we prove (2).
*Sufficiency:* Let $(x_n)_{n \in \mathbb{N}}$ be a Cauchy sequence in $E = \bigcap_{k=1}^{\infty} E_k$. Then $(x_n)_{n \in \mathbb{N}}$ is also Cauchy in each $E_k$, and for each $k$ there exists some point $y_k \in E_k$ such that

$$\lim_{n \to \infty} \|x_n - y_k\|_k = 0.$$

We may identify the elements $y_k$ with a single element $x$ in the intersection of all $E_k$. For, $y_{k+1} \in E_{k+1} \subset E_k$ and

$$\begin{aligned} \| y_{k+1} - y_k \|_k &\leq \| y_{k+1} - x_n \|_k + \| x_n - y_k \|_k \\ &\leq \| y_{k+1} - x_n \|_{k+1} + \| x_n - y_k \|_k \end{aligned}$$

from which it follows that $y_{k+1}$ coincides with $y_k$. This unique vector $x$ has the property that

$$\lim_{n \to \infty} \| x_n - x \|_k = 0$$

for every integer $k$; thus $E$ is complete.

*Necessity.* Let $E$ be complete and let $x$ belong to the intersection of all the $E_k$. Then for any $n$ there exists some $x_n \in E$ such that $\| x_n - x \|_n < 1/n$. For any $p \leq n$ we have

$$\| x - x_n \|_p \leq \| x - x_n \|_n < 1/n$$

so that $x_n \to x$ in $E_p$. It follows that $(x_n)_{n \in \mathbb{N}}$ is a Cauchy sequence in each $E_p$ and therefore in $E$. Write $\hat{x} \equiv \lim_{n \to \infty} x_n$ in $E$. Then, $\| x_n - \hat{x} \|_p \to 0$ for any $p$. This means that $\| x - \hat{x} \| = 0$ and therefore that $x = \hat{x} \in E$. □

**Linear metric space; Fréchet space** In any countably normed space $E$ it is always possible to introduce a metric given by

$$d(x,y) = \sum_{p=1}^{\infty} \frac{\| x - y \|_p}{2^p \cdot \left[ 1 + \| x - y \|_p \right]} \qquad (1.32)$$

and it can be shown that the topology of the resulting metric space is equivalent to the original topology of $E$ as a countably normed space. In addition to the usual metric axioms the distance function $d(\cdot, \cdot)$ will also satisfy the following additional conditions

**D4** $d(x,y) = d(x-y, 0)$,

**D5** (i) For any sequence $(\alpha_n)_{n \in \mathbb{N}}$ of scalars such that $\lim_{n \to \infty} \alpha_n = 0$ we have $\lim_{n \to \infty} d(\alpha_n x, 0) = 0$, for any $x \in E$.

(ii) For any sequence $(x_n)_{n \in \mathbb{N}} \subset E$ such that $x_n \to 0$ in $E$ and any scalar $\alpha$ we have $\lim_{n \to \infty} d(\alpha x_n, 0) = 0$.

A topological vector space $E$ which can be equipped with a metric which also satisfies D4 and D5 is sometimes called a **linear metric space**; if, in addition, $E$ is locally convex and complete then $E$ is said to be a **Fréchet space**. Thus every complete countably normed space is a Fréchet space. Note, however, that in practice it is easier to work in terms of the definition of the topology on $E$ as a countably normed space than in terms of the metric $d(x,y)$.

**Seminorms** In the general theory of topological vector spaces a more powerful and general method of constructing locally convex topologies is to work in terms of families of seminorms:

> A **seminorm** on a linear space $E$ is a real-valued function on $E$ which satisfies all the conditions for a norm save that we may have $\|x\| = 0$ for non-null vectors $x$. That is to say a seminorm on $E$ is a function $p : E \to \mathbb{R}$ such that
>
> **SN1.** $\forall x, y \in E : p(x+y) \leq p(x) + p(y)$,
>
> **SN2.** $\forall x \in E : p(\alpha x) = |\alpha| p(x)$ for every scalar $\alpha$.

A norm is a particular case of a seminorm in which the additional condition satisfied is that $p(x) = 0$ if and only if $x$ is the null vector of the linear space $E$ concerned. Thus, if $K$ is a fixed compact subset of $\mathbb{R}$ and if we define, as hitherto, the quantity $\| \varphi \|_{[p,K]}$ to be $\max_{0 \leq r \leq p} \{\sup_{x \in K} |\varphi^{(r)}(x)|\}$ then the result is a norm on the linear space $\mathcal{C}_K^p$ but only a semi-norm on the space $\mathcal{C}_0^p$ or the space $\mathcal{C}^p$. Taking the supremum over the whole of $\mathbb{R}$ in the definition would yield a true norm on both $\mathcal{C}_0^p$ and $\mathcal{C}^p$.

A family $\mathcal{P}$ of semi-norms on $E$ is said to be a **separating** family if for each $x \in E, x \neq 0$, we can find some semi-norm $p \in \mathcal{P}$ such that $p(x) \neq 0$. If $\mathcal{P}$ is such a family we denote by $\mathcal{B}(0)$ the collection of all finite intersections of sets of the form,

$$\{x \in E : \ p(x) < \varepsilon\}$$

where $p \in \mathcal{P}$ and $\varepsilon > 0$. Then it is easy to show that $\mathcal{B}$ is a basis of the neighbourhoods of 0 for a locally convex vector space topology on $E$. This topology is called the topology generated by the family $\mathcal{P}$ of seminorms.

On the other hand, suppose that $E$ is any locally convex topological space whatsoever, and let $M$ be any convex, balanced, and absorbing neighbourhood of 0. The so-called **Minkowski functional** of $M$ defined as

$$p_M(x) \equiv \inf\{t > 0 : x \in tM\}$$

is a seminorm. The topology of $E$ can then be shown to be generated by a family of such seminorms.

### 1.2.5 Inductive limits

There remains one further extension of the basic idea of a normed space which we will need in the sequel. Recall first the definition of induced, or relative, topology:

Let $E_1$ and $E_2$ be non-empty sets such that $E_1 \subset E_2$. If $E_2$ has a topology $\mathcal{T}_2$ then the relative topology induced on $E_1$ is the family of sets of the form $G \cap E_1$, where $G \in \mathcal{T}_2$. Suppose now that $E_1$ has a topology $\mathcal{T}_1$ in its own right. Then we say that the inclusion $E_1 \subset E_2$ extends to the topological spaces $(E_1, \mathcal{T}_1)$ and $(E_2, \mathcal{T}_2)$ if $\mathcal{T}_1$ agrees with the relative topology induced on $E_1$ as a subset of $E_2$.

For example, let $K$ and $L$ be two compact subsets of $\mathbb{R}$ such that $K$ is contained in (the interior of) $L$. Take for $E_1$ the space $\mathcal{C}_K^0$ which is equipped with the norm $\|\varphi\|_{[0,K]} = \sup_{x \in K} |\varphi(x)|$ and for $E_2$ the space $\mathcal{C}_L^0$ equipped with the norm $\|\varphi\|_{[0,L]} = \sup_{x \in L} |\varphi(x)|$. Then $\mathcal{C}_K^0 \subset \mathcal{C}_L^0$, and the topology induced by the norm $\|\cdot\|_{[0,K]}$ on $\mathcal{C}_K^0$ agrees with the relative topology induced by the norm $\|\cdot\|_{[0,L]}$.

Now let $E_1 \subset E_2 \subset \ldots \subset E_n \subset E_{n+1} \subset \ldots$ be an increasing sequence of complete countably normed spaces, with the inclusion satisfying this extended definition. Then the linear space

$$E \equiv \bigcup_{n=1}^{\infty} E_n \qquad (1.33)$$

is called the **strict inductive limit** of the family $\{E_n\}_{n \in \mathbb{N}}$. Let $\mathcal{B}$ be the collection of all sets $A \subset E$ such that for each $n$ the intersection $A \cap E_n$ is a set of the form $\{x \in A : \|x\|_n < \varepsilon\}$. Then $\mathcal{B}$ is a basis for a topology $\mathcal{T}$ on $E$ and so $E$ is a topological vector space. Any TVS defined in this manner is said to be a (strict) **countable union space**.

**Remark 1.1** *More generally suppose we have an increasing sequence $\{E_n\}_{n \in \mathbb{N}}$ of complete countable union spaces in which, for each $n$, the topology induced by $\mathcal{T}_{n+1}$ on $E_n$ is stronger than (but not necessarily equivalent to) the topology $\mathcal{T}_n$ of $E_n$. Then the union $E$ is said to be the inductive limit of the $E_n$, and is referred to simply as a countable union space.*

**Exercise 1.6 :**

1. Let $\{p_i\}_{1 \leq i \leq n}$ be a finite set of semi-norms on a linear space $E$. Show that

    (a) $p(x) = \max\{p_i(x) : 1 \leq i \leq n\}$ is a semi-norm on $E$;

    (b) $q(x) = \sum_{i=1}^n a_i p_i(x)$, where $a_i \geq 0$ for all $i = 1, \ldots, n$, is a semi-norm on $E$.

2. Let $\{E_j\}_{j \in J}$ be a family of Banach spaces (with respect to norms $p_j, j \in J$) with the following property: for every $j', j''$ in $J$ there exists $j \in J$ such that

    $$E_j \subset E_{j'} \cap E_{j''}, \; p_j \geq \max\{p_{j'}, p_{j''}\}.$$

    Consider $E = \bigcap_{j \in J} E_j$ with the topology defined by the family of norms $\{p_j\}_{j \in J}$. Show that $E$ is complete.

## 1.2.6 Schwartz's definition of distributions

We can now rephrase the definitions of test function spaces and distributions as follows. Let $K$ be any fixed compact subset of $\mathbb{R}$. For any non-negative integer $p$, $\mathcal{C}_K^p$ is the linear space of all complex-valued functions $\varphi$ on $\mathbb{R}$ which have continuous derivatives at least up to order $p$ and which have support contained in $K$. We now denote by $\mathcal{D}_{p,K}$ the space $\mathcal{C}_K^p$ equipped with the norm

$$\|\varphi\|_{[p,K]} = \max_{0 \leq k \leq p} \left\{ \sup_{t \in K} |\varphi^{(k)}(t)| \right\}. \tag{1.34}$$

(Note that with $t$ restricted to $K$ the quantity on the right-hand side of (1.34) defines a norm on $\mathcal{C}_K^p$ but only a semi-norm on $\mathcal{C}_0^p$.) A sequence $(\varphi_n)_{n \in \mathbb{N}}$ converges to zero in this space if and only if, for $k = 0, 1, 2, \ldots, p$ each of the sequences $(\varphi_n^{(k)})_{n \in \mathbb{N}}$ of the $k$th derivatives of the $\varphi_n$ converges uniformly to zero. In the terminology suggested in §1.1.2 this is $p$-uniform convergence. Each such space $\mathcal{D}_{p,K}$ is complete with respect to this mode of convergence, that is with respect to the corresponding norm $\| \cdot \|_{[p,K]}$, and is therefore a Banach space.

If we allow $p$ to tend to infinity then we obtain the linear space $\mathcal{C}_K^\infty$ of all complex-valued functions which are infinitely differentiable and which vanish outside $K$. This is the intersection of all spaces $\mathcal{C}_K^p$ for $p = 0, 1, 2, \ldots$, and is closed under the mode of convergence described in §1.1.2 as $\omega$-uniform convergence. However there is no corresponding norm which we can define on $\mathcal{C}_K^\infty$ which will give rise to this mode of convergence. Instead we must turn to the concept of a **countably normed space** introduced in §1.2.4 above. We denote by

$$\mathcal{D}_K \equiv \bigcap_{p=0}^{\infty} \mathcal{D}_{p,K}$$

the countably normed space generated by the family of norms $(\| \cdot \|_{[p,K]})_{p \in \mathbb{N}_0}$. A sequence $(\varphi_n)_{n \in \mathbb{N}}$ contained in $\mathcal{D}_K$ then converges to zero in the sense of the topology of the countably normed space if and only if each sequence $(\varphi_n^{(k)})_{n \in \mathbb{N}}$ for $k = 0, 1, 2, \ldots$, converges uniformly to zero (i.e. the sequence converges to zero $\omega$-uniformly). From Theorem 1.6 it follows that $\mathcal{D}_K$ is complete and therefore is a Fréchet space. If we now write

$$K_n = \{ x \in \mathbb{R} : |x| \leq n \}$$

for $n = 1, 2, 3, \ldots$, then $(K_n)_{n \in \mathbb{N}}$ is an increasing sequence of compacts whose union is $\mathbb{R}$. Accordingly we can define the fundamental space of test functions, $\mathcal{D}(\mathbb{R})$, as the strict inductive limit,

$$\mathcal{D}(\mathbb{R}) = \bigcup_{n=1}^{\infty} \mathcal{D}_{K_n}. \tag{1.35}$$

The members of $\mathcal{D}(\mathbb{R})$ are of course the infinitely smooth functions on $\mathbb{R}$ with compact support. The mode of convergence corresponding to the inductive limit topology on $\mathcal{D}(\mathbb{R})$ is such that a sequence $(\varphi_n)_{n\in\mathbb{N}}$ converges in $\mathcal{D}$ to zero if all the $\varphi_n$ vanish outside the same compact set $K \sqsubset \mathbb{R}$, and each sequence $(\varphi_n^{(k)})_{n\in\mathbb{N}}$ converges uniformly to zero for $k = 0, 1, 2, \ldots$.

The generalisation to $\mathcal{D}(\mathbb{R}^m)$ is straightforward. The norms $\|\cdot\|_{[p,K]}$ are defined as

$$\|\varphi\|_{[p,K]} = \max_{0 \le |k| \le p} \left\{ \sup_{t \in K} |D^k \varphi(t)| \right\}$$

where $k$ is a multi-index of dimension $m$; a sequence $(\varphi_n)_{n\in\mathbb{N}}$ converges to zero in $\mathcal{D}(\mathbb{R}^m)$ if and only if all the $\varphi_n$ have support contained in the same fixed compact subset $K \sqsubset \mathbb{R}^m$ and each sequence $(D^k \varphi_n)_{n\in\mathbb{N}}$ converges uniformly to zero. With this topology $\mathcal{D}(\mathbb{R}^m)$ is a Fréchet space.

**Definition 1.3** *A distribution on $\mathbb{R}^m$ is a linear functional $\mu$ on $\mathcal{D}(\mathbb{R}^m)$ which is continuous with respect to the topology on $\mathcal{D}(\mathbb{R}^m)$. The linear space of all distributions on $\mathbb{R}^m$ is denoted by $\mathcal{D}'(\mathbb{R}^m)$.*

Spaces of test functions other than $\mathcal{D}(\mathbb{R})$ or $\mathcal{D}(\mathbb{R}^m)$ are of importance. In general let $\Omega$ be the basic set on which all functions of interest are defined; usually $\Omega$ will be $\mathbb{R}^m$ or else some open subset of $\mathbb{R}^m$. A test function space will be a linear space $\Phi$ of complex-valued functions defined on $\Omega$ which are such that

**TFS1** $\Phi$ is either a complete countably normed space or is a countable union space of complete countably normed spaces,

**TFS2** if a sequence $(\varphi_n)_{n\in\mathbb{N}}$ converges to zero in the space $\Phi$ then $(\varphi_n)_{n\in\mathbb{N}}$ converges pointwise to zero in $\Omega$.

Then any continuous linear functional on $\Phi$ may be called a (linear) **generalised function** on $\Omega$. The collection of all continuous linear functionals on $\Phi$ is called the **dual** of $\Phi$ and is usually denoted by $\Phi'$. Equality, addition and scalar multiplication are defined in $\Phi'$ just as they are defined for distributions in section §1.3.

Now let $\Phi$ and $\Psi$ be two countably normed spaces (or countable union spaces) and suppose that $\Psi$ is a linear subspace of $\Phi$. Assume that the convergence concept for $\Psi$ is *stronger* than the one for $\Phi$ in the sense that the convergence in $\Psi$ of a sequence $(\varphi_n)_{n\in\mathbb{N}}$ to some limit $\varphi$ implies its convergence in $\Phi$ to the same limit.

If $\mu$ is a functional in $\Phi'$ its restriction to $\Psi$ is a unique functional $\mu_{|\Psi}$ for which $<\mu_{|\Psi},\varphi> = <\mu,\varphi>$ whenever $\varphi \in \Psi$. Then clearly $\mu_{|\Psi}$ is linear and continuous on $\Psi$ and therefore belongs to $\Psi'$. However we cannot conclude that $\Phi'$ is a subspace of $\Psi'$ because two different members of $\Phi'$ may have the same restriction to $\Psi$. Nevertheless, if we add the hypothesis that $\Psi$ is a dense subspace of $\Phi$ then more can be said as the following theorem shows.

**Theorem 1.9** *Let $\Phi$ and $\Psi \subset \Phi$ be two countably normed spaces (or two countable union spaces) such that $\Psi$ is a dense subspace of $\Phi$. If convergence in $\Psi$ is stronger than that in $\Phi$ in the sense described above, then $\Phi'$ is a subspace of $\Psi'$.*

**Proof:** Since $\Psi \subset \Phi$, it is clearly impossible for two different members of $\Psi'$ to correspond to the same member of $\Phi'$. Thus we need merely to prove the following: if $\mu \in \Phi'$, then the values of $<\mu,\varphi>$ as $\varphi$ traverses $\Psi$ uniquely determine the values of $<\mu,\varphi>$ as $\varphi$ traverses $\Phi$. This will show that no two different members of $\Phi'$ can have the same restriction on $\Psi$. Because $\Psi$ is dense in $\Phi$, we can find for each $\varphi \in \Phi$ a sequence $(\varphi_n)_{n\in\mathbb{N}}$ of elements of $\Psi$ which converges in $\Phi$ to $\varphi$. Thus for any $\mu \in \Phi'$, $<\mu,\varphi_n> \to <\mu,\varphi>$ as $n \to \infty$; hence, $<\mu,\varphi>$ is uniquely determined by the sequence $(<\mu,\varphi_n>)_{n\in\mathbb{N}}$. $\square$

Thus, continuous linear functionals on certain superspaces of $\mathcal{D}$ turn out to be distributions of various special kinds.

**Exercise 1.7 :**

1. For each $p \in \mathbb{N}_0$ define

$$\mathcal{D}_p = \bigcup_{n=1}^{\infty} \mathcal{D}_{p,K_n},$$

where $K_n = \{x \in \mathbb{R} : |x| \leq n\}$. Show that $\mathcal{D}$ is a dense subspace of each $\mathcal{D}_p$ (with respect to the topology of $\mathcal{D}_p$).

2. For each $p \in \mathbb{N}_0$ we have $\mathcal{D}'_p \subset \mathcal{D}'$. Show that $\mu \in \mathcal{D}'$ is a distribution of finite order if and only if $\mu \in \mathcal{D}'_p$ for some $p \in \mathbb{N}_0$. Then

$$\mathcal{D}'_{\text{fin}} = \bigcup_{p=0}^{\infty} \mathcal{D}'_p \subset \mathcal{D}'$$

is the space of all finite-order distributions, and for each $p \in \mathbb{N}_0$, $\mathcal{D}'_p$ is the linear space of $\mathcal{D}'_{\text{fin}}$ comprising all distributions of order $\leq p$.

## 1.3 CALCULUS OF DISTRIBUTIONS

The following definitions are perfectly general but for simplicity we work throughout in terms of distributions defined on $\mathbb{R}$, unless explicitly stated to the contrary.

### 1.3.1 Operations on distributions

Linear operations on distributions are defined as for functionals in general. Thus:

**Addition and scalar multiplication**  Given two distributions $\mu, \nu$ and any scalar $a \in \mathbb{C}$, define the distributions $(\mu + \nu)$ and $a\mu$ by

$$<\mu+\nu,\varphi> \;=\; <\mu,\varphi>+<\nu,\varphi>$$
$$<a\mu,\varphi> \;=\; a<\mu,\varphi>$$

for any test function $\varphi \in \mathcal{D}$.

**Equality**  The sense in which two distributions can be said to be equal has already been discussed in §1.1.6. Briefly we say that two distributions $\mu_1, \mu_2$ are equal if and only if

$$<\mu_1,\varphi> \;=\; <\mu_2,\varphi>$$

for every test function $\varphi \in \mathcal{D}$. In particular for regular distributions $\mu_{(h)}$ and $\mu_{(g)}$, defined by locally integrable functions $h$ and $g$ respectively, we will have $\mu_{(h)} = \mu_{(g)}$ if and only if $h(x) = g(x)$ almost everywhere.

**Null distribution**  The null, or zero, distribution **0** is defined to be that distribution for which we have $<\mathbf{0},\varphi> = 0$ for every function $\varphi \in \mathcal{D}$. It behaves as an additive unit since for any distribution $\mu$ we have

$$\mathbf{0}+\mu = \mu+\mathbf{0} \;;\; 0\mu = \mathbf{0}.$$

**Support of a distribution**  The support of a distribution $\mu$ is now defined as the complement of the largest open set on which $\mu$ is zero. Thus the support of a distribution is always a closed set. Every point of the support of $\mu$ is called an **essential** point for $\mu$; if the support of $\mu$ is contained in a set $A$ then $\mu$ is often said to be **concentrated** on $A$. For example the delta distribution $\delta_a$ is concentrated on the set $\{a\}$. If $\mu_{(h)}$ is a regular distribution generated by a continuous function $h$ then the support of the distribution $\mu_{(h)}$ coincides with the support of the function $h$. However note that a point $x$ may be an essential point for $\mu_{(h)}$ even though $h(x) = 0$ (for example if $h(x) = x$ then the support of $\mu_{(h)}$ is the whole of $\mathbb{R}$ although $h(0) = 0$).

**Translate** For $a \in \mathbb{R}$ the translation operator $\tau_a$ is defined for ordinary functions by $\tau_a \varphi(x) = \varphi(x-a)$. For a regular distribution $\mu_{(h)}$ this gives

$$\begin{aligned}
<\mu_{(h)}, \tau_a \varphi> &= \int_{-\infty}^{+\infty} \varphi(x-a) h(x)\, dx \\
&= \int_{-\infty}^{+\infty} \varphi(y) h(y+a)\, dy = <\mu_{(g)}, \varphi>
\end{aligned}$$

where $g = \tau_{-a} h$. Accordingly we define the translate $\tau_a \mu$ of a general distribution $\mu$ by

$$<\tau_a \mu, \varphi> = <\mu, \tau_{-a} \varphi> \tag{1.36}$$

for any $\varphi \in \mathcal{D}$.

### 1.3.2 Differentiation of distributions

If $\mu$ is any distribution then its derivative is the distribution $D\mu$ defined by

$$<D\mu, \varphi> = <\mu, -\varphi'> \tag{1.37}$$

for all $\varphi \in \mathcal{D}$, where $\varphi'$ denotes the ordinary classical derivative of the (infinitely smooth) function $\varphi$. It is immediate that every distribution is infinitely differentiable in this sense. This definition of the distributional derivative is chosen so as to agree with that of the ordinary derivative of a smooth function. Thus if $\mu_{(h)}$ is a regular distribution defined by a continuously differentiable function $h$ then

$$\begin{aligned}
<D\mu_{(h)}, \varphi> &\equiv <\mu_{(h)}, -\varphi'> = -\int_{-\infty}^{+\infty} \varphi'(x) h(x) dx \\
&= +\int_{-\infty}^{+\infty} \varphi(x) h'(x) dx \equiv <\mu_{(h')}, \varphi>
\end{aligned}$$

for any test function $\varphi \in \mathcal{D}$. As already discussed informally in §1.1.1, we have

$$\begin{aligned}
<DH, \varphi> &= <H, -\varphi'> \\
&= -\int_0^{+\infty} \varphi'(x) dx = \varphi(0) = <\delta, \varphi>
\end{aligned}$$

so that $\delta$ is the distributional derivative of the Heaviside distribution $H$. Similarly the distribution $\delta'$, which maps $\varphi \in \mathcal{D}$ into the number $-\varphi'(0)$, is seen to be the (first) distributional derivative of $\delta$ itself. For we have

$$<D\delta, \varphi> = <\delta, -\varphi'> = -\varphi'(x)|_{x=0} = -\varphi'(0) \equiv <\delta', \varphi>$$

and in general the $p$th derivative, $\delta^{(p)}$, of the delta distribution is given by the mapping

$$\varphi \rightsquigarrow <\delta^{(p)}, \varphi> = (-1)^p \varphi^{(p)}(0) \tag{1.38}$$

where $\varphi$ is any test function in $\mathcal{D}$.
It is clear that differentiation is a linear operation for distributions:

$$D(\mu + \nu) = D\mu + D\nu \quad ; \quad D(a\mu) = aD\mu . \tag{1.39}$$

**Primitive of a distribution** A primitive of a continuous function $h$ is by definition any (continuously differentiable) function $g$ such that $g'(x) = h(x)$ for all $x$. By analogy we would wish to define a primitive of a given distribution $\mu$ to be a distribution $\nu$ such that $D\nu = \mu$. In that case $\nu$ should satisfy

$$<\nu, -\varphi'> \equiv <\mu, \varphi> \tag{1.40}$$

for all $\varphi \in \mathcal{D}$. Every distribution $\mu$ has a primitive in this sense; in fact $\mu$ will have infinitely many primitives, any two of which will differ by a constant distribution. However we cannot use the right-hand side of (1.40) as a definition of $\nu$ as a distribution since it only defines a linear functional on a certain subspace of $\mathcal{D}$, viz. on the subspace $\partial \mathcal{D}$ of all functions which are the derivatives of test functions in $\mathcal{D}$. It is necessary to show that this functional $\nu \equiv D^{-1}\mu$ can be extended to the whole space $\mathcal{D}$ as a linear functional. First note that if $\varphi_0$ is some fixed function in $\mathcal{D}$ such that

$$\int_{-\infty}^{+\infty} \varphi_0(x) dx = 1 \tag{1.41}$$

then every function $\varphi \in \mathcal{D}$ can be expressed in the form

$$\varphi(x) = k\varphi_0(x) + \psi(x) \tag{1.42}$$

where $k = \int_{-\infty}^{+\infty} \varphi(x) dx$ and $\psi \in \partial \mathcal{D}$. In view of (1.41) the function $\varphi_0$ itself cannot belong to $\partial \mathcal{D}$, and so $<D^{-1}\mu, \varphi_0>$ is not defined by (1.40). If we assign an arbitrary value $c$ to $<D^{-1}\mu, \varphi_0>$ then we can extend $\nu = D^{-1}\mu$ to the whole of $\mathcal{D}$ by defining it as that functional given by

$$\begin{aligned} <\nu, \varphi> &= kc + <D^{-1}\mu, \psi> \\ &= <\mu_{(c)}, \varphi> - <\mu, \theta> \end{aligned} \tag{1.43}$$

where $\theta$ is the function in $\mathcal{D}$ such that $\theta' = \psi = \varphi - k\varphi_0$ and $\mu_{(c)}$ is the constant distribution given by

$$<\mu_{(c)}, \varphi> = c \int_{-\infty}^{+\infty} \varphi(x) \, dx . \tag{1.44}$$

We leave it to the reader to confirm that $\nu \equiv D^{-1}\mu$ is linear and continuous on $\mathcal{D}$ and is therefore a distribution. Thus, a primitive of $\delta$ will be any distribution of the form $H + \mu_{(c)}$, where $H$ is the Heaviside distribution. We usually denote a constant distribution such as $\mu_{(c)}$ by the same symbol $c$ used to denote the number $c \in \mathbb{C}$ and the constant function which generates $\mu_{(c)}$. Similarly from now on we often adopt the common practice of using the same symbol $h$ to denote both the locally integrable function and the regular distribution $\mu_{(h)}$ generated by it.

**Exercise 1.8 :**

1. Show that if $f \in \mathcal{C}(\mathbb{R})$ then its support is identical with that of the regular distribution $\mu_{(f)}$ defined by
$$<\mu_{(f)},\varphi> = \int_{-\infty}^{+\infty} \varphi(x)f(x)dx$$
for all $\varphi \in \mathcal{D}$. Is this generally true for $f \in L^1_{\text{loc}}(\mathbb{R})$?

2. For all $\varphi \in \mathcal{D}$ set
$$<\mu,\varphi> = \int_{-\infty}^{+\infty}\int_{-\infty}^{+\infty} \exp(-x^2 - y^2)\varphi(\sin(xy))dxdy.$$
Show that $\mu$ is a distribution and determine its support.

3. Show that the function $\log(|x|)$ is locally integrable and therefore defines a regular distribution, and that the distributional derivative of this is the singular distribution defined by the principal value $\text{Pv}\{x^{-1}\}$ as given in example 1.1.3.

4. Let $h \in L^1_{\text{loc}}(\mathbb{R})$ and let $a, b \in \mathbb{R}, a \neq 0$. Show that for any $\varphi$ in $\mathcal{D}$ we have
$$<\mu_{(h)}(ax), \varphi(x)> \; = \; <\mu_{(h)}(x), |a|^{-1}\varphi(x/a)>.$$
Use this result to determine the distributions defined by the expressions $\delta(ax + b)$ and $\delta'(ax + b)$.

5. Formulate appropriate definitions of odd and even distributions, and prove that the derivative of an even distribution is an odd distribution.

### 1.3.3 Structure theorems for distributions

Every function $f$ which is continuous on $\mathbb{R}$ is locally integrable and therefore generates a regular distribution $\mu_{(f)}$. In such a case the identification of the function $f$ with the distribution $\mu_{(f)}$ is a particularly harmless abuse of notation since no other *continuous* function will generate the same distribution $\mu_{(f)}$. Hence we may consider $\mathcal{C}(\mathbb{R})$ as a linear subspace of $\mathcal{D}'$. Each member $f$ of $\mathcal{C}(\mathbb{R})$ is differentiable to any order in the distributional sense and, for each $k \in \mathbb{N}_0$, $D^k f$ represents a (generally singular) distribution. For example, the function

$$x_+(x) = xH_0(x) = \begin{cases} x & \text{for } x > 0 \\ 0 & \text{for } x \geq 0 \end{cases} \quad (1.45)$$

belongs to $C(\mathbb{R})$ and defines a regular distribution given by

$$<x_+, \varphi> = \int_0^{+\infty} x\varphi(x)dx$$

for all $\varphi \in \mathcal{D}$. The second derivative of this distribution is the delta function. There is an important converse in the fact that *every* distribution $\mu$ in $\mathcal{D}'$ can be expressed, at least locally, as a finite order derivative of a continuous function. We consider first the case when $\mu$ is a distribution of order 0.

**Theorem 1.10** *Let $\mu$ be a distribution of order 0 and let $K = [a,b]$ be a compact subset of $\mathbb{R}$. Then there exists a continuous function $h$ (which vanishes identically outside $[a,b]$) such that for any function $\varphi \in \mathcal{D}_K$*

$$<\mu, \varphi> = \int_a^b h(t)\varphi''(t)dt \equiv <D^2 h, \varphi>.$$

**Proof:** Since $\mu$ is a distribution of order 0 there exists a constant $M = M(K)$ such that for all functions $\varphi$ belonging to $\mathcal{D}_K$ we have

$$|<\mu,\varphi>| \leq M \cdot \sup_{x \in K} |\varphi(x)| \equiv M \cdot \|\varphi\|_{[0,K]}.$$

We can extend $\mu$ as a linear functional on the space $C_K^0$ of continuous functions with support contained in $K$ such that $\mu$ is continuous with respect to the norm $\|\cdot\|_{[0,K]}$. Following Zemanian [117] we can define a function $u(x,t)$ on $\mathbb{R}^2$ as follows:

$$u(x,t) = \begin{cases} (t-b)(x-a)/(b-a) & \text{for } a \leq x \leq t \leq b, \\ (x-b)(t-a)/(b-a) & \text{for } a \leq t \leq x \leq b, \\ 0 & \text{otherwise.} \end{cases} \quad (1.46)$$

Then $u(x,t)$ belongs to $C_K^0$ for each fixed $t$ and

$$h(t) = <\mu(x), u(x,t)> \quad (1.47)$$

is well defined as a continuous function of $t$ which vanishes identically outside $[a,b]$. Further, given any function $\varphi \in \mathcal{D}_K$ we have

$$\varphi(x) = \int_a^b u(x,t)\varphi''(t)dt \quad (1.48)$$

**Sec. 1.3]**                                                                      **Calculus of Distributions**

as can be shown by breaking the interval of integration into $a \leq t \leq x$ and $x \leq t \leq b$ and then integrating by parts twice in both of the resulting integrals.

Given $\varepsilon > 0$ we can choose a partition $\{t_i\}_{1 \leq i \leq n}$ of $[a,b]$ such that

$$|\varphi(x) - \sum_{i=1}^n u(x,\tau_i)\varphi''(\tau_i)(t_i - t_{i-1})| \leq \varepsilon(b-a)$$

for any choice of the points $\tau_i$ in the sub-intervals $[t_{i-1}, t_i]$. Hence, from (1.47), it follows that

$$|<\mu,\varphi> - \sum_{i=1}^n h(\tau_i)\varphi''(\tau_i)(t_i - t_{i-1})| \leq |<\mu(x),\varphi(x) - \sum_{i=1}^n u(x,\tau_i)\varphi''(\tau_i)(t_i - t_{i-1})>|$$

$$\leq M \cdot \sup_{x \in K} |\varphi(x) - \sum_{i=1}^n u(x,\tau_i)\varphi''(\tau_i)(t_i - t_{i-1})| \leq M\varepsilon(b-a)$$

and since, for an appropriate choice of partition, we also have

$$|\int_a^b h(t)\varphi''(t)dt - \sum_{i=1}^n h(\tau_i)\varphi''(\tau_i)(t_i - t_{i-1})| \leq \varepsilon(b-a)$$

the result follows.                                                                                                                      □

Using this theorem we can now prove the general result:

**Theorem 1.11 (Structure Theorem)** *Every distribution $\mu \in \mathcal{D}'(\mathbb{R})$ is locally a finite order derivative of a continuous function.*

**Proof:** Let $\mu$ be a distribution which is defined at least over a neighbourhood of some fixed compact $K \subset \mathbb{R}$; without loss of generality we may take $K$ to be the finite closed interval $[a,b]$. From the local boundedness property of distributions it follows that there must exist some $p \in \mathbb{N}_0$ such that for all $\varphi \in \mathcal{D}_K$ we will have

$$|<\mu,\varphi>| \leq M \cdot \sup_{x \in K} |\varphi^{(p)}(x)| \qquad (1.49)$$

where $M$ depends on $[a,b]$ but is independent of $\varphi$. Let $\partial \mathcal{D}_K$ denote the space of all derivatives of functions in $\mathcal{D}_K$, and let $D^{-1}\mu$ be some particular primitive of $\mu$. Recalling the definition of $D^{-1}\mu$ as given in (1.43) and using the fact that, for any $\varphi \in \mathcal{D}_K$,

$$\begin{aligned}
|k| &= |\int_a^b \varphi(x)dx| \\
&\leq (b-a)\left\{\sup_{a \leq x \leq b} |\varphi(x)|\right\} = (b-a)\left\{\sup_{a \leq x \leq b} \left|\int_a^x \varphi'(t)dt\right|\right\} \\
&\leq (b-a)^2 \left\{\sup_{a \leq x \leq b} |\varphi'(x)|\right\} \leq \ldots \leq (b-a)^p \left\{\sup_{a \leq x \leq b} |\varphi^{(p-1)}(x)|\right\}
\end{aligned}$$

we have

$$\begin{aligned}
|<D^{-1}\mu,\varphi>| &= |<D^{-1}\mu, k\varphi_0+\psi>| = |ck+<D^{-1}\mu,\psi>| \\
&\leq |ck| + M\cdot \sup_{a\leq x\leq b}|\psi^{(p-1)}(x)| \\
&\leq |ck| + M\cdot \sup_{a\leq x\leq b}|\varphi^{(p-1)}(x)-k\varphi_0^{(p-1)}(x)| \\
&\leq |c|(b-a)^p\cdot \sup_{a\leq x\leq b}|\varphi^{(p-1)}(x)| + M\sup_{a\leq x\leq b}|\varphi^{(p-1)}(x)| \\
&\quad + M(b-a)^p\cdot \sup_{a\leq x\leq b}|\varphi^{(p-1)}(x)|\cdot \sup_{a\leq x\leq b}|\varphi_0^{(p-1)}(x)| \\
&= M_1\cdot \sup_{a\leq x\leq b}|\varphi^{(p-1)}(x)|
\end{aligned} \qquad (1.50)$$

where $M_1$ is independent of $\varphi$. If (1.49) holds for $p=1$ then (1.50) also holds for $p=1$ and from Theorem 1.10 it follows that

$$<\mu,\varphi> = <D^{-1}\mu,-\varphi'> = <h'',-\varphi'> = <h''',\varphi> \qquad (1.51)$$

where $h$ is a continuous function. The general result now follows by induction. □

Thus for an arbitrary distribution $\mu \in \mathcal{D}'$ we can say that for each compact subset $K$ of $\mathbb{R}$ there exists a non-negative integer $p$ and a continuous function $f$ such that, for every $\varphi \in \mathcal{D}_K$ we have

$$\begin{aligned}
<\mu,\varphi> &= <D^p f,\varphi> \\
&= (-1)^p<f,\varphi^{(p)}> = (-1)^p\int_{-\infty}^{+\infty}f(x)\varphi^{(p)}(x)dx.
\end{aligned}$$

Suppose, in particular, that $\mu$ has compact support $K=[a,b]$. Let $I_\varepsilon$ denote the closed interval $[a-\varepsilon, b+\varepsilon]$, and let $I$ be any finite open interval containing $I_\varepsilon$. If $\lambda$ is a function in $\mathcal{D}$ such that $\lambda(x)=1$ on $I_\varepsilon$ and $\lambda(x)=0$ outside $I$ then for any $\varphi \in \mathcal{D}$ we have

$$<\mu,\varphi> \equiv <\mu,\lambda\varphi> = <D^p f,\lambda\varphi> = (-1)^p\int_{-\infty}^{+\infty}f(x)D^p\{\lambda(x)\varphi(x)\}dx.$$

By the Leibniz differentiation formula for products we can write

$$D^p\{\lambda\varphi\} = \sum_{q=0}^{p}\binom{p}{q}D^{p-q}\lambda D^q\varphi$$

and it follows that

$$\begin{aligned}
<\mu,\varphi> &= <f,(-1)^p\sum_{q=0}^{p}\binom{p}{q}D^{p-q}\lambda D^q\varphi> \\
&= <\sum_{q=0}^{p}D^q\{(-1)^{p+q}\binom{p}{q}fD^{p-q}\lambda\},\varphi>.
\end{aligned}$$

This means that we can set $\mu \equiv \sum_{q=0}^{p} D^q f_q$, where for each $q = 0, 1, \ldots, p$ the function $f_q(x) \equiv (-1)^{p+q} \binom{p}{q} f(x) D^{p-q} \lambda(x)$ is continuous and has support in $I$. Thus we have the following corollary to theorem 1.11:

**Corollary 1.3.1** *If $\mu \in \mathcal{D}'$ has compact support $K$ then $\mu$ can be expressed (non-uniquely) as the sum of finitely many finite-order derivatives of continuous functions, each with support contained in a neighbourhood $I$ of $K$.*

## 1.4 THE J.S.SILVA THEORY OF DISTRIBUTIONS

### 1.4.1 Distributions of finite differential order

Let $I$ denote an arbitrary interval of $\mathbb{R}$ (open or closed, but in any case with more than one point), and let $\{I_\alpha\}_{\alpha \in A}$ denote the family of all compact subsets of $I$. Then for every distribution $\mu \in \mathcal{D}'(I)$ and any fixed $\alpha \in A$ there exists, according to theorem 1.11, a continuous function $f_\alpha$ and a non-negative integer $r_\alpha$ such that

$$<\mu, \varphi> = <D^{r_\alpha} f_\alpha, \varphi> \qquad (1.52)$$

for all $\varphi \in \mathcal{D}_{I_\alpha}$. This fact can be expressed by saying that the restriction of the distribution $\mu \in \mathcal{D}'(I)$ to $\mathcal{D}_{I_\alpha}$ is of **differential order** $\leq r_\alpha$. The number $r_\alpha \in \mathbb{N}_0$ is by no means unique; in fact if we denote by $\Im_a f_\alpha$ the function defined by

$$\Im_a f_\alpha(x) = \int_a^x f_\alpha(t) dt \quad, \quad x \in I_\alpha$$

where $a$ is a fixed point in $I_\alpha$, then for all $\varphi \in \mathcal{D}_{I_\alpha}$,

$$<\mu, \varphi> = <D^{r_\alpha} f_\alpha, \varphi> = <D^{r_\alpha+1} \Im_a f_\alpha, \varphi>$$

where $\Im_a f_\alpha$ is again a continuous function. From now on we take $r_\alpha$ to be the smallest possible number for which (1.52) holds.
Let $I_{\alpha_1}$ and $I_{\alpha_2}$ be two compact subsets of $I$ such that $I_{\alpha_1} \subset I_{\alpha_2}$; then $r_{\alpha_1} \leq r_{\alpha_2}$. If $I$ is itself compact then $I$ belongs to the family $\{I_\alpha\}_{\alpha \in A}$ and moreover, for all $\alpha \in A$, we have $I_\alpha \subset I$. Consequently the family $\{r_\alpha\}_{\alpha \in A}$ has a maximum (corresponding to the compact $I$). We say then that $\mu$ is a distribution of finite differential order and that its (differential) order is $r \equiv \max\{r_\alpha : \alpha \in A\}$.
Suppose on the other hand that $I$ is not compact. Then the family $\{r_\alpha\}_{\alpha \in A}$ may or may not be bounded above, depending on the particular distribution $\mu \in \mathcal{D}'(I)$ concerned. If there exists some $r \in \mathbb{N}_0$ such that $r_\alpha \leq r$ for all $\alpha \in A$ then $\mu \in \mathcal{D}'(I)$

is said to be a distribution of finite differential order $\leq r$. If no such $r$ exists then $\mu$ is said to be of infinite differential order.

Note that the concept of differential order defined here differs from the Schwartz order of a distribution given earlier in §1.1.3 (cf. Definition 1.2). We have seen, for example, that the delta function is a linear functional on $\mathcal{D}$ which is bounded with respect to the uniform norm $\|\varphi\|_{[0,K]} = \sup_{x \in K} |\varphi(x)|$ and is therefore a distribution of Schwartz order 0. Since it is the second order derivative of the continuous function $x_+$, however, it is a distribution of differential order 2. This distinction between Schwartz order and differential order can sometimes cause confusion, particularly as it is common practice to refer to "the order" of a distribution without any further qualification. On the other hand it remains true that a distribution is of finite differential order if and only if it is of finite order in the sense of Schwartz. Therefore, without any ambiguity, we can continue to use the notation $\mathcal{D}'_{\text{fin}}(\mathbb{R})$ to denote the space of all distributions of finite order on $\mathbb{R}$. We are often interested mainly in knowing whether or not a distribution is of finite or of infinite order, and are less concerned with the value of the order (in either the differential or the Schwartz sense). Thus it is of immediate interest that $\sum_{n=-\infty}^{+\infty} \delta(t-n)$ is of finite order (whether we understand this in the Schwartz or the differential sense) and that $\sigma(x) \equiv \sum_{n=0}^{\infty} n \cdot \delta^{(n)}(x-n)$, defined in §1.1.4, is of infinite order (again in either sense).

### 1.4.2 The J. S. Silva axioms

Theorem 1.11 and the corollary 1.3.1 stated above suggest an alternative definition of distributions to the Schwartz theory already described. This was developed in the 1950's by the Portuguese mathematician José Sebastião e Silva. J. S. Silva's idea was to define distributions from the outset as formal derivatives, $D^r f$, of continuous functions establishing beforehand a compatible system of axioms which these idealised objects should satisfy. These axioms should obviously embody the general abstract properties of Schwartz distributions so that the latter are seen to constitute one of the possible models of the axiom system itself. The principle of developing a theory as the logical consequence of a (preferably) small set of axioms always has a strong intrinsic appeal from the mathematical point of view. In addition it often allows a much clearer understanding of precisely what has been achieved by the theory in question.

The basic fact that any distribution is, locally, a finite order derivative of a continuous function allowed J. S. Silva to construct one particular model following closely the same process as that used to obtain the rational numbers from the integers. This has the advantage of a development of a strictly algebraic character without the need for the underlying structure of topological vector spaces which the Schwartz approach properly demands. The J. S. Silva axiomatic definition of a general distribution includes as an axiom the partition of unity given in §1.1.6 under Theorem 1.6. For the sake of simplicity we follow the example of the treatment given in one of J. S. Silva's original papers [102] and restrict attention here to the axiomatic definition of distributions of finite order on $\mathbb{R}$. The treatment of distributions of arbitrary order from an axiomatic point of view is given in J. S. Silva's earlier paper [101].

Let $\mathcal{C}(\mathbb{R})$ be the linear space of all complex-valued continuous functions defined on $\mathbb{R}$ and, as usual, denote by $\mathcal{C}^1(\mathbb{R})$ the subspace of $\mathcal{C}(\mathbb{R})$ constituted by all continuously differentiable functions on $\mathbb{R}$. Taking as primitive notions those of *continuous function* and *derivative*, distributions on $\mathbb{R}$ may therefore be characterised as elements of a linear space $E$ for which two linear maps are defined

$$\iota : \mathcal{C}(\mathbb{R}) \to E \quad , \quad D : E \to E$$

so that

**Axiom 1** *$\iota$ is injective, that is every function in $\mathcal{C}(\mathbb{R})$ is a distribution on $\mathbb{R}$.*

**Axiom 2** *To every distribution $\mu$ on $\mathbb{R}$ there corresponds a distribution $D\mu$, called the derivative of $\mu$, such that if $\mu = \iota(f)$ with $f \in \mathcal{C}^1(\mathbb{R})$ then $D\mu = \iota(f')$.*

**Axiom 3** *If $\mu$ is a distribution on $\mathbb{R}$ then there exists a continuous function $f \in \mathcal{C}(\mathbb{R})$ and a natural number $r \in \mathbb{N}_0$ such that we have $\mu = D^r \iota(f)$.*

**Axiom 4** *Given two functions $f, g \in \mathcal{C}(\mathbb{R})$, and a natural number $r$, the equality $D^r \iota(f) = D^r \iota(g)$ holds if and only if $f - g$ is a polynomial function of degree $< r$.*

Axiom 2 extends the usual concept of derivative of a function differentiable on $\mathbb{R}$ to the whole space of all continuous functions on $\mathbb{R}$. This means that in some sense every continuous function on $\mathbb{R}$ becomes infinitely differentiable on $\mathbb{R}$. For every $r \in \mathbb{N}_0$ the iterated indefinite integral operator $\Im^r$ transforms $\mathcal{C}(\mathbb{R})$ into itself; on the other hand the iterated differential operator $D^r$, introduced by the above axiom system, transforms $\mathcal{C}(\mathbb{R})$ into a superset. The members of this superset are called **(finite differential order) distributions** on $\mathbb{R}$.

In view of the injective map $\iota$ we may write $\mathcal{C}(\mathbb{R}) \subset E$ instead of (the more rigorous statement) $\iota(\mathcal{C}(\mathbb{R})) \subset E$. Hence we freely write $f \in E$ for any $f \in \mathcal{C}(\mathbb{R})$.

### 1.4.3 Consistency of the J. S. Silva axioms

To show that the given system of axioms is consistent J. S. Silva constructed a model himself. Although it may be of comparatively little use for applications of distribution theory it is of interest because of its intrinsic simplicity and wholly algebraic character. Consider the Cartesian product $\mathbb{N}_0 \times \mathcal{C}(\mathbb{R})$ and introduce the relation $\sim$ defined by

$$\forall (r,f), (s,g) \in \mathbb{N}_0 \times \mathcal{C}(\mathbb{R}) : [(r,f) \sim (s,g) \Leftrightarrow$$
$$\exists a \in \mathbb{R}, m \in \mathbb{N}_0 : [(m \geq r,s) \Rightarrow (\Im_a^{m-r} f - \Im_a^{m-s} g \in \Pi_m)]\,],$$

where $\Pi_m \subset \mathcal{C}(\mathbb{R})$ denotes the set of all complex-valued polynomial functions of degree $< m$ (for consistency we take $\Pi_0 = \{0\}$), and $\Im_a^k$ denotes the $k$th iterated indefinite integral operator with origin at the point $a \in \mathbb{R}$. Then $\sim$ is clearly an equivalence relation. The quotient set comprising all equivalence classes

$$[r,f] \equiv \{(s,g) \in \mathbb{N}_0 \times \mathcal{C}(\mathbb{R}) : (s,g) \sim (r,f)\}$$

will be denoted by

$$\mathcal{C}_\infty(\mathbb{R}) \equiv \mathbb{N}_0 \times \mathcal{C}(\mathbb{R})/\sim .$$

We now show that the set $\mathcal{C}_\infty(\mathbb{R})$, equipped with a suitable derivation operator, satisfies the J. S. Silva axiomatic system. The map

$$\iota : \mathcal{C}(\mathbb{R}) \to \mathcal{C}_\infty(\mathbb{R})$$

defined by $\iota(f) = [0, f]$ is clearly injective since we have that $\iota(f) = \iota(g)$ implies that $(0, f) \sim (0, g)$, which means that $f - g = 0$. Thus Axiom 1 is satisfied. Now define the mapping

$$D : \mathcal{C}_\infty(\mathbb{R}) \to \mathcal{C}_\infty(\mathbb{R})$$

by setting

$$D[r, f] = [r+1, f], \ \forall r \in \mathbb{N}_0, \ \forall f \in \mathcal{C}(\mathbb{R}).$$

This operation does not depend on the pair $(r, f)$ chosen to represent the equivalence class $[r, f]$. If $[r, f] = [s, g]$ then $(r, f) \sim (s, g)$ and therefore there exists $a \in \mathbb{R}$ and $m \in \mathbb{N}_0$ (with $m \geq r, s$) such that

$$\Im_a^{m-r} f - \Im_a^{m-s} g \in \Pi_m.$$

Since this is equivalent to

$$\Im_a^{(m+1)-(r+1)} f - \Im_a^{(m+1)-(s+1)} g \in \Pi_m \subset \Pi_{m+1}$$

it follows that $[r+1, f] = [s+1, g]$, which shows that

$$D[r, f] = D[s, g]$$

as asserted. The class $[r+1, f]$ is called the derivative of the class $[r, f]$ and $D$ is the derivative operator defined on $\mathcal{C}_\infty(\mathbb{R})$. If $f$ is a differentiable function on $\mathbb{R}$ then since $f - \Im_a f'$ is a constant polynomial we have

$$D[0, f] = [1, f] = [0, f'] = \iota(f')$$

and so Axiom 2 holds.

Next, for every $f \in \mathcal{C}(\mathbb{R})$, we make the identification $\iota(f) = f$. Then, for any distribution $[r, f] \in \mathcal{C}_\infty(\mathbb{R})$, we obtain

$$[r, f] = D[r-1, f] = D^2[r-2, f] = \ldots = D^r[0, f] = D^r f,$$

which shows that Axiom 3 holds.

Finally if $D^r f = D^r g$ then $(r, f) \sim (r, g)$ which means that $f - g$ belongs to $\Pi_r$ and so Axiom 4 holds.

Thus the set $\mathcal{C}_\infty(\mathbb{R})$ with the derivation operator $D$ defined as above is a model for the Silva set of axioms. Moreover it is possible to establish the result:

**Theorem 1.12** *Every model for the J.S.Silva axiomatic system for finite order distributions on $\mathbb{R}$ is isomorphic to $\mathcal{C}_\infty(\mathbb{R})$.*

### 1.4.4 Schwartz distributions of finite order

We show now that $\mathcal{D}'_{\text{fin}}(\mathbb{R})$ also is a model for the J.S.Silva axioms given above. Let $f$ be a function in $\mathcal{C}(\mathbb{R})$. For any $p \in \mathbb{N}_0$ the functional $\mu_{(f)} : \mathcal{C}_0^p(\mathbb{R}) \to \mathbb{C}$ defined by

$$<\mu_{(f)}, \varphi> = \int_{-\infty}^{+\infty} f(t)\varphi(t)\, dt$$

is linear and continuous, and hence defines a distribution in $\mathcal{D}'_{\text{fin}}(\mathbb{R})$. Moreover the correspondence $\iota : \mathcal{C}(\mathbb{R}) \to \mathcal{D}'_{\text{fin}}(\mathbb{R})$ which associates each function $f$ with the distribution $\mu_{(f)}$ is clearly injective. Therefore we may write $\mathcal{C}(\mathbb{R}) \subset \mathcal{D}'_{\text{fin}}(\mathbb{R})$, and so Axiom 1 is satisfied.

If $\mu \in \mathcal{D}'_{\text{fin}}(\mathbb{R})$ then $\mu$ is a continuous linear functional in $\mathcal{D}'_p$ for some $p \in \mathbb{N}_0$. Hence the functional $D\mu$, mapping $\mathcal{C}_0^{p+1}(\mathbb{R})$ into $\mathbb{C}$, defined by

$$<D\mu, \varphi> = <\mu, -\varphi'>$$

is a continuous linear functional in $\mathcal{D}'_{p+1}$, and so it must belong to $\mathcal{D}'_{\text{fin}}(\mathbb{R})$. Furthermore, if $\mu = \mu_{(f)}$ is generated by a continuously differentiable function $f$ on $\mathbb{R}$, then we have

$$<D\mu_{(f)}, \varphi> = <\mu_{(f)}, -\varphi'> = -\int_{-\infty}^{+\infty} f(t)\varphi'(t)\, dt$$
$$= +\int_{-\infty}^{+\infty} f'(t)\varphi(t)\, dt = <\mu_{(f')}, \varphi>$$

that is, $D\mu_{(f)}$ may be identified with $f'$, the ordinary classical derivative of $f$. This shows that $\mathcal{D}'_{\text{fin}}(\mathbb{R})$ satisfies Axiom 2.

Axiom 3 is obviously satisfied, from Corollary 1.3.1, and so we need only to show that $\mathcal{D}'_{\text{fin}}(\mathbb{R})$ satisfies Axiom 4. Let $\mu$ be any distribution in $\mathcal{D}'_{\text{fin}}(\mathbb{R})$. Then $\mu = D^r f$ for some continuous function $f$ on $\mathbb{R}$ and for some non-negative integer $r$. Hence $\mu$ has a primitive, say $P\mu = D^{r-1}f$, which is also a distribution. (Note that if $r = 0$ then $D^{-1}f$ is the usual primitive of a function on $\mathbb{R}$.) Moreover there are infinitely many such primitives of $\mu$, any two of which differ by a constant distribution. Hence, using induction over $r \in \mathbb{N}$, it is easy to see that if $\mu, \nu \in \mathcal{D}'_{\text{fin}}(\mathbb{R})$ are such that $\mu \equiv D^r f, \nu \equiv D^r g$ for some $f, g \in \mathcal{C}(\mathbb{C})$ and $r \in \mathbb{N}$ then $\mu = \nu$ implies $f - g \in \Pi_r$. That is, Axiom 4 holds.

## 1.5 GENERALISED POWERS AND PSEUDO-FUNCTIONS

### 1.5.1 Distributions of the form $x_+^\lambda, x_-^\lambda$

In many elementary applications a relatively small class of generalised functions is sufficient. This consists of the regular distributions (those which are generated by, and which may be identified with, locally integrable functions), delta functions, derivatives of delta functions and linear combinations of these. This may give the impression that the delta function and its derivatives are the only singular distributions which really matter. However there are many examples of singular distributions other than these which are of immediate practical interest. The definition of distributions corresponding to arbitrary powers of $x$ is a case in point.

For any real value of $\lambda > -1$ the function $|x|^\lambda$ is locally integrable on $\mathbb{R}$, and we can define functions $x_+^\lambda, x_-^\lambda$ by the equations

$$x_+^\lambda = |x|^\lambda H_0(x) = \begin{cases} |x|^\lambda & \text{for } x > 0 \\ 0 & \text{for } x \leq 0 \end{cases} \quad (1.53)$$

$$x_-^\lambda = |x|^\lambda H_0(-x) = \begin{cases} 0 & \text{for } x \geq 0 \\ |x|^\lambda & \text{for } x < 0 \end{cases} \quad (1.54)$$

These are themselves locally integrable in $\mathbb{R}$ and generate regular distributions $x_+^\lambda, x_-^\lambda$ according to

$$<x_+^\lambda, \varphi(x)> = \int_{-\infty}^{+\infty} x_+^\lambda(x)\varphi(x)dx \equiv \int_0^{+\infty} x^\lambda \varphi(x)\, dx \quad (1.55)$$

$$<x_-^\lambda, \varphi(x)> = \int_{-\infty}^{+\infty} x_-^\lambda(x)\varphi(x)dx \equiv \int_0^{+\infty} x^\lambda \varphi(-x)dx \quad (1.56)$$

for any $\varphi \in \mathcal{D}$. It follows readily that

$$x_+^\lambda = \frac{1}{\lambda+1}\left\{\frac{d}{dx}x_+^{\lambda+1}\right\} \; ; \; x_-^\lambda = -\frac{1}{\lambda+1}\left\{\frac{d}{dx}x_-^{\lambda+1}\right\} \quad (1.57)$$

both in the classical sense (that is, in the sense of the formula 1.5 of integration by parts) and as distributions. However, for $\lambda \leq -1$ the integrals in (1.55) and (1.56) will generally diverge for arbitrary $\varphi \in \mathcal{D}$ and we cannot use them to define distributions. Instead, for $\lambda < -1$ and $\lambda \neq -2, -3, \ldots$, we define the (singular) distributions $x_+^\lambda, x_-^\lambda$ inductively using (1.57).

We can obtain explicit formulas for the action of such distributions on an arbitrary test function $\varphi$. First, let $\lambda = -1 - \alpha$ where $0 < \alpha < 1$. Then for all $x \neq 0$ the ordinary function $x^{-\alpha-1}H_0(x)$ is well defined as the classical derivative of the locally integrable function $-\frac{1}{\alpha}x^{-\alpha}H_0(x)$. Consider instead the singular distribution $x_+^{-1-\alpha}$ which is, by definition, the distributional derivative of $-\frac{1}{\alpha}x_+^{-\alpha}$:

$$<x_+^{-1-\alpha}, \varphi> \equiv <D\left\{-\frac{1}{\alpha}x_+^{-\alpha}\right\}, \varphi> = <-\frac{1}{\alpha}x_+^{-\alpha}, -\varphi'>$$

$$= \lim_{\varepsilon \downarrow 0}\int_\varepsilon^{+\infty}\left\{-\frac{1}{\alpha}x^{-\alpha}\right\}\{-\varphi'(x)\}dx$$

$$= \lim_{\varepsilon \downarrow 0}\int_\varepsilon^{+\infty}\frac{\varphi'(x)}{\alpha x^\alpha}dx \quad (1.58)$$

Integrating by parts, and using the fact that $\varphi$ has compact support, we have

$$\int_\varepsilon^{+\infty}\frac{\varphi'(x)}{\alpha x^\alpha}dx = \int_\varepsilon^{+\infty}\frac{\varphi(x)}{x^{\alpha+1}}dx - \frac{\varphi(\varepsilon)}{\alpha\varepsilon^\alpha}$$

$$\equiv \int_\varepsilon^{+\infty}\frac{\varphi(x) - \varphi(0)}{x^{\alpha+1}}dx + \frac{\varphi(0) - \varphi(\varepsilon)}{\alpha\varepsilon^\alpha}. \quad (1.59)$$

Now $\varphi(x) - \varphi(0) = x\varphi'(\theta x)$, where $0 < \theta < 1$. This shows on the one hand that the integral on the right-hand side of (1.59) converges absolutely as $\varepsilon \downarrow 0$ (since $\alpha > 0$), and on the other that

$$\frac{\varphi(\varepsilon) - \varphi(0)}{\alpha \varepsilon^{\alpha}} = \varepsilon^{1-\alpha} \frac{\varphi'(\theta \varepsilon)}{\alpha}$$

which tends to 0 with $\varepsilon$ (since $\alpha < 1$). Hence we have

$$\begin{aligned} <x_+^{-1-\alpha}, \varphi> &= \int_0^{+\infty} \frac{\varphi(x) - \varphi(0)}{x^{\alpha+1}} dx \\ &\equiv \int_{-\infty}^{+\infty} [\varphi(x) - \varphi(0)] \left\{ \frac{H_0(x)}{x^{\alpha+1}} \right\} dx. \end{aligned} \quad (1.60)$$

Conventionally this can be written as

$$<x_+^{-1-\alpha}, \varphi> \equiv <\mathrm{Pf}\left\{\frac{H_0(x)}{x^{\alpha+1}}\right\}, \varphi> = \int_{-\infty}^{+\infty} \varphi(x) \mathrm{Pf}\left\{\frac{H_0(x)}{x^{\alpha+1}}\right\} dx \quad (1.61)$$

where the notation $\mathrm{Pf}\{H_0(x)/x^{\alpha+1}\}$ indicates that the (symbolic) integral of (1.61) is to be evaluated according to the modified form given in (1.60). We refer to $\mathrm{Pf}\{H_0(x)/x^{\alpha+1}\}$ as a **pseudo-function** to distinguish it from the ordinary function $H_0(x)/x^{\alpha+1}$.

A similar analysis for the case $\lambda = -2 - \alpha$, where $0 < \alpha < 1$, shows that the action of the singular distribution $x_+^{-2-\alpha}$ on an arbitrary test function $\varphi$ is given by

$$\begin{aligned} <x_+^{-2-\alpha}, \varphi> &\equiv <D\left\{-\frac{1}{\alpha+1}x_+^{-1-\alpha}\right\}, \varphi> \\ &= \int_0^{+\infty} \frac{\varphi(x) - \varphi(0) - x\varphi'(0)}{x^{\alpha+2}} dx \\ &\equiv \int_{-\infty}^{+\infty} [\varphi(x) - \varphi(0) - x\varphi'(0)] \left\{\frac{H_0(x)}{x^{\alpha+2}}\right\} dx. \end{aligned}$$

Once again, it is convenient to write this in the conventional form

$$<x_+^{-2-\alpha}, \varphi> \equiv <\mathrm{Pf}\left\{\frac{H_0(x)}{x^{\alpha+2}}\right\}, \varphi> = \int_{-\infty}^{+\infty} \varphi(x) \mathrm{Pf}\left\{\frac{H_0(x)}{x^{\alpha+2}}\right\} dx$$

where the notation $\mathrm{Pf}\{H_0(x)/x^{\alpha+2}\}$ distinguishes the distribution in question from the ordinary function $H_0(x)/x^{\alpha+2}$.

In general, for $-n - 1 < \lambda < -n$, it can be shown that

$$<x_+^{\lambda}, \varphi> = \int_0^{+\infty} x^{\lambda} \left\{\varphi(x) - \sum_{j=0}^{n-1} \frac{\varphi^{(j)}(0)}{j!} x^j\right\} dx.$$

Similarly, for $-n-1 < \lambda < -n$ we have

$$< x_-^\lambda, \varphi > = \int_0^{+\infty} x^\lambda \left\{ \varphi(-x) - \sum_{j=0}^{n-1} \frac{\varphi^{(j)}(0)}{j!} (-x)^j \right\} dx.$$

The singular distributions $x_+^\lambda$ and $x_-^\lambda$ coincide with the ordinary functions $H_0(x)|x|^\lambda$ and $H_0(-x)|x|^\lambda$ respectively on $\mathbb{R}\backslash\{0\}$, but need to be distinguished from them. This is done by using the pseudo-function notation and writing $x_+^\lambda \equiv \text{Pf}\{H_0(x)|x|^\lambda\}$ and $x_-^\lambda \equiv \text{Pf}\{H_0(-x)|x|^\lambda\}$.

It remains to discuss the definition of distributions of the form $x_+^{-n}, x_-^{-n}$, where $n = 1, 2, \ldots$. First we define the locally integrable function, and regular distribution,

$$\log(x_+) = \begin{cases} \log|x| & \text{, for } x > 0 \\ 0 & \text{, for } x \leq 0 \end{cases} \tag{1.62}$$

and then set

$$x_+^{-1} = D\{\log(x_+)\}, \tag{1.63}$$

with, for $n = 2, 3, \ldots$,

$$x_+^{-n} = \frac{1}{1-n} D x_+^{1-n}. \tag{1.64}$$

Then,

$$\begin{aligned}< x_+^{-1}, \varphi > &= \; < D\log(x_+), \varphi > \\ &= \; < \log(x_+), -\varphi' > = \lim_{\varepsilon \downarrow 0} \int_\varepsilon^{+\infty} \{-\varphi'(x)\} \log(x) dx \end{aligned}$$

and an integration by parts shows that

$$\begin{aligned}\int_\varepsilon^{+\infty} \{-\varphi'(x)\} \log(x) dx &= \varphi(\varepsilon) \log(\varepsilon) + \left\{ \int_\varepsilon^1 \frac{\varphi(x)}{x} dx + \int_1^{+\infty} \frac{\varphi(x)}{x} dx \right\} \\ &= [\varphi(\varepsilon) - \varphi(0)] \log(\varepsilon) + \int_\varepsilon^1 \frac{\varphi(x) - \varphi(0)}{x} dx + \int_1^{+\infty} \frac{\varphi(x)}{x} dx.\end{aligned}$$

Now $\{\varphi(\varepsilon) - \varphi(0)\}\log(\varepsilon) = \{\varepsilon \log(\varepsilon)\}\varphi'(\theta\varepsilon)$ for some $\theta \in (0,1)$, and $\varphi'(x)$ is bounded. Hence $\{\varphi(\varepsilon) - \varphi(0)\} \log(\varepsilon)$ tends to 0 with $\varepsilon$ and so

$$\begin{aligned}< x_+^{-1}, \varphi > &= \int_0^1 \frac{\varphi(x) - \varphi(0)}{x} dx + \int_1^{+\infty} \frac{\varphi(x)}{x} dx \\ &\equiv \int_0^{+\infty} \frac{\varphi(x) - \varphi(0) H_0(1-x)}{x} dx.\end{aligned}$$

For $n = 2$ we have

$$<x_+^{-2}, \varphi> \;=\; <-Dx_+^{-1}, \varphi> \;=\; <x_+^{-1}, \varphi'>$$
$$=\; \lim_{\varepsilon \downarrow 0} \int_\varepsilon^1 \frac{\varphi'(x) - \varphi'(0)}{x}\,dx + \int_1^{+\infty} \frac{\varphi'(x)}{x}\,dx.$$

Integration by parts shows that

$$\int_\varepsilon^1 \frac{\varphi'(x) - \varphi'(0)}{x}\,dx \;=\; \varphi(1) - \frac{\varphi(\varepsilon)}{\varepsilon} + \int_\varepsilon^1 \frac{\varphi(x) - x\varphi'(0)}{x^2}\,dx$$
$$\equiv \varphi(1) - \varphi(0) - \frac{\varphi(\varepsilon) - \varphi(0)}{\varepsilon} + \int_\varepsilon^1 \frac{\varphi(x) - \varphi(0) - x\varphi'(0)}{x^2}\,dx$$

and

$$\int_1^{+\infty} \frac{\varphi'(x)}{x}\,dx \;=\; -\varphi(1) + \int_1^{+\infty} \frac{\varphi(x)}{x^2}\,dx$$
$$\equiv -\varphi(1) + \varphi(0) + \int_1^{+\infty} \frac{\varphi(x) - \varphi(0)}{x^2}\,dx.$$

Hence, it follows that

$$<x_+^{-2}, \varphi> \;=\; \int_0^1 \frac{\varphi(x) - \varphi(0) - x\varphi'(0)}{x^2}\,dx + \int_1^{+\infty} \frac{\varphi(x) - \varphi(0)}{x^2}\,dx - \varphi'(0).$$

In general, for $n = 2, 3, \ldots$, we can derive the following result:

$$<x_+^{-n}, \varphi> \;=\; \int_0^{+\infty} x^{-n}\left\{\varphi(x) - \sum_{j=0}^{n-2} \frac{\varphi^{(j)}(0)}{j!}x^j - \frac{\varphi^{(n-1)}(0)}{(n-1)!}x^{n-1}H_0(1-x)\right\}dx$$
$$- \frac{\varphi^{(n-1)}(0)}{(n-1)!}\left\{\sum_{r=1}^{n-1} r^{-1}\right\}. \tag{1.65}$$

In the same way we can define distributions of the form $x_-^{-n}$ and obtain explicit formulas for their action on test functions. Thus we first define $x_-^{-1}$ by

$$<x_-^{-1}, \varphi> \;=\; <-D\log(x_-), \varphi> \;=\; <\log(x_-), \varphi'(x)>$$

where we write

$$\log(x_-) = \begin{cases} 0 & \text{, for } x \geq 0 \\ \log|x| & \text{, for } x < 0 \end{cases} \tag{1.66}$$

Then for $n = 2, 3, \ldots$, we define $x_-^{-n}$ inductively from the equation

$$x_-^{-n} \;=\; \frac{1}{n-1}Dx_-^{1-n}. \tag{1.67}$$

These definitions give

$$<x_-^{-1},\varphi> = \int_0^{+\infty} \frac{\varphi(-x) - \varphi(0)H_0(1-x)}{x}\,dx,$$

and in general for $n = 2, 3, \ldots$,

$$<x_-^{-n},\varphi> = \int_0^{+\infty} x^{-n}\left\{\varphi(-x) - \sum_{j=0}^{n-2}(-1)^j\frac{\varphi^{(j)}(0)}{j!}x^j\right.$$

$$\left. - (-1)^{n-1}\frac{\varphi^{(n-1)}(0)}{(n-1)!}x^{n-1}H_0(1-x)\right\}dx - (-1)^{n-1}\frac{\varphi^{(n-1)}(0)}{(n-1)!}\left\{\sum_{r=1}^{n-1}r^{-1}\right\}.$$

**Exercise 1.9 :**

1. Confirm directly that if $x_+^{-3}$ is defined as $-\frac{1}{2}Dx_+^{-2}$ then for any $\varphi \in \mathcal{D}$ we have

$$<x_+^{-3},\varphi> = \int_0^{+\infty} x^{-3}\left[\varphi(x) - \varphi(0) - x\varphi'(0) - \frac{x^2}{2}\varphi''(0)H_0(1-x)\right]dx - \frac{3}{4}\varphi''(0)$$

2. Give an inductive proof of the general formula (1.65) for $<x_+^{-n},\varphi>, n = 2, 3, \ldots$.

### 1.5.2 Pseudo-functions and Hadamard finite parts

The definitions given in §1.5.1 of the distributions $x_+^\lambda, x_-^\lambda$ are the most straightforward in the sense that the simple differential relationships

$$Dx_+^{\lambda+1} = +(\lambda+1)x_+^\lambda \;;\; Dx_-^{\lambda+1} = -(\lambda+1)x_-^\lambda \qquad (1.68)$$

are preserved throughout, even when $\lambda$ takes negative integral values. They seem first to have been given in Fisher [21], while older texts give alternative definitions for generalised powers of $x$. This, together with the use of pseudo-function notation by some authors (e.g. Zemanian) and formal products $x^\lambda H(x)$ by others, can give rise to considerable confusion. We consider, in particular, the treatment given in Gel'fand and Shilov [26]. Let $\lambda$ be a complex parameter such that $\Re(\lambda) > -n-1$ and $\lambda \neq -1, -2, \ldots, -n$. Then for any $\varphi \in \mathcal{D}$ and any $\varepsilon > 0$ we have

$$\int_\varepsilon^{+\infty} x^\lambda \varphi(x)dx = \int_\varepsilon^1 x^\lambda \left\{\varphi(x) - \sum_{k=1}^n \frac{\varphi^{(k-1)}(0)}{(k-1)!}x^{k-1}\right\}dx$$

$$+ \int_1^{+\infty} x^\lambda \varphi(x)dx + \sum_{k=1}^n \frac{\varphi^{(k-1)}(0)}{(k-1)!(\lambda+k)}(1-\varepsilon^{\lambda+k}) \quad (1.69)$$

If we allow $\varepsilon$ to tend to zero then for values of $\lambda$ such that $\Re(\lambda) < -1$ the classical integral $\int_0^{+\infty} x^\lambda \varphi(x)dx$ will generally diverge since the right-hand side of (1.69) contains negative powers of $\varepsilon$. We could obtain the finite value $< x_+^\lambda, \varphi >$ by simply agreeing to discard from the right-hand side of (1.69) just those terms in $\varepsilon$ which become infinite as $\varepsilon \downarrow 0$. This technique of neglecting appropriately defined infinite quantities was devised by J. Hadamard [28] and the resulting finite value extracted from the divergent integral is usually referred to as the **Hadamard Finite Part**. It is common practice to use it in conjunction with the pseudo-function notation introduced in §1.5.1, and to write:

$$\text{Fp} \int_0^{+\infty} x^\lambda \varphi(x) dx = \lim_{\varepsilon \downarrow 0} \left\{ \int_\varepsilon^{+\infty} x^\lambda \varphi(x) dx + \sum_{k=1}^n \frac{\varphi^{(k-1)}(0)}{(k-1)!(\lambda+k)} \varepsilon^{\lambda+k} \right\}$$

$$= \int_{-\infty}^{+\infty} \varphi(x) \text{Pf}\{H_0(x)|x|^\lambda\} dx = < x_+^\lambda, \varphi > \quad (1.70)$$

for $\Re(\lambda) > -n-1, \lambda \neq -1, -2, \ldots, -n$. For $-n-1 < \Re(\lambda) < -n+1$ Gel'fand and Shilov define the distribution $F_{-n}(x_+, \lambda)$ by

$$< F_{-n}(x_+, \lambda), \varphi > = \int_0^{+\infty} x^\lambda \left\{ \varphi(x) - \sum_{k=1}^{n-1} \frac{\varphi^{(k-1)}(0)}{(k-1)!} x^{k-1} \right.$$

$$\left. - \frac{\varphi^{(n-1)}(0)}{(n-1)!} x^{n-1} H_0(1-x) \right\} dx. \quad (1.71)$$

For each fixed $\varphi \in \mathcal{D}$ the expression $< F_{-n}(x_+, \lambda), \varphi >$ is an analytic function of $\lambda$ in $|\Re(\lambda) + n| < 1$. In particular it is well defined when $\lambda = -n$, and it is this which Gel'fand and Shilov define as the value of the expression $< x_+^{-n}, \varphi >$. However, we shall retain the definition of $x_+^{-n}$ already given in §1.5.1, and we use the pseudo-function notation $< \text{Pf}\{H_0(x)|x|^{-n}\}, \varphi(x) >$ to denote $< F_{-n}(x_+, -n), \varphi >$. Thus, while for $\lambda \neq -1, -2, \ldots$, the pseudo-function $\text{Pf}\{H_0(x)|x|^\lambda\}$ is taken to be identical with the distribution $x_+^\lambda$, for negative integer values of $\lambda$ we have the following relations:

$$x_+^{-1} = D \log(x_+) = \text{Pf}\{H_0(x)/x\}$$
$$x_+^{-2} = -Dx_+^{-1} = \text{Pf}\{H_0(x)/x^2\} + \delta'(x)$$
$$x_+^{-3} = -\tfrac{1}{2}Dx_+^{-2} = \text{Pf}\{H_0(x)/x^3\} - \tfrac{1}{2}\{1 + \tfrac{1}{2!}\}\delta''(x)$$

and in general for $n = 2, 3, \ldots$, we have

$$x_+^{-n} = \frac{1}{-n+1} Dx_+^{-n+1} = \text{Pf}\{H_0(x)/x^n\} - \frac{(-1)^{n-1}}{(n-1)!} \left\{ \sum_{r=1}^{n-1} r^{-1} \right\} \delta^{(n-1)}(x).$$

As is easily confirmed the following differential relationship holds for pseudo-functions $\text{Pf}\{H_0(x)/x^n\}$ where $n = 1, 2, \ldots$,

$$D\text{Pf}\left\{\frac{H_0(x)}{x^n}\right\} = -n \, \text{Pf}\left\{\frac{H_0(x)}{x^{n+1}}\right\} + \frac{(-1)^n}{n!} \delta^{(n)}(x). \quad (1.72)$$

In the convention adopted by Gel'fand and Shilov the distribution $x_+^\lambda$ is identified with the pseudo-function $\mathrm{Pf}\{H_0(x)|x|^\lambda\}$ for all $\lambda$, including the negative integral values $\lambda = -1, -2, \ldots$. This is perfectly consistent and has the advantage that the set of formulas for the action of distributions $x_+^\lambda$ on arbitrary test functions $\varphi$ consists of relatively simple and coherent integral expressions. On the other hand it obliges us to use the differentiation formula of (1.72) in the negative integer case instead of the straightforward (1.68). Because of this we prefer here and in what follows to retain the distinction between the distribution $x_+^{-n}$ and the pseudo-function $\mathrm{Pf}\{H_0(x)|x|^{-n}\}$ as given above.

We can define the pseudo-functions $\mathrm{Pf}\{H_0(-x)|x|^{-n}\}$ by

$$< \mathrm{Pf}\left\{\frac{H_0(-x)}{|x|^n}\right\}, \varphi(x) > \; = \; < \mathrm{Pf}\left\{\frac{H_0(x)}{|x|^n}\right\}, \varphi(-x) > . \qquad (1.73)$$

This gives

$$x_-^{-1} = -D \, \log(x_-) = \mathrm{Pf}\left\{\frac{H_0(-x)}{|x|}\right\}$$

and for $n = 2, 3, \ldots$,

$$x_-^{-n} = \frac{1}{-n+1} Dx_-^{-n+1} = \mathrm{Pf}\left\{\frac{H_0(-x)}{|x|^n}\right\} - \frac{1}{(n-1)!}\left\{\sum_{r=1}^{n-1} r^{-1}\right\} \delta^{(n-1)}(x).$$

The pseudo-functions $\mathrm{Pf}\{H_0(-x)|x|^{-n}\}$ satisfy the following differential relationship:

$$D \, \mathrm{Pf}\left\{\frac{H_0(-x)}{|x|^n}\right\} = n \, \mathrm{Pf}\left\{\frac{H_0(-x)}{|x|^{n+1}}\right\} - \frac{1}{n!} \delta^{(n)}(x) \qquad (1.74)$$

where $n = 1, 2, \ldots$. Note that for $n = 1, 2, \ldots$, we can write without any ambiguity

$$x^{-n} = x_+^{-n} + (-1)^n x_-^{-n} \equiv \mathrm{Pf}\left\{\frac{H_0(x)}{|x|^n}\right\} + (-1)^n \, \mathrm{Pf}\left\{\frac{H_0(-x)}{|x|^n}\right\} \qquad (1.75)$$

and we have always $Dx^{-n} = -nx^{-n-1}$.

**Exercise 1.10 :**

1. Show that

$$< \mathrm{Pf}\left\{\frac{H_0(x)}{x^{3/2}}\right\}, \varphi(x) > \; = \; 2 \int_0^{+\infty} \frac{\varphi'(x)}{x^{1/2}} \, dx$$

for any $\varphi \in \mathcal{C}_0^1(\mathbb{R})$.

2. Express as an absolutely convergent integral the Hadamard finite part of the generally divergent integral

$$\int_{-\infty}^{+\infty} \varphi(x) x_+^{-4/3} dx,$$

where $\varphi$ is an arbitrary continuously differentiable function of compact support.

**3.** Show that $h(x) = H_0(x) \log(x)/x$ is not locally integrable and therefore does not define a regular distribution. Show that by extracting an appropriate finite part of the (generally divergent) integral

$$\int_0^{+\infty} \varphi(x) \frac{\log(x)}{x} dx$$

we can define a pseudo-function $\text{Pf}\{h(x)\}$ such that

$$< \text{Pf}\{h(x)\}, \varphi(x) > = \lim_{\varepsilon \downarrow 0} \left\{ \int_\varepsilon^{+\infty} \frac{\log(x)}{x} \varphi(x) dx + \frac{1}{2} \varphi(0) \log^2(\varepsilon) \right\}.$$

Confirm that this pseudo-function is a distribution.

### 1.5.3 Distributions of the form $(x \pm i0)^\lambda$

The function $(x + iy)^\lambda$ is defined as

$$(x + iy)^\lambda = \exp\{\lambda \text{Log}(x + iy)\} = \exp\{\lambda[\log |x + iy| + i\text{Arg}(x + iy)]\}$$

where we adopt the common convention of lower case letters to denote principal values of logarithm and argument, with upper case indicating the many-valued functions. If we choose the principal value $\arg(x + iy)$ then

$$(x + iy)^\lambda = \exp\{\lambda[\log |x + iy| + i \arg(x + iy)]\}$$

is a single-valued analytic function of the variable $z = x + iy$ in each of the half-planes $\{z \in \mathbb{C} : y > 0\}$ and $\{z \in \mathbb{C} : y < 0\}$. The following (ordinary) functions are defined for all complex $\lambda$:

$$\begin{aligned}
(x + i0)^\lambda &= \lim_{y \downarrow 0}(x + iy)^\lambda = \lim_{y \downarrow 0}\{|x + iy|^\lambda \exp[i\lambda \arg(x + iy)]\} \\
&= \lim_{y \downarrow 0}(x^2 + y^2)^{\lambda/2} \exp[i\lambda \arg(x + iy)] \\
&= \begin{cases} |x|^\lambda & \text{for } x > 0 \\ |x|^\lambda \exp(i\lambda\pi) & \text{for } x < 0 \end{cases}
\end{aligned}$$

$$\begin{aligned}
(x - i0)^\lambda &= \lim_{y \uparrow 0}(x + iy)^\lambda = \lim_{y \uparrow 0}\{|x + iy|^\lambda \exp[i\lambda \arg(x + iy)]\} \\
&= \lim_{y \uparrow 0}(x^2 + y^2)^{\lambda/2} \exp[i\lambda \arg(x + iy)] \\
&= \begin{cases} |x|^\lambda & \text{for } x > 0 \\ |x|^\lambda \exp(-i\lambda\pi) & \text{for } x < 0 \end{cases}
\end{aligned}$$

For $\Re(\lambda) > -1$ we have the identities

$$(x+i0)^\lambda \equiv x_+^\lambda + \exp(+i\lambda\pi)\, x_-^\lambda \tag{1.76}$$

$$(x-i0)^\lambda \equiv x_+^\lambda + \exp(-i\lambda\pi)\, x_-^\lambda \tag{1.77}$$

and we can use these, together with the definitions of the generalised functions $x_+^\lambda, x_-^\lambda$ for $\lambda \neq -1, -2, \ldots$, to define corresponding distributions $(x+i0)^\lambda, (x-i0)^\lambda$ for all values of $\lambda$ apart from the negative integers. To complete the definitions we need to allow $\lambda$ to tend to the value $-n$ and to take the appropriate limiting values of the distributions concerned. Suppose once again that $\Re(\lambda) > -n-1$ and that $\lambda \neq -1, -2, \ldots, -n$. If we impose the additional constraint $\Re(\lambda) < -n+1$ then we can express $<x_+^\lambda, \varphi>$ in the form

$$<x_+^\lambda, \varphi> = \int_0^1 x^\lambda \left\{ \varphi(x) - \sum_{k=1}^n \frac{\varphi^{(k-1)}(0)}{(k-1)!} x^{k-1} \right\} dx$$

$$+ \int_1^{+\infty} x^\lambda \left\{ \varphi(x) - \sum_{k=1}^{n-1} \frac{\varphi^{(k-1)}(0)}{(k-1)!} x^{k-1} \right\} dx + \frac{\varphi^{(n-1)}(0)}{(n-1)!(\lambda+n)} \tag{1.78}$$

The final term on the right-hand side of (1.78) is the only part of the expression which fails to converge as $\lambda$ tends to $-n$. The sum of the integrals, using the Gel'fand and Shilov notation introduced above, is $<F_{-n}(x_+, \lambda), \varphi>$ and is an analytic function of $\lambda$ in the strip $|\Re(\lambda)+n| < 1$. In fact, within this strip we may write

$$x_+^\lambda = \frac{(-1)^{n-1}\delta^{(n-1)}(x)}{(n-1)!(\lambda+n)} + F_{-n}(x_+, \lambda) \tag{1.79}$$

and, similarly,

$$x_-^\lambda = \frac{\delta^{(n-1)}(x)}{(n-1)!(\lambda+n)} + F_{-n}(x_-, \lambda). \tag{1.80}$$

As the detailed analysis given by Gel'fand and Shilov makes clear the representations (1.79) and (1.80) may be interpreted as Laurent expansions for the generalised functions $x_+^\lambda, x_-^\lambda$ respectively about the point $\lambda = -n$. In this sense $F_{-n}(x_+, \lambda)$ and $F_{-n}(x_-, \lambda)$ are the regular parts of these expressions and, in the notation adopted in the present text, they assume the values $\mathrm{Pf}\{H_0(x)/|x|^n\}$ and $\mathrm{Pf}\{H_0(-x)/|x|^n\}$ respectively at the point $\lambda = -n$ itself. Using the fact that

$$\exp(\pm i\lambda\pi) = (-1)^n \exp[\pm i\pi(\lambda+n)] = (-1)^n\{1 \pm i(\lambda+n)\pi + \cdots\}$$

we can substitute in the expressions (1.76) and (1.77) defining the generalised functions $(x+i0)^\lambda, (x-i0)^\lambda$ and allow $\lambda \to -n$ to get

$$(x \pm i0)^{-n} = \text{Pf}\left\{\frac{H_0(x)}{|x|^n}\right\} + (-1)^n \text{Pf}\left\{\frac{H_0(-x)}{|x|^n}\right\} \mp \frac{(-1)^{n-1} i\pi}{(n-1)!} \delta^{(n-1)}(x)$$

$$\equiv x_+^{-n} + (-1)^n x_-^{-n} \mp \frac{(-1)^{n-1} i\pi}{(n-1)!} \delta^{(n-1)}(x). \qquad (1.81)$$

We may write this in the equivalent form

$$(x+i0)^{-n} = x^{-n} - \frac{(-1)^{n-1}}{(n-1)!} i\pi \delta^{(n-1)}(x) \qquad (1.82)$$

$$(x-i0)^{-n} = x^{-n} + \frac{(-1)^{n-1}}{(n-1)!} i\pi \delta^{(n-1)}(x) \qquad (1.83)$$

using the definition of the generalised function $x^{-n}$ given in (1.75). It follows immediately from the definitions that we always have

$$D(x+i0)^\lambda = \lambda (x+i0)^{\lambda-1} \; ; \; D(x-i0)^\lambda = \lambda(x-i0)^{\lambda-1}.$$

**Remark 1.2** *An alternative definition of $(x+i0)^{-n}$ and $(x-i0)^{-n}$ can be given as follows. The locally integrable function $\log(x+i0)$ is defined as*

$$\log(x+i0) \equiv \lim_{y\downarrow 0} \log(x+iy) = \begin{cases} \log|x| + i\pi &, \text{ for } x<0 \\ \log|x| &, \text{ for } x>0. \end{cases}$$

*For an arbitrary $\varphi \in \mathcal{D}$ we have*

$$\int_{-\infty}^{+\infty} \varphi(x) \log(x+iy) dx = \int_{-\infty}^{+\infty} \varphi(x) \frac{1}{2} \log(x^2+y^2) dx + i \int_{-\infty}^{+\infty} \varphi(x) \arctan(y/x) dx$$

*for any fixed $y > 0$. As $y \to 0$ through positive values we have,*

(a) *$\frac{1}{2} \log(x^2+y^2)$ decreases monotonically to $\log|x|$, and*

(b) *$|\arctan(y/x)| \le \pi$ and $\arctan(y/x)$ converges to $\pi$ for $x<0$ and to $0$ for $x>0$.*

*Hence, taking the limit in the distributional sense (see §2.2), we may write*

$$\log(x+i0) = \lim_{y\downarrow 0} \log(x+iy) = \lim_{y\downarrow 0}\{\log|x+iy| + i\arg(x+iy)\}$$
$$= \log|x| + i\pi H(-x)$$
$$= \begin{cases} \log|x| + i\pi &, \text{ for } x<0 \\ \log|x| &, \text{ for } x>0 \end{cases}$$

and it follows that
$$D\log(x+i0) = D\log|x| + i\pi DH(-x)$$
$$= Pf\left\{\frac{1}{x}\right\} - i\pi\delta(x). \quad (1.84)$$

*Similarly*
$$\log(x-i0) = \lim_{y\downarrow 0}\log(x+iy) = \log|x| - i\pi H(-x)$$
*so that*
$$D\log(x-i0) = Pf\left\{\frac{1}{x}\right\} + i\pi\delta(x).$$

*Thus we have*
$$D\log(x+i0) = (x+i0)^{-1} \; ; \; D\log(x-i0) = (x-i0)^{-1}$$

*which could have been used as definitions of the generalised functions* $(x+i0)^{-1}$ *and* $(x-i0)^{-1}$. *The functions* $(x\pm i0)^{-n}$ *for* $n = 2, 3, \ldots$, *could then be defined by successive differentiation.*

### 1.5.4 Neutrices and neutrix limits

The definition of pseudo-functions, such as those discussed above in §1.5.2, by the technique of extracting appropriate finite parts is due, as the name suggests, to Hadamard. In fact Hadamard's method can be regarded as a particular application of the so-called **neutrix calculus** developed by J. van der Corput [13]. This is a very general principle for the discarding of unwanted infinite quantities from asymptotic expansions and has been widely exploited in the context of generalised functions, notably by Brian Fisher [22] [23] in connection with the problem of distributional multiplication. The essential idea of a neutrix limit can be summarised as follows.

Let $X$ be some non-empty set and let $G$ be an additive commutative group. We consider functions $f$ defined on $X$ and taking values in $G$; with the usual natural definition of addition of functions, $G^X$ itself forms an additive commutative group. The object of the neutrix calculus is to define a subgroup of $G^X$ which consists of those functions which (in an appropriate context) it is desirable to consider as **negligible**. This subgroup $\mathcal{N}$ is said to be a **neutrix** if the only constant function which it contains is the zero map of $X$ into $G$.

Suppose, in particular, that the domain $X$ is a subspace of a topological space $Y$, and suppose that $y$ is a limit point of $X$ which does not belong to $X$. Take for $G$ the real number system $\mathbb{R}$ or, more generally, the complex number system $\mathbb{C}$. Now let $\mathcal{N}$ be a commutative group of functions mapping $X$ into $G$ such that, if

$\mathcal{N}$ contains any function $g(x)$ for which we have $\lim_{x \to y} g(x) = c$ (a constant), then we must have $c = 0$. Then $\mathcal{N}$ is a neutrix, for if $g$ belongs to $\mathcal{N}$ and $g(x) \equiv c$ then it is surely true that $g(x) \to c$ as $x \to y$, and so $c = 0$.

If $f$ is defined on $X$ and if there exists some number $c$ such that the function $g(x) = f(x) - c$ is a member of $\mathcal{N}$ (briefly, if $f(x) - c$ is a negligible function) then we define the **neutrix limit** of $f(x)$ as $x$ tends to $y$ to be the number $c$:

$$\mathcal{N} - \lim_{x \to y} f(x) = c.$$

This limit, whenever it does exist, is certainly unique. For if $g_1(x) \equiv f(x) - c_1$ and $g_2(x) \equiv f(x) - c_2$ both belong to $\mathcal{N}$, then the constant function $(g_1 - g_2)(x) = c_2 - c_1$ also belongs to $\mathcal{N}$ and so $c_2 - c_1 = 0$. Moreover it is easy to see that the definition of neutrix limit agrees with that of the usual sense of $\lim_{x \to y} f(x)$ whenever the latter exists.

**Example 1.5.1** Let $X = \mathbb{N}$ and take $G$ to be the real number system $\mathbb{R}$. The functions belonging to $G^X$ are then simply real sequences, and the neutrix calculus can be used to generalise the concept of sequential convergence. Thus, for example, we could take $\mathcal{N}$ to be the set of all linear sums of the functions

$$n^\lambda \cdot \log^{r-1}(n) \; ; \; \log^r(n)$$

where $\lambda > 0$ and $r = 1, 2, \ldots$, together with all functions (sequences) which converge to 0 in the usual sense as $n \to \infty$.

Then the function $f(n) = n^3 + 2n^2 + 5 - 1/n$, for example, would have the neutrix limit 5. Again, if we take $f_1(n) = n^3 \log(n) + 7n^2$ then $f_1(n)$ is a negligible function in the sense of the neutrix. The function $f_2(n) = 6 + 1/n$ has the limit 6 in the usual sense as $n \to \infty$, and we can conclude that the sum $f_1(n) + f_2(n)$ has the neutrix limit 6.

**Example 1.5.2** Let $X = \mathbb{R}^+ \equiv \{x \in \mathbb{R} : 0 < x < +\infty\}$, $G = \mathbb{R}$, and $\mathcal{N}$ the set of all finite linear combinations of functions of the form

$$\varepsilon^\lambda \log^{r-1}(\varepsilon) \; ; \; \log^r(\varepsilon)$$

where $\lambda < 0, r = 1, 2, \ldots$, and $\varepsilon \in \mathbb{R}^+$, together with all functions which converge to zero as we approach the origin.

Then for $\lambda > -n - 1$ and $\lambda \neq -1, -2, \ldots, -n$, the identity (1.69)

$$\int_\varepsilon^{+\infty} x^\lambda \varphi(x) dx = \int_\varepsilon^1 x^\lambda \left\{ \varphi(x) - \sum_{k=1}^n \frac{\varphi^{(k-1)}(0)}{(k-1)!} x^{k-1} \right\} dx$$

$$+ \int_1^{+\infty} x^\lambda \varphi(x) dx + \sum_{k=1}^n \frac{\varphi^{(k-1)}(0)}{(k-1)!(\lambda+k)} \left(1 - \varepsilon^{\lambda+k}\right)$$

shows immediately that the pseudo-functions $\mathrm{Pf}\{H_0(x)|x|^\lambda\}$ admit straightforward definitions as neutrix limits:

$$< \mathrm{Pf}\{H_0(x)|x|^\lambda\}, \varphi(x) > \equiv < x_+^\lambda, \varphi > = \mathcal{N}-\lim_{\varepsilon \downarrow 0} \int_\varepsilon^{+\infty} x^\lambda \varphi(x) dx.$$

Moreover the definition as a neutrix limit continues to work when $\lambda$ takes (negative) integer values. Thus for $n = 1, 2, \ldots$, we have

$$\begin{aligned}
\int_\varepsilon^{+\infty} x^{-n} \varphi(x) dx &= \int_\varepsilon^{+\infty} x^{-n} \left\{ \varphi(x) - \sum_{k=0}^{n-2} \frac{\varphi^{(k)}(0)}{k!} x^k \right. \\
&\qquad \left. - \frac{\varphi^{(n-1)}(0)}{(n-1)!} x^{n-1} H_0(1-x) \right\} dx \\
&\quad + \sum_{k=0}^{n-2} \frac{\varphi^{(k)}(0)}{k!} \int_\varepsilon^{+\infty} x^{k-n} dx + \frac{\varphi^{(n-1)}(0)}{(n-1)!} \int_\varepsilon^1 x^{-1} dx \\
&\equiv \int_\varepsilon^{+\infty} x^{-n} \left\{ \varphi(x) - \sum_{k=0}^{n-2} \frac{\varphi^{(k)}(0)}{k!} x^k \right. \\
&\qquad \left. - \frac{\varphi^{(n-1)}(0)}{(n-1)!} x^{n-1} H_0(1-x) \right\} dx \\
&\quad + \sum_{k=0}^{n-2} \frac{\varphi^{(k)}(0)}{k!(n-k-1)} \varepsilon^{k-n+1} - \frac{\varphi^{(n-1)}(0)}{(n-1)!} \log(\varepsilon)
\end{aligned}$$

and it follows at once that

$$< \mathrm{Pf}\left\{ \frac{H_0(x)}{|x|^n} \right\}, \varphi(x) > = \mathcal{N}-\lim_{\varepsilon \downarrow 0} \int_\varepsilon^{+\infty} x^{-n} \varphi(x) dx.$$

# 2

# Further Properties of Distributions

## 2.1 CONVOLUTION AND DIRECT PRODUCTS

### 2.1.1 The convolution integral and its generalisation

If $f(x), g(x)$ are continuous functions with compact supports then they each define a regular distribution $\mu_{(f)}, \mu_{(g)}$. The classical convolution of $f$ and $g$ is defined as the integral

$$h(x) \equiv (f * g)(x) = \int_{-\infty}^{+\infty} f(y)g(x-y)dy \equiv \int_{-\infty}^{+\infty} f(x-y)g(y)dy \qquad (2.1)$$

and is again a continuous function with compact support. The function $h$ itself defines a regular distribution $\mu_{(h)}$ which we would clearly wish to call the convolution of the regular distributions $\mu_{(f)}$ and $\mu_{(g)}$,

$$<\mu_{(h)}, \varphi> \equiv <\mu_{(f*g)}, \varphi> = \int_{-\infty}^{+\infty} \varphi(x) \left\{ \int_{-\infty}^{+\infty} f(y)g(x-y)dy \right\} dx. \qquad (2.2)$$

We can easily extend this definition to apply to wider classes of regular distributions. For example we can discard the conditions that $f$ and $g$ should be continuous and have compact supports, and suppose merely that they both belong to $L^1(\mathbb{R})$. Then the function $h(x)$ is well defined for almost all $x \in \mathbb{R}$ and is itself a member of

$L^1(\mathbb{R})$. Provided we impose sufficiently strong conditions on one of the factors we can even extend the definition to include arbitrary singular distributions. In particular if $\varphi$ is a function in $\mathcal{D}$ and $\mu$ any distribution in $\mathcal{D}'$ then for each $x \in \mathbb{R}$ the expression

$$h(x) = <\mu(y), \varphi(x-y)> \qquad (2.3)$$

is well defined and is a natural generalisation of the convolution integral 2.1. Accordingly we define the convolution $\mu * \varphi$ to be the regular distribution generated by this function $h$. It is not difficult to show that this definition does make sense. In the first place recall that on any given neighbourhood of the support of $\varphi$ we can always represent $\mu$ as a finite-order derivative of a continuous function $f$. Hence, for any sufficiently large $a \in \mathbb{R}^+$ we can always find a continuous function $f$ and an integer $m \in \mathbb{N}_0$ such that, for all $x \in [-a, +a]$, we have

$$\begin{aligned} h(x) &= <\mu(y), \varphi(x-y)> = <D^m f(y), \varphi(x-y)> \\ &= (-1)^m <f(y), \frac{\partial^m}{\partial y^m} \varphi(x-y)> = <f(y), \varphi^{(m)}(x-y)> \\ &= \int_{-a}^{+a} f(y) \varphi^{(m)}(x-y) dy \equiv \left(f * \varphi^{(m)}\right)(x). \end{aligned}$$

We may differentiate under the integral sign as often as we wish, and so $h$ must be an infinitely differentiable function. In particular $h$ generates a regular distribution.

Although the definition of the convolution $\mu * \varphi$ represents a special case of particular importance it does not help us to develop a satisfactory broad definition which applies to a sufficiently large class of distributions. We cannot use the classical integral (2.1) as a model for a more general definition of convolution, involving possibly singular distributions for both factors, since the product of arbitrary distributions is presently undefined. We shall need to examine this deficiency in more detail at a later stage (see §2.4 at the end of the present chapter). However for the moment we return to the convolution integral (2.1) where the factors $f$ and $g$ are continuous functions of compact support, and consider its role as a (regular) distribution. For an arbitrary test function $\varphi \in \mathcal{D}$ we can carry out the following re-formulation of the action of $\mu_{(h)}$ on $\varphi$:

$$\begin{aligned} <\mu_{(h)}(x), \varphi(x)> &= \int_{-\infty}^{+\infty} \varphi(x) \left\{ \int_{-\infty}^{+\infty} f(y) g(x-y) dy \right\} dx \\ &= \int_{-\infty}^{+\infty} \int_{-\infty}^{+\infty} f(y) g(x-y) \varphi(x) dy dx \\ &= \int_{-\infty}^{+\infty} \int_{-\infty}^{+\infty} f(x) g(y) \varphi(x+y) dy dx \qquad (2.4) \end{aligned}$$

This suggests that a satisfactory definition of convolution might be obtained if we attempt to generalise (2.4) for arbitrary distributions in place of the functions (regular distributions) $f$ and $g$. The required generalisation involves what is usually called the **direct product** (or the **tensor product**) of distributions. Although we shall continue to restrict attention to one-dimensional distributions the formulation of the direct product does require us to work initially in terms of distributions on $\mathbb{R}^2$, and some notational conventions will be helpful. Following Zemanian [117] we shall often use the notation $\mathbb{R}^2_{x,y}$ to denote Euclidean space $\mathbb{R}^2$ comprising ordered pairs $(x,y)$ of reals. Similarly we shall sometimes find it convenient to distinguish between $\mathcal{D}(\mathbb{R}_y)$ as the space of test functions $\varphi(y)$ defined on the space $\mathbb{R}_y$ from the corresponding space of test functions $\varphi(x)$ defined on the space $\mathbb{R}_x$.

### 2.1.2 Direct products of distributions

Given a distribution $\mu(x)$ in $\mathcal{D}'(\mathbb{R}_x)$ and a distribution $\nu(y)$ in $\mathcal{D}'(\mathbb{R}_y)$ we can define a certain two-dimensional distribution (i.e. a member of $\mathcal{D}'(\mathbb{R}^2_{x,y})$) which we denote by $\mu(x) \otimes \nu(y)$:

Let $\varphi(x,y) \in \mathcal{D}(\mathbb{R}^2_{x,y})$; for each $x \in \mathbb{R}_x$ the function $\varphi(x,y)$ is infinitely differentiable and of compact support in $\mathbb{R}_y$ and hence $<\nu(y), \varphi(x,y)>$ is well defined for every $x$. Moreover this function of $x$ is certainly of compact support in $\mathbb{R}_x$, and

$$\begin{aligned}\frac{d}{dx}<\nu(y),\varphi(x,y)> &= \lim_{h\to 0}\left\{\frac{1}{h}[<\nu(y),\varphi(x+h,y)> - <\nu(y),\varphi(x,y)>]\right\} \\ &= \lim_{h\to 0}<\nu(y),[\varphi(x+h,y)-\varphi(x,y)]/h> \\ &= <\nu(y),\frac{\partial}{\partial x}\varphi(x,y)> \end{aligned} \qquad (2.5)$$

where the second step is justified by linearity and the third by continuity. A similar argument shows that the function $<\nu(y),\varphi(x,y)>$ is actually infinitely differentiable in $x$ and therefore belongs to $\mathcal{D}(\mathbb{R}_x)$. Accordingly we can define the two-dimensional distribution $\mu(x) \otimes \nu(y)$, which is called the **direct product** of $\mu$ and $\nu$, by writing

$$<\mu(x)\otimes\nu(y),\varphi(x,y)> \,=\, <\mu(x),<\nu(y),\varphi(x,y)>> \qquad (2.6)$$

The proof that this does define a distribution on $\mathbb{R}^2_{x,y}$ is left as an exercise.

Now if $\varphi(x) \in \mathcal{D}(\mathbb{R}_x)$ and $\psi(y) \in \mathcal{D}(\mathbb{R}_y)$, the product $\varphi(x)\psi(y)$ belongs to $\mathcal{D}(\mathbb{R}^2_{x,y})$. Further, since

$$<\mu(x)\otimes\nu(y),\varphi(x)\psi(y)> \,=\, <\mu(x),\varphi(x)<\nu(y),\psi(y)>>$$

we obtain

$$< \mu(x) \otimes \nu(y), \varphi(x)\psi(y) > = (< \nu(y), \psi(y) >) \cdot (< \mu(x), \varphi(x) >)$$

and similarly

$$< \nu(y) \otimes \mu(x), \varphi(x)\psi(y) > = (< \nu(y), \psi(y) >) \cdot (< \mu(x), \varphi(x) >).$$

It can be shown that test functions of the form

$$\theta(x,y) = \sum_k \varphi_k(x)\psi_k(y) \qquad (2.7)$$

where the $\varphi_k$ belong to $\mathcal{D}(\mathbb{R}_x)$ and the $\psi_k$ belong to $\mathcal{D}(\mathbb{R}_y)$, and the summation over $k$ is finite, form a dense subspace of $\mathcal{D}(\mathbb{R}^2_{x,y})$. As a result it follows readily that the direct product is commutative. The direct product is also associative. Given three distributions $\mu(x), \nu(y)$, and $\vartheta(t)$, we can define a three-dimensional distribution, $\mu(x) \otimes \nu(y) \otimes \vartheta(t)$ whose value at $\varphi(x,y,t) \in \mathcal{D}(\mathbb{R}^3_{x,y,t})$ is given uniquely by

$$\begin{aligned}
< [\mu(x) \otimes \nu(y)] \otimes \vartheta(t), \varphi(x,y,t) > &\equiv\, < \mu(x) \otimes \nu(y), < \vartheta(t), \varphi(x,y,t) >> \\
&\equiv\, < \mu(x), < \nu(y), < \vartheta(t), \varphi(x,y,t) >>> \\
&=\, < \mu(x), < \nu(y) \otimes \vartheta(t), \varphi(x,y,t) >> \\
&=\, < \mu(x) \otimes [\nu(y) \otimes \vartheta(t)], \varphi(x,y,t) > .
\end{aligned}$$

Clearly the process could be extended to define direct products of any order $n$.

**Example 2.1.1** Let $h_1(x), h_2(y)$ be locally integrable functions on $\mathbb{R}_x$ and $\mathbb{R}_y$ respectively. Then the pointwise product $h_1(x) \cdot h_2(y)$ is a locally integrable function on $\mathbb{R}^2_{x,y}$. The regular distribution defined by this product function is the direct product of the regular distributions defined by the factor functions $h_1(x)$ and $h_2(y)$ respectively. For we have

$$\begin{aligned}
< h_1(x) \otimes h_2(y), \varphi(x,y) > &= < h_1(x), < h_2(y), \varphi(x,y) >> \\
&= \int_{-\infty}^{+\infty} h_1(x) \left\{ \int_{-\infty}^{+\infty} h_2(y)\varphi(x,y)\, dy \right\} dx
\end{aligned}$$

and since $\varphi(x,y)h_1(x)h_2(y)$ is locally integrable and of compact support in $\mathbb{R}^2_{x,y}$ we can write the iterated integral as a double integral to get

$$\begin{aligned}
< h_1(x) \otimes h_2(y), \varphi(x,y) > &= \int_{-\infty}^{+\infty}\int_{-\infty}^{+\infty} h_1(x)h_2(y)\varphi(x,y)dydx \\
&= < h_1(x)h_2(y), \varphi(x,y) >
\end{aligned}$$

**Example 2.1.2** The direct product of the singular distribution $\delta$ with the regular distribution $H$ is defined as a simple integration process:

$$<\delta(x) \otimes H(y), \varphi(x,y)> = <\delta(x), \int_0^{+\infty} \varphi(x,y)dy> = \int_0^{+\infty} \varphi(0,y)dy.$$

More generally let $h(y,t)$ be a locally integrable function on $\mathbb{R}_y \times \mathbb{R}_t$. Then

$$\begin{aligned}<\delta_a(x) \otimes h(y,t), \varphi(x,y,t)> &= <\delta_a(x), <h(y,t), \varphi(x,y,t)>> \\ &= <\delta_a(x), \int_{-\infty}^{+\infty}\int_{-\infty}^{+\infty} \varphi(x,y,t)h(y,t)dtdy> \\ &= \int_{-\infty}^{+\infty}\int_{-\infty}^{+\infty} \varphi(a,y,t)h(y,t)dtdy.\end{aligned}$$

**Example 2.1.3** As an example of the direct product of two singular distributions consider $\delta_a \otimes \delta_b$:

$$\begin{aligned}<\delta_a(x) \otimes \delta_b(y), \varphi(x,y)> &= <\delta_a(x), <\delta_b(y), \varphi(x,y)>> \\ &= <\delta_a(x), \varphi(x,b)> = \varphi(a,b).\end{aligned}$$

This direct product of two one-dimensional distributions could equally well be defined directly as the two-dimensional distribution $\delta(x-a, y-b)$ characterised by the sampling property $\varphi(x,y) \to \varphi(a,b)$.

If $h(x)$ is a locally integrable function then the regular distribution $\mu_{(h)}$ generated by $h(x)$ can be identified with a distribution of mass or charge along the $x$-axis, for which $h(x)$ is the density function. On this basis $\delta_a(x)$ represents a discrete point mass, of unit value, concentrated at the point $x = a$. In the same way we can interpret $h(x,y)$ as a continuous (as opposed to discrete) planar distribution of mass in $\mathbb{R}^2_{x,y}$. The direct product $\delta_a(x) \otimes h(y)$ represents a one-dimensional distribution of mass in $\mathbb{R}^2_{x,y}$ which is spread along the line $x = a$ with density $h(y)$. The direct product $\delta_a(x) \otimes \delta_b(y)$ similarly corresponds to a unit point mass located in the plane at $(a,b)$. Finally, the three-dimensional distribution $\delta_a(x) \otimes h(y,t)$ can be interpreted as a surface layer of mass spread over the plane $x = a$ with surface density given by $h(y,t)$.

**Exercise 2.1 :**

1. If $\mu, \nu, \sigma$, are distributions show that

$$\mu \otimes (a\nu + b\sigma) = a(\mu \otimes \nu) + b(\mu \otimes \sigma),$$

where $a, b$ are any complex numbers.

2. If $\varphi(x,y)$ is an arbitrary test function in $\mathcal{D}(\mathbb{R}^2)$ show that

$$< \text{Pf}\left\{\frac{H_0(x)}{x}\right\} \otimes \text{Pf}\left\{\frac{H_0(y)}{y}\right\}, \varphi(x,y) >$$

can be expressed as a finite sum of absolutely convergent double integrals.

3. For arbitrary distributions $\mu, \nu \in \mathcal{D}'$, and any $r = (j,k) \in \mathbb{N}_0^2$ show that

$$D^j\mu \otimes D^k\nu = D^r(\mu \otimes \nu).$$

### 2.1.3 The convolution of distributions

The expression (2.4) for the action of the convolution of continuous functions $f$ and $g$ on a test function $\varphi$ suggests an extended definition of convolution for distributions based on the direct product. Formally we would like to write

$$< (\mu * \nu)(x), \varphi(x) > \;=\; < \mu(x) \otimes \nu(y), \varphi(x+y) >$$
$$=\; < \mu(x), < \nu(y), \varphi(x+y) >>,$$

but $\varphi(x+y)$, although infinitely differentiable, does not have compact support in $\mathbb{R}^2_{x,y}$. The most we can say is that its support is contained in some infinite strip of finite width running parallel to the line $x+y=0$. However, a meaning can be assigned to this expression if the supports of $\mu$ and $\nu$ are suitably restricted. In particular, suppose that the support of $\mu(x) \otimes \nu(y)$ intersects the support of $\varphi(x+y)$ in a bounded set, say $\Omega$. Let $\lambda(x,y)$ be some test function in $\mathcal{D}(\mathbb{R}^2_{x,y})$ which is equal to unity over a neighbourhood of $\Omega$. Then the function $\lambda(x,y)\varphi(x+y)$ is a test function which coincides with $\varphi(x+y)$ on $\Omega$. Hence we may advance the definition

$$< (\mu * \nu)(x), \varphi(x) > \;=\; < \mu(x) \otimes \nu(y), \lambda(x,y)\varphi(x+y) > \qquad (2.8)$$

On the basis of this definition the convolution of two distributions $\mu$ and $\nu$ certainly exists whenever any one of the following conditions is satisfied:

(1) either $\mu$ or $\nu$ has compact support;
(2) both $\mu$ and $\nu$ have supports bounded on the left;
(3) both $\mu$ and $\nu$ have supports bounded on the right.

**Example 2.1.4** Let $f$ and $g$ be locally integrable functions with supports satisfying one or other of the conditions listed above. Then

$$< h, \varphi > \;=\; < f * g, \varphi > \;=\; \int_{-\infty}^{+\infty}\int_{-\infty}^{+\infty} f(x)g(y)\varphi(x+y)dxdy$$

where the double integral converges absolutely since the integrand is a locally integrable function and has bounded support in the $(x, y)$-plane. Hence, with $x = \tau$ and $y = t - \tau$, we can use the Fubini theorem to transform the double integral into an iterated integral to get

$$<h, \varphi> = \int_{-\infty}^{+\infty} \varphi(t) \left\{ \int_{-\infty}^{+\infty} f(\tau) g(t - \tau) \, d\tau \right\} dt$$

The function $h(t)$ is given by the integral in braces; it is defined almost everywhere and is locally integrable.

**Example 2.1.5** For any distribution $\mu$ and any non-negative integer $m$,

$$\begin{aligned} <\delta^{(m)} * \mu, \varphi> &= <\mu * \delta^{(m)}, \varphi> \\ &= <\mu(x), <\delta^{(m)}(y), \varphi(x+y)>> \\ &= <\mu(x), (-1)^m \varphi^{(m)}(x)> = <D^m \mu(x), \varphi(x)> \end{aligned}$$

and therefore $\delta^{(m)} * \mu = D^m \mu$ in $\mathcal{D}'$. In particular, $\delta * \mu = \mu$ for every $\mu \in \mathcal{D}'$. More generally we get

$$\delta^{(m)}(x - a) * \mu(x) = D^m \mu(x - a).$$

Although convolution is a commutative operation for distributions (whenever it is well defined) it is not generally associative. Even if $\mu * (\nu * \vartheta)$ and $(\mu * \nu) * \vartheta$ are both well defined, they need not be equal. For example we have the following situation: Let 1 and 0 denote the regular distributions generated by the constant functions $f(x) \equiv 1$ and $g(x) \equiv 0$ respectively. Then

$$1 * (\delta' * H(x)) = 1 * \delta = 1,$$

whereas

$$(1 * \delta') * H(x) = 0 * H(x) = 0.$$

It can be shown that $\mu * (\nu * \vartheta)$ and $(\mu * \nu) * \vartheta$ both exist and are equal if any one of the following conditions is fulfilled:

**(1)** all the supports are bounded on the left,

**(2)** all the supports are bounded on the right,

**(3)** at least two of the supports are bounded,

(see, for example, Zemanian [117]).

**Exercise 2.2 :**

1. If $\sigma$ is well defined as the convolution of the distributions $\mu$ and $\nu$, so that $\sigma = \mu * \nu$, show that

   (a) $\sigma(x-a) = \mu(x-a) * \nu(x) = \mu(x) * \nu(x-a)$,
   (b) $D\sigma = (D\mu) * \nu = \mu * (D\nu)$,
   (c) $D^m \sigma = (D^p \mu) * (D^q \nu)$, where $p + q = m$, and $p, q \geq 0$.

2. If $h$ is a locally integrable function such that $h(x) = 0$ for all $x < 0$, show that

$$\left\{ h(x) * \frac{H_0(x)}{\sqrt{\pi x}} \right\} * \frac{H_0(x)}{\sqrt{\pi x}} = \left\{ \int_0^x h(y) dy \right\} H_0(x).$$

3. Show that

$$D^2 \left\{ H_0(x) \int_{-\infty}^{+\infty} \log(y) \cdot \log(x-y) dy \right\} = \text{Pf}\left\{ \frac{H_0(x)}{x} \right\} * \text{Pf}\left\{ \frac{H_0(x)}{x} \right\}.$$

## 2.2 CONVERGENCE OF DISTRIBUTIONS

### 2.2.1 Sequence of distributions

In §1.5.3 the distributions $(x \pm i0)^{-n}, n \in \mathbb{N}$, were introduced in an ad hoc manner as the limits of distributions $(x \pm i0)^\lambda$ as the parameter $\lambda$ approached the value $-n$. The idea of limit and of convergence in the sense of $\mathcal{D}'$ can be formalised in the following way.

A sequence $(\mu_n)_{n \in \mathbb{N}}$ of distributions is said to converge to the distribution $\mu$ as its limit if and only if

$$\forall \varphi \in \mathcal{D} : <\mu, \varphi> = \lim_{n \to \infty} <\mu_n, \varphi> . \qquad (2.9)$$

As it stands this definition tacitly assumes that the functional $\mu$ defined as the limit of the $\mu_n$ is necessarily a distribution. In fact it is possible to establish the following theorem:

**Theorem 2.1** *Let $(\mu_n)_{n \in \mathbb{N}}$ be a sequence of distributions which is such that the limit $\lim_{n \to \infty} <\mu_n, \varphi>$ is well defined for each $\varphi \in \mathcal{D}$. Then the resulting functional on $\mathcal{D}$ is a distribution; in other words the space $\mathcal{D}'$ is closed under sequential convergence.*

**Proof:** A direct proof, due to M.S.Brodskii, can be found in Gel'fand and Shilov [26], or in Zemanian [117]. (The result is actually a consequence of the fact that the Banach-Steinhaus Theorem can be generalised so as to apply to $\mathcal{D}$.) □

Let $(h_n)_{n \in \mathbb{N}}$ be a sequence of locally integrable functions which converges almost everywhere to a limit function $h$. If there exists a locally integrable function $g$ such that $|h_n(x)| \leq g(x)$ everywhere for each $n$ then the Lebesgue Dominated Convergence theorem ensures that $h(x)$ is locally integrable and that

$$\int_a^b h(x)dx = \lim_{n \to \infty} \int_a^b h_n(x)dx$$

for every interval $[a, b]$. If $\varphi \in \mathcal{D}$ then the product $\varphi(x)h_n(x)$ is certainly locally integrable (and in fact is absolutely integrable over $(-\infty, +\infty)$). Also the sequence $(\varphi h_n)_{n \in \mathbb{N}}$ converges pointwise almost everywhere to the limit $\varphi(x)h(x)$, and we have

$$|\varphi(x)h_n(x)| \leq \|\varphi\|_{[0,\text{supp}(\varphi)]} \cdot |h_n(x)| \leq \|\varphi\|_{[0,\text{supp}(\varphi)]} \cdot g(x).$$

Therefore each function $\varphi(x)h_n(x)$ is dominated by the locally integrable function $\|\varphi\|_{[0,\text{supp}(\varphi)]} \cdot g(x)$, and so the Lebesgue theorem gives

$$\int_{-\infty}^{+\infty} \varphi(x)h(x)dx = \lim_{n \to \infty} \int_{-\infty}^{+\infty} \varphi(x)h_n(x)dx \equiv <h, \varphi>.$$

The sequence of regular distributions generated by the $(h_n)_{n \in \mathbb{N}}$ thus converges in the distributional sense to the (regular) distribution generated by the pointwise limit $h$. By contrast consider the case when, for $n = 1, 2, \ldots$, we let $d_n$ denote the function given by

$$d_n(x) \equiv \begin{cases} n & \text{for } 1/n < x < 2/n, \\ 0 & \text{otherwise.} \end{cases} \quad (2.10)$$

This is a sequence of locally integrable functions which converges pointwise to zero everywhere on $\mathbb{R}$. However, for any test function $\varphi \in \mathcal{D}$ we get

$$\int_{-\infty}^{+\infty} \varphi(x)d_n(x)\,dx = \int_{1/n}^{2/n} \varphi(x)d_n(x)\,dx = \varphi(\theta_n) \quad (2.11)$$

where $1/n < \theta_n < 2/n$. Since $\varphi$ is continuous at the origin it follows that

$$\lim_{n \to \infty} \int_{-\infty}^{+\infty} \varphi(x)d_n(x)\,dx = \varphi(0) \quad (2.12)$$

so that the sequence of regular distributions generated by $(d_n)_{n \in \mathbb{N}}$ converges in $\mathcal{D}'$ to the singular distribution $\delta$. Thus the distributional limit of a sequence of functions need not coincide with the pointwise limit. In fact the distributional limit may be well defined when the pointwise limit fails to exist (even almost everywhere). For example, suppose that for $n = 1, 2, 3, \ldots$, we take $f_n(x) = \sin(n!x)$. The pointwise limit of this sequence of (locally integrable) functions exists only at points $x = p\pi/q$, where $p$ and $q$ are integers (since for all integral $n \geq q$, $nx$ will be a multiple of $\pi$). It converges for no other value of $x$ and therefore does not converge almost everywhere. However the distributional limit is well defined as the zero distribution since we have

$$<f_n,\varphi> = \int_{-\infty}^{+\infty} \varphi(x)\sin(n!x)dx = \frac{1}{n!}\int_{-\infty}^{+\infty} \varphi'(x)\cos(n!x)dx$$

and so

$$|<f_n,\varphi>| \leq \frac{1}{n!}\int_{-\infty}^{+\infty}|\varphi'(x)|dx \to 0 \quad \text{as} \quad n \to \infty.$$

Even when a pointwise limit does exist its behaviour with respect to differentiation is far from simple, as is well known. If a sequence $(f_n)_{n\in\mathbb{N}}$ of differentiable functions converges in the pointwise sense to a differentiable limit $f$ it may or may not be true that the sequence $(f'_n)_{n\in\mathbb{N}}$ converges to $f'$. The situation with regard to distributional convergence is more straightforward since we have the result:

**Theorem 2.2** *Differentiation of distributions is a continuous operation with respect to convergence in $\mathcal{D}'$.*

**Proof:** Whenever a sequence $(\mu_n)_{n\in\mathbb{N}} \subset \mathcal{D}'$ converges to a limit $\mu$ in the sense of $\mathcal{D}'$ we have that, for any test function $\varphi$, $<D\mu_n,\varphi> \equiv <\mu_n,-\varphi'>$ converges to $<\mu,-\varphi'> \equiv <D\mu,\varphi>$ as $n \to \infty$. □

### 2.2.1.1 Infinite series in $\mathcal{D}'$

The concept of distributional limit introduced above allows us to give a proper definition of the sum of an infinite series of distributions. We say that the infinite series

$$\sum_{k=1}^{\infty} \mu_k$$

converges to $\mu$ as its sum if and only if the sequence of corresponding partial sums converges to $\mu$ as its limit in the above sense. That is to say

$$\mu = \sum_{k=1}^{\infty} \mu_k \Leftrightarrow \mu = \lim_{m\to\infty}\left\{\sum_{k=1}^{m} \mu_k\right\}. \qquad (2.13)$$

Similarly we can give meaning to expressions like $\sum_{k=-\infty}^{+\infty} \mu_k$. For example, the expression $\sum_{m=-\infty}^{+\infty} \delta(x-mp)$, where $p > 0$, represents the distributional limit of the sequence $(\mu_n)_{n\in\mathbb{N}}$ where for each $n$ we have

$$\mu_n(x) = \sum_{m=-n}^{+n} \delta(x-mp).$$

Sec. 2.2] **Convergence of Distributions** 73

This is a distribution given by

$$< \sum_{m=-\infty}^{+\infty} \delta(x-mp), \varphi(x) > \equiv \sum_{m=-\infty}^{+\infty} \varphi(mp)$$

for every $\varphi \in \mathcal{D}$. (Since $\varphi$ always vanishes outside some finite interval there will be only finitely many non-zero values $\varphi(mp)$ to be summed and so the series on the right-hand side turns out to be a finite sum.)

Theorem 2.2 on the continuity of differentiation with respect to convergence in $\mathcal{D}'$ implies that term-by-term differentiation of an infinite series of distributions is always permissible. Hence we have, for example,

$$D\left\{\sum_{m=-\infty}^{+\infty} \delta(x-mp)\right\} = \sum_{m=-\infty}^{+\infty} D\delta(x-mp) = \sum_{m=-\infty}^{+\infty} \delta'(x-mp)$$

where

$$< \sum_{m=-\infty}^{+\infty} \delta'(x-mp), \varphi(x) > = - \sum_{m=-\infty}^{+\infty} \varphi'(mp).$$

### 2.2.1.2 Periodic impulse trains

In general a distribution $\mu$ is said to be periodic, with period $2\pi$, if and only if

$$< \mu(x), \varphi(x) > = < \tau_{2\pi}\mu(x), \varphi(x) > = < \mu(x), \varphi(x+2\pi) >, \forall \varphi \in \mathcal{D}.$$

The distribution given by $\sum_{m=-\infty}^{+\infty} \delta(x-2m\pi)$ is an example of such a periodic distribution and as such admits a representation as a generalised Fourier series. A simple non-rigorous demonstration of this can be given as follows.

Let $\rho(x) = [H_0(x+d/2) - H_0(x-d/2)]/d$, where $x \in \mathbb{R}$ and $d$ is a fixed number such that $0 < d < 2\pi$. Denote by $T_{2\pi}[\rho](x)$ the $2\pi$-periodic extension of $\rho(x)$, given by

$$T_{2\pi}[\rho](x) \equiv \sum_{n=-\infty}^{+\infty} \rho(x-2n\pi)$$

The classical Fourier expansion of the periodic function $T_{2\pi}[\rho](x)$ is

$$T_{2\pi}[\rho](x) \sim \frac{1}{2\pi} \sum_{n=-\infty}^{+\infty} \frac{\sin(nd/2)}{n/2} \exp(inx)$$

$$\equiv \frac{1}{2\pi}\left[1 + 2\sum_{n=1}^{+\infty} \frac{\sin(nd/2)}{nd/2} \cos(nx)\right]$$

where the Fourier series converges to $T_{2\pi}[\rho](x)$ at all points other than $x = 2n\pi + d/2$ or $x = 2n\pi - d/2$. As $d$ tends to zero the distributional limit of $\rho(x)$ is $\delta(x)$, while that of $T_{2\pi}[\rho](x)$ is $\sum_{n=-\infty}^{+\infty} \delta(x - 2n\pi)$. Hence, formally at least, we get

$$\sum_{n=-\infty}^{+\infty} \delta(x - 2n\pi) \sim \frac{1}{2\pi} \sum_{n=-\infty}^{+\infty} \exp(inx)$$

$$= \frac{1}{2\pi} \left[ 1 + 2 \sum_{n=1}^{+\infty} \cos(nx) \right].$$

For a rigorous derivation of this result we can appeal to the following theorem:

**Theorem 2.3** *A trigonometric series $\sum_{n=-\infty}^{+\infty} b_n \exp(inx)$ converges in $\mathcal{D}'$ if there exist numbers $M > 0$ and $k \in \mathbb{N}_0$ such that $|b_n| < M \cdot |n|^k$ for all $n \in \mathbb{N}_0$. Its sum can then be written as*

$$\sum_{n=-\infty}^{+\infty} b_n \exp(inx) = b_0 + D^r f(x) \tag{2.14}$$

*where $r$ is a non-negative integer $\geq k + 2$, and $f(x)$ is the continuous function given by the series*

$$f(x) = \sum_{n=-\infty, n \neq 0}^{+\infty} \frac{b_n}{(in)^r} \exp(inx). \tag{2.15}$$

**Proof:** Since $|b_n/(in)^r| \leq M|n|^{-2}$ for all $n$ the series (2.15) converges uniformly and so $f(x)$ is well defined as a continuous periodic function. Further the partial sums of this series define a sequence of regular distributions converging in $\mathcal{D}'$ to the regular distribution $f(x)$ as its limit. Hence we may differentiate it distributionally term by term $r$ times; adding the constant $b_0$ then gives (2.14). □

By this theorem the series $\sum_{n=-\infty}^{+\infty} \exp(inx)$ converges in $\mathcal{D}'$ to the distribution $1 + D^2 f(x)$, where $f(x)$ is the continuous periodic function of period $2\pi$ given by

$$f(x) = -\sum_{n=-\infty, n \neq 0}^{+\infty} \frac{\exp(inx)}{n^2}.$$

This is the classical Fourier series expansion of a periodic function given in $(0, 2\pi)$ by $\frac{\pi^2}{6} - \frac{1}{2}(x - \pi)^2$. Differentiating twice we get

$$D^2 f(x) = -1 + \sum_{n=-\infty}^{+\infty} 2\pi \delta(x - 2n\pi)$$

and so we may conclude that

$$\sum_{n=-\infty}^{+\infty} \exp(inx) = \sum_{n=-\infty}^{+\infty} 2\pi \delta(x - 2n\pi). \tag{2.16}$$

**Exercise 2.3 :**

1. Let $(f_n)_{n\in\mathbb{N}}$ be a sequence of functions in $L^2(\mathbb{R})$ which converges in the mean-square sense to a limit function $f$ on every finite interval $[a,b]$. Prove that the corresponding regular distributions $f_n$ converge to the regular distribution $f$ in the sense of $\mathcal{D}'$.

2. Show that in the sense of convergence in $\mathcal{D}'$ we have
$$\lim_{h\to 0} \frac{\mu - \tau_h \mu}{h} = D\mu,$$
where $\mu$ is any distribution in $\mathcal{D}'$ and $\tau_h$ is the translation operator.

3. Let $a \in \mathbb{R}$ and for each $n \in \mathbb{N}$ define $f_n(x) = n^a \sin(nx)$. Prove that the sequence $(f_n)_{n\in\mathbb{N}}$ converges in $\mathcal{D}'$ to the zero distribution.

### 2.2.2 Delta sequences

A **delta sequence** is a sequence of ordinary functions (strictly, of regular distributions generated by ordinary functions) which converges to the singular distribution $\delta(x)$. The sequence $(d_n)_{n\in\mathbb{N}}$ discussed in §2.2.1 is a particularly simple example of such a delta sequence. More generally let $\rho(x)$ be some fixed, non-negative function such that
$$\int_{-\infty}^{+\infty} \rho(x)\, dx = 1,$$
and write
$$\rho_n(x) \equiv n\rho(nx). \tag{2.17}$$
Then if $f$ is an arbitrary continuous function we have
$$\int_{-\infty}^{+\infty} f(x)\rho_n(x)\, dx = \int_{-\infty}^{+\infty} f(t/n)\rho(t)\, dt$$
$$= f(0) + \int_{-\infty}^{+\infty} [f(t/n) - f(0)]\,\rho(t)\, dt.$$
Also,
$$\left| \int_0^{+\infty} [f(t/n) - f(0)]\,\rho(t)\, dt \right| \le \left\{ \int_0^T + \int_T^{+\infty} \right\} |f(t/n) - f(0)| \cdot |\rho(t)|\, dt = I_1 + I_2$$
and given any $\varepsilon > 0$ we can always choose $T$ such that
$$I_2 \le 2 \cdot \left\{ \sup_{t \ge T} |f(t)| \right\} \cdot \int_T^{+\infty} |\rho(t)|\, dt < \varepsilon/2\,.$$

Again, $0 \leq t \leq T$ implies that $0 \leq t/n \leq T/n$. Hence having chosen $T$ we can then always choose $n$ large enough to ensure that

$$\sup_{0 \leq t \leq T} |f(t/n) - f(0)| < \varepsilon/2$$

and therefore such that $I_1 < \varepsilon/2$. A similar argument applies to the integral from $-\infty$ to $0$ and so we have the result

$$\lim_{n \to \infty} \int_{-\infty}^{+\infty} f(x) \rho_n(x) \, dx = f(0).$$

In particular this holds for any $f \in \mathcal{D}$. Therefore the distributional limit of the sequence of regular distributions generated by the functions $\rho(x)$ is the delta function $\delta(x)$. For example let $\rho(x) = x$ for $0 \leq x \leq 1$, $\rho(x) = 2 - x$ for $1 \leq x \leq 2$, and $\rho(x) = 0$ otherwise. Then the above conditions are satisfied and so the sequence $(\rho_n)_{n \in \mathbb{N}}$ of regular distributions converges to $\delta(x)$. In this case the pointwise limit of the sequence of functions $(\rho_n)_{n \in \mathbb{N}}$ is identically zero. More typically, taking

$$\rho(x) = \frac{1}{\pi(1+x^2)} \quad \text{and} \quad \rho(x) = \frac{1}{\pi}\left\{\frac{\sin(x)}{x}\right\}^2,$$

we get the well known results

$$\delta(x) = \lim_{n \to \infty} \frac{n}{\pi(1+n^2x^2)} \quad \text{and} \quad \delta(x) = \lim_{n \to \infty} \frac{\sin^2(nx)}{n\pi x^2},$$

the limits being understood in the distributional sense.

In Gel'fand and Shilov [26] a more general form of this result is given by defining a sequence $(\rho_n)_{n \in \mathbb{N}}$ of locally integrable functions to be a **delta-convergent sequence** if it satisfies the conditions

(1) for any $M > 0$ and for any $a, b$ such that $|a| \leq M$, $|b| \leq M$, the quantity

$$\left| \int_a^b \rho_n(x) dx \right|$$

is bounded by a constant independent of $a$, $b$, and $n$, and depending only on $M$, and

(2) for any fixed, non-zero numbers $a$ and $b$ we have

$$\lim_{n \to \infty} \int_a^b \rho_n(x) dx = \begin{cases} 0 & \text{for } a < b < 0 \text{ and } 0 < a < b, \\ 1 & \text{for } a < 0 < b. \end{cases}$$

If we write
$$h_n(x) \equiv \int_{-1}^{x} \rho_n(t)dt,$$
then we have
$$\lim_{n\to\infty} h_n(x) = \begin{cases} 1 & \text{for } x > 0 \\ 0 & \text{for } x < 0 \end{cases}$$
and therefore $\lim_{n\to\infty} h_n(x) \equiv H(x)$. Hence the functions $h_n$ are uniformly bounded over every finite interval. It follows that for any $\varphi$ in $\mathcal{D}$ we will have
$$\lim_{n\to\infty} \int_{-\infty}^{+\infty} h_n(x)\varphi(x)dx = \int_{-\infty}^{+\infty} \left\{\lim_{n\to\infty} h_n(x)\right\}\varphi(x)dx = \int_{0}^{+\infty} \varphi(x)dx$$
so that the $(h_n)_{n\in\mathbb{N}}$ converge distributionally to the Heaviside distribution $H$ and the sequence of the derivatives converges distributionally to the delta distribution $\delta$. For example let
$$\rho_n(x) = \frac{\sin(nx)}{\pi x}. \tag{2.18}$$
Then for each $n$ we have a locally integrable function and
$$\int_{-\infty}^{+\infty} \rho_n(x)dx = \frac{1}{\pi} \int_{-\infty}^{+\infty} \frac{\sin(nx)}{x} dx = 1.$$
Also, if $b > a > 0$, then
$$\int_{a}^{b} \rho_n(x)dx = \frac{1}{\pi} \int_{a}^{b} \frac{\sin(nx)}{x} dx = \frac{1}{\pi} \int_{na}^{nb} \frac{\sin(y)}{y} dy,$$
and the existence of the infinite integral
$$\int_{0}^{+\infty} \frac{\sin(y)}{y} dy$$
ensures that the integral $\int_a^b \rho_n(x)dx$ tends to 0 as $n \to \infty$. Further,
$$\left| \int_{a}^{b} \frac{\sin(nx)}{\pi x} dx \right| = \frac{1}{\pi} \left| \int_{na}^{nb} \frac{\sin(y)}{y} dy \right| \leq \frac{1}{\pi} \log(b/a),$$
which shows that the integral is uniformly bounded. Hence the sequence $(\rho_n)_{n\in\mathbb{N}}$ is a delta-convergent sequence and
$$\lim_{n\to\infty} \frac{\sin(nx)}{\pi x} = \delta(x) \tag{2.19}$$
in the sense of $\mathcal{D}'$.

If we set
$$g_n(x) = \frac{1}{\pi} \int_{0}^{n} \cos(tx)dt \equiv \frac{\sin(nx)}{\pi x}$$

then we have

$$< g_n(x), \varphi(x) > = \int_{-\infty}^{+\infty} \varphi(x) \left\{ \frac{1}{\pi} \int_0^n \cos(tx) dt \right\} dx$$

which converges to $\varphi(0)$ as $n$ tends to $\infty$. Hence, in the sense of distributions, we can write

$$\lim_{n \to \infty} \frac{1}{\pi} \int_0^n \cos(tx) dt = \delta(x). \qquad (2.20)$$

This result is often interpreted as the assignment of a (symbolic) meaning to an integral which is divergent in the classical sense:

$$\frac{1}{\pi} \int_0^{+\infty} \cos(tx) dt = \delta(x). \qquad (2.21)$$

This is consistent with the formal use of the Fourier inversion integral applied to the constant function 1:

$$\mathcal{F}^{-1}[1] = \frac{1}{2\pi} \int_{-\infty}^{+\infty} \exp(i\omega x) d\omega = \frac{1}{\pi} \int_0^{+\infty} \cos(\omega x) d\omega = \delta(x).$$

### 2.2.2.1 Classification of delta-sequences

There is an extensive literature both of theoretical aspects of distributions and of applications in which delta sequences of various types play a central role. In a number of important situations the results obtained depend to a greater or lesser extent on particular properties of the delta sequences used, and it is helpful to have a classification of the main types of such sequences. Often the work in question is phrased in terms of **delta nets** rather than **delta sequences**; that is to say, in terms of families $\{\rho_{(\varepsilon)}\}_{0 < \varepsilon \leq 1}$ of functions such that $\lim_{\varepsilon \downarrow 0} \rho_{(\varepsilon)}(x) = \delta(x)$ in the sense of $\mathcal{D}'$. Although the use of delta nets is superficially more general, all results of importance can be achieved equally well in terms of delta sequences and for the most part we work in terms of these.

Note first that if $(\rho_n)_{n \in \mathbb{N}}$ is any delta sequence in which the functions $\rho_n$ are constrained to be infinitely smooth, or even to be members of $\mathcal{D}$, then the continuity of differentiation shows that, for any $m \in \mathbb{N}$ and any $a \in \mathbb{R}$ we have

$$\delta^{(m)}(x - a) = \lim_{n \to \infty} \rho_n^{(m)}(x - a).$$

**Strict delta sequences** Let $(\rho_n)_{n \in \mathbb{N}}$ be any sequence of functions in $\mathcal{D}$ which satisfies the conditions

  A   $\rho_n(x) \geq 0$ for all $x \in \mathbb{R}$ and for each $n \in \mathbb{N}$,

B $\int_{-\infty}^{+\infty} \rho_n(x)dx = 1$ for each $n \in \mathbb{N}$,

$C_1$ $\text{supp}(\rho_n) \to \{0\}$ as $n \to \infty$.

Then $(\rho_n)_{n \in \mathbb{N}}$ will be called a **strict** delta sequence. Such sequences were introduced by Hirata-Ogata [31] and by Mikusinski [74]. In some contexts the condition $C_1$ is replaced by one or other of the following stronger conditions:

$C_2$ $\text{supp}(\rho_n) \to \{0\}$ as $n \to \infty$, and for each $p \in \mathbb{N}_0$ there exists $M_p > 0$ such that
$$\int_{-\infty}^{+\infty} |x|^p |\rho_n^{(p)}(x)| dx \leq M_p$$
for all $n \in \mathbb{N}$;

$C_3$ $\text{supp}(\rho_n) \subset [-1/n, +1/n]$, and for each $p \in \mathbb{N}_0$ there exists $M_p > 0$ such that
$$\int_{-\infty}^{+\infty} |x|^p |\rho_n^{(p)}(x)| dx \leq M_p$$
for all $n \in \mathbb{N}$.

**Model delta sequences** A particularly important type of strict delta sequence is defined by taking

$C_4$ $\rho_n(x) = n\rho(nx)$, for each $n \in \mathbb{N}$, where $\rho$ is some fixed function in $\mathcal{D}$ such that
$$\rho(x) \geq 0, \, \forall x \in \mathbb{R} \text{ and } \int_{-\infty}^{+\infty} \rho(x)dx = 1.$$

These are often referred to as **model** delta sequences and occur very frequently throughout the literature, often with additional qualifying restrictions (such as, for example, the requirement that $\rho(x)$ should be an even function).
Strict delta sequences have been conveniently classified by Oberguggenberger [82] as $C_i$-delta sequences, $(i = 1, 2, 3, 4)$, according to the particular condition $C_i$ which is satisfied.

**Exercise 2.4 :**

1. Find the limit in $\mathcal{D}'$ as $n \to \infty$ of each of the following sequences
    (a) $f_{1n}(x) = n^3 x / [\pi(n^2 x^2 + 1)^2]$,
    (b) $f_{2n}(x) = \frac{1}{n} \exp(-x^2/n)$,
    (c) $f_{3n}(x) = \sqrt{n} \exp(-nx^2)$,
    (d) $f_{4n}(x) = [nx \cos(nx) - \sin(nx)]/x^2$.

2. Give an example of a strict delta sequence $(\rho_n)_{n \in \mathbb{N}}$ which is such that $(\rho_n^2)_{n \in \mathbb{N}}$ does not converge in $\mathcal{D}'$.

## 2.3 SEQUENTIAL THEORIES OF DISTRIBUTIONS

For $\mu \in \mathcal{D}'$ and any function $\rho \in \mathcal{D}$ the convolution $\mu * \rho$ is well defined as

$$(\mu * \rho)(x) = <\mu(y), \rho(x-y)> \tag{2.22}$$

and is an infinitely smooth function. For $n = 1, 2, \ldots$, let

$$\rho_n(x) \equiv n\rho(nx),$$

so that $(\rho_n)_{n \in \mathbb{N}}$ is a delta sequence of infinitely smooth functions, and define

$$\mu_n(x) \equiv \mu * \rho_n(x) \equiv <\mu(y), \rho_n(x-y)>.$$

Then the sequence $(\mu_n)_{n \in \mathbb{N}}$ of regular distributions, each generated by a $\mathcal{C}^\infty$ function, converges in $\mathcal{D}'$ to $\mu$:

$$<\mu, \varphi> = \lim_{n \to \infty} <\mu_n, \varphi> \equiv \lim_{n \to \infty} \int_{-\infty}^{+\infty} \varphi(x)\mu_n(x)dx \tag{2.23}$$

Thus each distribution $\mu$ in $\mathcal{D}'(\mathbb{R})$ admits a (non-unique) representation as the limit, in the sense of $\mathcal{D}'$, of a sequence of ordinary functions - in this case of infinitely differentiable functions. This suggests the possibility of *defining* distributions in terms of such sequences, the non-uniqueness of the representation being taken care of by some appropriate equivalence relation for the sequences. The idea first appears in Schwartz's work and has since been exploited from several different points of view. Most recently it has been used in certain nonstandard approaches to the theory of distributions (and to other kinds of generalised functions) as will be discussed in the last chapter of the present text. We give first an outline of one of the earliest comprehensive sequential treatments of distribution theory.

### 2.3.1 The Mikusinski sequential theory of distributions

Mikusinski's sequential treatment of distributions stems from a paper published as early as 1948 [72], but the ideas are most comprehensively developed in the 1973 text by Antosik, Mikusinski and Sikorski [1]. For one-dimensional distributions a genuinely elementary theory is presented which uses equivalence classes of continuous functions rather than infinitely smooth ones. As in the case of the J.S.Silva theory there are some advantages in limiting the discussion initially to finite-order distributions.

Consider the linear space $\mathcal{C}(I)$ of all functions defined and continuous on an open (possibly infinite) interval $I \subset \mathbb{R}$. A sequence $(f_n)_{n \in \mathbb{N}}$ of functions belonging to $\mathcal{C}(I)$ is said to be **fundamental** if and only if

(1) there exists an integer $p$ and a sequence $(F_n)_{n\in\mathbb{N}} \subset \mathcal{C}^p(I)$ such that we have $F_n^{(p)}(x) = f_n(x)$ for all $x \in I$, and for all $n \in \mathbb{N}$, and,

(2) the sequence $(F_n)_{n\in\mathbb{N}}$ converges uniformly on every compact subset of $I$. (This is usually referred to as *almost uniform convergence* on the open subset $I$ of $\mathbb{R}$.)

Two fundamental sequences $(f_{1n})_{n\in\mathbb{N}}, (f_{2n})_{n\in\mathbb{N}}$, are said to be **equivalent**, and we write $(f_{1n})_{n\in\mathbb{N}} \sim (f_{2n})_{n\in\mathbb{N}}$, if and only if there exists an integer $p$ and sequences $(F_{1n})_{n\in\mathbb{N}}, (F_{2n})_{n\in\mathbb{N}}$ such that

(1) $F_{1n}^{(p)}(x) = f_{1n}(x)$ and $F_{2n}^{(p)}(x) = f_{2n}(x)$ for all $x \in I$ and for all $n \in \mathbb{N}$, and

(2) both sequences $(F_{1n})_{n\in\mathbb{N}}$ and $(F_{2n})_{n\in\mathbb{N}}$ converge almost uniformly to the same limit $F$.

A (finite-order) distribution in the sense of Mikusinski is then defined to be an equivalence class $[(f_n)_{n\in\mathbb{N}}]$ of fundamental sequences with respect to the equivalence relation $\sim$ defined above. The space $\mathcal{C}(I)$ itself is embedded in the set of such Mikusinski distributions by identifying each $f \in \mathcal{C}(I)$ with the equivalence class $[(f_n)_{n\in\mathbb{N}}]$ defined by the constant sequence $f_n(x) = f(x)$ for all $n \in \mathbb{N}$. Let $(K_n)_{n\in\mathbb{N}}$ be a sequence of compacts of $I$ so that $I = \lim_{n\to\infty} K_n$. From the Weierstrass approximation theorem for continuous functions it follows that for every function $f$ in $\mathcal{C}(I)$ there exists a sequence $(P_n)_{n\in\mathbb{N}}$ of polynomials such that

$$|f(x) - P_n(x)| < \frac{1}{n}$$

for all $x \in K_n$ and for all $n \in \mathbb{N}$. Thus $(P_n)_{n\in\mathbb{N}}$ converges almost uniformly to $f$. Hence each such Mikusinski distribution actually admits a representation as an equivalence class $[(P_n)_{n\in\mathbb{N}}]$ where the $P_n$ are polynomials on $I$. If we now define the distributional differentiation operator $D$ by

$$D[(P_n)_{n\in\mathbb{N}}] \equiv [(P_n')_{n\in\mathbb{N}}] \qquad (2.24)$$

then every Mikusinski distribution is seen to be infinitely differentiable in this sense. Moreover it is now easy to show (using Corollary 1.3.1) that the Mikusinski distributions of finite order may be identified with Schwartz distributions of finite order.

Distributions of infinite order can be brought into the Mikusinski theory by a simple extension of the definitions of fundamental sequence and equivalence. In general we say that a sequence $(f_n)_{n \in \mathbb{N}} \subset \mathcal{C}(I)$ is fundamental if and only if for each compact subset $K$ of $I$ there exists some $p = p(K)$ in $\mathbb{N}$ and a sequence $(F_n)_{n \in \mathbb{N}} \subset \mathcal{C}^p(I)$ such that $F_n^{(p)} = f_n$ on $K$ for all $n \in \mathbb{N}$ and $(F_n)_{n \in \mathbb{N}}$ converges uniformly on $K$. Two such fundamental sequences $(f_{1n})_{n \in \mathbb{N}}, (f_{2n})_{n \in \mathbb{N}}$, are equivalent if for every compact $K \subset I$ there exists some $p = p(K)$ in $\mathbb{N}$ and sequences $(F_{1n})_{n \in \mathbb{N}}, (F_{2n})_{n \in \mathbb{N}}$, such that $F_{1n}^{(p)} = f_{1n}, F_{2n}^{(p)} = f_{2n}$ on $K$, and both sequences $(F_{1n})_{n \in \mathbb{N}}, (F_{2n})_{n \in \mathbb{N}}$ converge uniformly to the same limit on $K$. A distribution in general is then defined as an equivalence class of fundamental sequences.

In the second part of the text by Antosik, Mikusinski and Sikorski the definition of distributions in $\mathbb{R}^m$, or more generally in an open subset $\Omega$ of $\mathbb{R}^m$, the same basic approach is used. However instead of fundamental sequences of continuous functions the theory is founded on fundamental sequences of infinitely differentiable functions. This avoids the need for the intermediate definition of a generalised derivation operator for continuous functions.

### 2.3.2 The generalised functions of Temple, Lighthill and Jones

George Temple [106] followed the general idea of Mikusinski and developed a sequential theory of distributions which was particularly designed to be suitable for a broadly based audience. The theory was presented, in a succinct and somewhat simplified form, in the well known text by James Lighthill [65]. Subsequently the approach was greatly extended and developed by D.S.Jones in a comprehensive text [42] which was later superseded by a revised second edition [46]. The sequential treatment allows much of the background of functional analysis and topology associated with the Schwartz theory to be avoided, or at least side-stepped. Nevertheless the classification and the properties of the various test functions used remain of crucial importance. In the approach initiated by Temple and developed by Lighthill and Jones a useful and evocative terminology is introduced which we reproduce here:

**Fine functions**: a function $\varphi : \mathbb{R} \to \mathbb{C}$ is said to be fine if it is infinitely differentiable and, together with each of its derivatives, vanishes identically outside some finite interval. (The fine functions are therefore precisely the members of the space $\mathcal{D}(\mathbb{R})$ of the Schwartz test functions.)

**Good functions**: a function $\gamma : \mathbb{R} \to \mathbb{C}$ is said to be good if it is infinitely differentiable and such that

$$\lim_{|x| \to \infty} |x^r \gamma^{(k)}(x)| = 0$$

for every pair of integers $r, k \geq 0$. (The good functions constitute the larger space $\mathcal{S}(\mathbb{R})$ of test functions which is needed to develop the theory of the distributional Fourier transform. See Chapter 3)

**Fairly good functions**: a function $\psi : \mathbb{R} \to \mathbb{C}$ is said to be fairly good if it is infinitely differentiable and if there exists some fixed integer N such that
$$\lim_{|x| \to \infty} |x^{-N} \gamma^{(k)}(x)| = 0$$
for every non-negative integer $k$.

**Moderately good functions**: a function $\theta : \mathbb{R} \to \mathbb{C}$ is said to be moderately good if it is infinitely differentiable and if, for each non-negative integer $k$, there exists a finite $N_k$ such that
$$\lim_{|x| \to \infty} |x^{-N_k} \theta^{(k)}(x)| = 0$$

In both of the texts by Jones the main concern is not with distributions in general but with an important subclass for which there is a natural and straightforward definition of a (generalised) Fourier transform. This is the subclass of the so-called **tempered distributions** (or distributions of slow growth) which appear in the Schwartz theory as linear continuous functionals defined on the space $\mathcal{S} \equiv \mathcal{S}(\mathbb{R})$ of good functions. Somewhat confusingly (in the light of later developments) they are referred to by Jones as "*generalised functions*".

**Generalised Functions** A sequence $(\gamma_n)_{n \in \mathbb{N}}$ of good functions is said to be regular if, for every good function $f$, the limit
$$\lim_{n \to \infty} \int_{-\infty}^{+\infty} f(x) \gamma_n(x) dx \qquad (2.25)$$
is well defined. Two regular sequences are said to be equivalent if and only if they have the same limit in this sense. An equivalence class $[(\gamma_n)_{n \in \mathbb{N}}]$ of such regular sequences is then called a *(Jones) generalised function*.

If we write $\gamma$ for the generalised function defined by a sequence $(\gamma_n)_{n \in \mathbb{N}}$ of good functions then the action of $\gamma$ on an arbitrary good function $f$ can be conveniently denoted by a symbolic integration sign:
$$\fint_{-\infty}^{+\infty} \gamma(x) f(x) dx = \lim_{n \to \infty} \int_{-\infty}^{+\infty} \gamma_n(x) f(x) dx. \qquad (2.26)$$

Every function $\varphi \in \mathcal{D}$ is a fine function and therefore certainly a good function, and so the symbolic integral $f_{-\infty}^{+\infty} \gamma(x)\varphi(x)dx$ is well defined by (2.26). In this way the Jones generalised function $\gamma$ defines a certain functional on $\mathcal{D}$, and it is not difficult to confirm that this functional is linear and continuous on $\mathcal{D}$. That is to say each generalised function $\gamma$, in the sense of Jones, defines a certain distribution in $\mathcal{D}'$ and may be identified with that distribution. On the other hand not every distribution in $\mathcal{D}'$ may be identified with a Jones generalised function; as remarked above the generalised functions correspond to a proper sub-class of $\mathcal{D}'$. In the terminology used by Jones in the second version of his text [46] arbitrary distributions, which are defined as equivalence classes of sequences of fine functions, are described as **weak functions**. The actual definition is as follows:

**Weak Functions** A sequence $(\varphi_n)_{n \in \mathbb{N}}$ of fine functions is said to be regular if, for every fine function $\varphi$, the limit

$$\lim_{n \to \infty} \int_{-\infty}^{+\infty} \varphi(x)\varphi_n(x)dx$$

is well defined. Two regular sequences are said to be equivalent if and only if they have the same limit in this sense for every fine function $\varphi$. An equivalence class of such regular sequences is then called a weak function.

Once again it is not difficult to show that each weak function defines a certain linear and continuous functional on $\mathcal{D}$. This time however the converse is also true; to each distribution in $\mathcal{D}'$ there corresponds a (unique) weak function with which it may be identified.

There are undeniable advantages in presenting the theory of distributions in terms of a sequential framework like this. For the most part one needs only the relatively familiar apparatus of elementary analysis to develop the theory and can avoid excursions into functional analysis and topological vector space theory. Operations on distributions (weak functions) appear in a particularly simple light. Thus the sum of two weak functions $\gamma = [(\gamma_n)_{n \in \mathbb{N}}]$ and $\xi = [(\xi_n)_{n \in \mathbb{N}}]$ is simply the equivalence class $\sigma = [(\sigma_n)_{n \in \mathbb{N}}]$ where for each $n \in \mathbb{N}$ the (fine) function $\sigma_n$ is the ordinary pointwise sum of the functions $\gamma_n$ and $\xi_n$. Similarly the derivative of the weak function $\gamma = [(\gamma_n)_{n \in \mathbb{N}}]$ is the weak function $D\gamma$ defined by the equivalence class $[(\gamma'_n)_{n \in \mathbb{N}}]$. Finally the fundamental result that every distribution is locally a finite-order derivative of a continuous function takes the following simple and straightforward form:

**Theorem 2.4** *(Jones [46, Theorem 11.7]) Any weak function $\gamma(x)$ can be expressed in the form*

$$\gamma(x) = \sum_{m=-\infty}^{+\infty} f_m^{(r_m)}(x) \qquad (2.27)$$

*where $f_m$ is continuous and vanishes outside an interval of the form $m - 1 - \varepsilon \leq x \leq m + 1 + \varepsilon$, $\varepsilon$ being an arbitrary positive real number. (Differentiation is here understood in the distributional, or weak function, sense on the right-hand side of (2.27)).*

Nevertheless the sequential approach has not been as generally accepted as an appropriate basis on which to teach the theory of distributions as these advantages might lead one to suppose. There remains a strong conviction that the natural significance of distributions (and of other types of generalised functions) is to be found in their role as functionals, that is as members of certain dual spaces. Recent developments in nonlinear theories of generalised functions and in the application of Nonstandard Analysis, however, suggests that there is much to be gained from sequential treatments. We return to this theme in the last two chapters.

## 2.4 THE MULTIPLICATION PROBLEM

The theory of distributions is a linear theory. As a result not all the operations which we can carry out on ordinary functions will extend to $\mathcal{D}'$, or at least will not do so without modification in some respects. Certain operations on smooth functions (such as addition, and multiplication by scalars) can be extended without difficulty to arbitrary distributions; these are called **regular** operations. Others (such as convolution, multiplication and change of variable) can be defined only for particular distributions or for certain restricted subclasses of distributions; these are called **irregular** operations. We have seen already in the case of convolution that we are obliged to impose certain restrictions on the distributions concerned before a satisfactory definition of the operation can be obtained. And even then the convolution may fail to exhibit such a simple and desirable property as associativity. Similar limitations arise when we try to define a multiplicative operation for distributions which generalises the pointwise product of ordinary functions, and it is this problem which has attracted the greatest interest since the original Schwartz definition of distribution.

### 2.4.1 The elementary Schwartz product

The first systematic examination of the problem of defining a satisfactory product for general distributions is given in Schwartz [97]. There is an immediate difficulty. It is reasonable to require that the definition of multiplication for arbitrary distributions should be consistent with the ordinary pointwise product of functions. However given arbitrary regular distributions generated by locally integrable functions $h_1(x), h_2(x)$, it does not necessarily follow that the pointwise product $h_1(x)h_2(x)$ will be locally integrable; hence the product $h_1 h_2$ may not generate a (regular) distribution. Nevertheless there is a natural definition for the product if one of the factors is a regular distribution generated by an infinitely differentiable function. If $\theta$ is an infinitely differentiable function ($\theta \in \mathcal{C}^\infty$) and $\mu$ is an arbitrary distribution ($\mu \in \mathcal{D}'$) then we define the **Schwartz Product**, or **S-product**, to be the distribution $\theta \mu$ given by

$$< \theta\mu, \varphi > \; = \; < \mu, \theta\varphi > \tag{2.28}$$

for all $\varphi \in \mathcal{D}$. This is certainly well defined as a distribution since the (pointwise) product $\theta\varphi$ is always a function in $\mathcal{D}$. Also, if $\mu_{(g)}$ is a regular distribution, generated by a locally integrable function $g$, then it is immediately clear that the Schwartz product does agree with the ordinary pointwise product. For we have

$$\begin{aligned}< \theta\mu_{(g)}, \varphi > &= \int_{-\infty}^{+\infty} \varphi(x)\left[\theta(x)g(x)\right] dx \\ &= \int_{-\infty}^{+\infty} \left[\varphi(x)\theta(x)\right] g(x) dx \; = \; < \mu_{(g)}, \theta\varphi >\end{aligned}$$

Accordingly we shall continue to indicate the S-product by simple juxtaposition of the factors, in the same way as for ordinary multiplication.

#### 2.4.1.1 Differentiation of the Schwartz product

If $\theta \in \mathcal{C}^\infty$ and $\mu \in \mathcal{D}'$ then we have

$$\begin{aligned}< D(\theta\mu), \varphi > &= \; < \theta\mu, -\varphi' > \; = \; < \mu, -\theta\varphi' > \\ &= \; < \mu, -D(\theta\varphi) > \; + \; < \mu, \theta'\varphi > \\ &= \; < D\mu, \theta\varphi > \; + \; < \mu, \theta'\varphi > \; = \; < \theta D\mu, \varphi > \; + \; < \theta'\mu, \varphi ></aligned>$$

and thus the usual product rule for differentiation applies to the S-product:

$$D(\theta\mu) \; = \; \theta D\mu + \theta'\mu. \tag{2.29}$$

More generally the Leibniz property holds:

$$< \theta(D^r \mu), \varphi > = < \sum_{k=0}^{r} \binom{r}{k}(-1)^k D^{r-k}[\mu \theta^{(k)}], \varphi > . \qquad (2.30)$$

A mild generalisation of the S-product can be obtained as follows. Given a space $E$ of test functions we say that a function $\theta$ is a *multiplier* for $E$ if

(1) $\theta\varphi$ belongs to $E$ for every member $\varphi$ of $E$,

(2) if the sequence $(\varphi_n)_{n \in \mathbb{N}}$ converges in $E$ to $\varphi$, then the sequence $(\theta\varphi_n)_{n \in \mathbb{N}}$ converges in $E$ to $\theta\varphi$.

Now let $\mu$ be a distribution defined for all $\varphi \in E$ (i.e. $\mu \in E'$) and let $\theta$ be a multiplier for $E$. Then $\theta\mu$ is well defined as that distribution on $E$ given by the mapping

$$\theta\mu : \varphi \rightsquigarrow < \theta\mu, \varphi > = < \mu, \theta\varphi > . \qquad (2.31)$$

**Example 2.4.1** $\delta_a$ is well defined on the test function space $\mathcal{C}_0$ of all continuous functions of compact support. If $\theta$ is any function continuous on $\mathbb{R}$ then $\theta$ is certainly a multiplier for $\mathcal{C}_0$ and we have

$$< \theta\delta_a, \varphi > \equiv < \delta_a(x), \theta(x)\varphi(x) > = \theta(a)\varphi(a).$$

Thus $\theta\delta_a$ is well defined as the distribution $\theta(a)\delta_a$.

**Example 2.4.2** $\delta'_a$ is well defined on the test function space $\mathcal{C}_0^1(\mathbb{R})$ of all continuously differentiable functions of compact support. If $\theta$ is any continuously differentiable function then $\theta$ is a multiplier for $\mathcal{C}_0^1(\mathbb{R})$ and we have

$$< \theta\delta'_a, \varphi > = < \delta'_a(x), \theta(x)\varphi(x) >$$
$$= -\frac{d}{dx}[\theta(x)\varphi(x)]\bigg|_{x=a} = -\theta(a)\varphi'(a) - \theta'(a)\varphi(a) .$$

Thus $\theta\delta'_a$ is well defined as the distribution $\theta(a)\delta'_a - \theta'(a)\delta_a$.

**Example 2.4.3** $x^{-1} \equiv \text{Pv}\{x^{-1}\}$ is a distribution of order 1, and $\theta(x) = x$ is a multiplier on $\mathcal{C}_0^1(\mathbb{R})$. Hence we get,

$$< x\text{Pv}\{x^{-1}\}, \varphi(x) > = < \text{Pv}\{x^{-1}\}, x\varphi(x) >$$
$$= \int_0^{+\infty} \frac{x\varphi(x) + x\varphi(-x)}{x} dx$$
$$= \int_0^{+\infty} [\varphi(x) + \varphi(-x)] dx$$
$$= \int_{-\infty}^{+\infty} \varphi(x) dx = < 1, \varphi(x) >$$

and so, $xx^{-1} \equiv x\text{Pv}\{x^{-1}\} = 1$.

The domain of definition of the S-product can be further extended by relaxing the differentiability conditions on the function $\theta$ provided that corresponding restrictions are placed on the degree of singularity of the other factor $\mu$. For example, suppose that $\mu$ is the $r^{th}$ derivative of a function $g$ belonging to $L^p(a,b)$ for some $p \geq 1$. Then if $\theta$ is a function whose $r^{th}$ (classical) derivative belongs to $L^q(a,b)$, where $p^{-1} + q^{-1} = 1$, writing

$$\theta\mu = \sum_{j=0}^{r} \binom{r}{j} (-1)^j D^{r-j}[\theta^{(j)} g]$$

yields a distribution well defined on the interval $(a,b)$.

**Exercise 2.5 :**

1. Find simpler equivalent expressions for each of the following:

   (a) $\cos(x) + \delta(x)\sin(x)$,
   
   (b) $\sin(x) + \delta(x)\cos(x)$,
   
   (c) $\delta''(x)\cos(2x)$.

2. Prove that, if $0 < \alpha < 1$, then the Schwartz product

$$x \operatorname{Pf}\left\{\frac{H_0(x)}{x^{\alpha+1}}\right\}$$

is equal to the regular distribution $H_0(x)/x^\alpha$.

3. If $m, n \in \mathbb{N}$ show that

$$x^n \delta^{(m)}(x) = \begin{cases} 0 & \text{if } m < n \\ (-1)^n n! \delta(x) & \text{if } m = n \\ \frac{(-1)^n m!}{(m-n)!} \delta^{(m-n)}(x) & \text{if } m > n. \end{cases}$$

4. Since $\delta(x) = DH(x)$ the chain rule suggests that the expression $\delta(\sin(x))$ should satisfy

$$DH(\sin(x)) = \delta(\sin(x)) D(\sin(x)).$$

Show that this means that $\delta(\sin(x))$ is an infinite sum of delta functions. Find a similar interpretation for $\delta(\cos(\pi x/2))$.

### 2.4.2 The division problem for the Schwartz product

The division problem can be formulated as follows:

> Given a distribution $\sigma \in \mathcal{D}'(\mathbb{R})$ and a function $f \in \mathcal{C}^\infty(\mathbb{R})$, to find a distribution $\mu \in \mathcal{D}'(\mathbb{R})$ such that $f\mu = \sigma$.

If $f(x) \neq 0$ for all $x \in \mathbb{R}$ then $g(x) = [f(x)]^{-1}$ belongs to $\mathcal{C}^\infty(\mathbb{R})$ and there is a unique solution to the problem, namely $\mu = g\sigma$. Otherwise the division problem is not an elementary one. We consider the simplest case in which the set

$$f^{-1}(\{0\}) = \{x \in \mathbb{R} : f(x) = 0\}$$

consists of isolated zeros of finite order. By partition of unity and translation of the origin it is enough to consider only the case when $f(x) = x^m$ for some $m \in \mathbb{N}$.

**Theorem 2.5** *(a) If $\mu \in \mathcal{D}'(\mathbb{R})$ is such that $x^m \mu = 0$ then*

$$\mu = \sum_{j=0}^{m-1} c_j \delta^{(j)} \tag{2.32}$$

*where the $c_j$, $(j = 0, 1, \ldots, m-1)$, are arbitrary constants.*

*(b) For any given distribution $\sigma \in \mathcal{D}'(\mathbb{R})$ there exists a distribution $\mu \in \mathcal{D}'(\mathbb{R})$ such that*

$$x^m \mu = \sigma. \tag{2.33}$$

**Proof:** Let $\theta$ be an arbitrary function in $\mathcal{D}$ such that $\theta(x) = 1$ on some neighbourhood of the origin; then $\theta^{(r)}(0) = 0$ for all $r \geq 1$. Any function $\varphi \in \mathcal{D}$ may be written as

$$\varphi(x) = \theta(x) \cdot \sum_{j=0}^{m-1} \frac{\varphi^{(j)}(0)}{j!} x^j + \eta(x)$$

where $\eta \in \mathcal{D}$ is a function such that $\eta^{(s)}(0) = 0$ for $s = 0, \ldots, m-1$. Hence there exists $\psi \in \mathcal{D}$ such that $\eta(x) = x^m \psi(x)$. In fact, from Taylor's theorem, with the integral form of remainder, we have

$$\eta(x) = \sum_{k=0}^{m-1} \frac{\eta^{(k)}(0)}{k!} x^k + \frac{1}{(m-1)!} \int_0^x (x-t)^{m-1} \eta^{(m)}(t)\, dt$$

$$= \frac{1}{(m-1)!} \int_0^x (x-t)^{m-1} \eta^{(m)}(t)\, dt$$

$$= \frac{1}{(m-1)!} \int_0^1 (x - x\xi)^{m-1} \eta^{(m)}(x\xi)\, x\, d\xi = x^m \psi(x)$$

where
$$\psi(x) = \frac{1}{(m-1)!} \int_0^1 (1-\xi)^{m-1} \eta^{(m)}(x\xi)\, d\xi \;.$$
Since $\eta \in \mathcal{D}$ and we can freely differentiate under the integral sign it follows that $\psi$ is a function in $\mathcal{D}$. Hence
$$\varphi(x) \;=\; \theta(x) \cdot \sum_{j=0}^{m-1} \frac{\varphi^{(j)}(0)}{j!} x^j + x^m \psi(x) \tag{2.34}$$

(a) If $x^m \mu = 0$ then
$$<\mu,\varphi> \;=\; <\mu(x), \theta(x) \cdot \sum_{j=0}^{m-1} \frac{\varphi^{(j)}(0)}{j!} x^j + x^m \psi(x) >$$
$$= \sum_{j=0}^{m-1} \frac{1}{j!} <\mu(x), x^j \theta(x)>\; \varphi^{(j)}(0)$$
$$= <\sum_{j=0}^{m-1} c_j \delta^{(j)}, \varphi>$$
where
$$c_j = \frac{(-1)^j}{j!} <\mu(x), x^j \theta(x)>\;, \quad j=0,1,\ldots,m-1.$$

(b) Equation (2.33) means that
$$<\mu(x), x^m \psi(x)> \;=\; <\sigma(x), \psi(x)>\;, \quad \forall \psi \in \mathcal{D}(\mathbb{R})$$
and this surely defines a continuous linear functional on
$$\{\eta \in \mathcal{D} : [\psi \in \mathcal{D} : \eta(x) = x^m \psi(x)]\} \subset \mathcal{D}.$$
It remains to show that $\mu$ extends to a continuous linear functional on the whole of $\mathcal{D}$. Now for any $\varphi \in \mathcal{D}$ we have
$$<\mu,\varphi> \;=\; <\mu(x), \theta(x) \cdot \sum_{j=0}^{m-1} \frac{\varphi^{(j)}(0)}{j!} x^j + x^m \psi(x) >$$
$$= <\sum_{j=0}^{m-1} c_j \delta^{(j)}, \varphi> + <\sigma(x), \psi(x)> \tag{2.35}$$
which shows that $<\mu,\varphi>$ is a well defined number for every $\varphi \in \mathcal{D}$. It follows easily from (2.35) that $\mu$ is linear and continuous on $\mathcal{D}(\mathbb{R})$. $\square$

**Corollary 2.4.1** *If $\mu_0$ is any particular solution of $x^m \mu = \sigma$, where $\sigma \in \mathcal{D}'(\mathbb{R})$, then the general solution is given by*
$$\mu = \mu_0 + \sum_{j=0}^{m-1} c_j \delta^{(j)}$$
*where the $c_j, (j=0,1,\ldots,m-1)$, are arbitrary constants.*

### 2.4.3 Problems associated with multiplication

The form of the S-product applies, at best, only to some restricted class of distributions. Even so there appear to be undesirable aspects of distributional multiplication at the most elementary levels. The most celebrated counter-example was exhibited at the outset by Schwartz himself, in connection with the associativity property. It is not difficult to show that in any Schwartz product involving three factors, say $\theta, \psi$ and $\mu$, associativity will hold provided at least two of the factors are regular distributions generated by infinitely differentiable functions. In contrast, recall that the Schwartz product of the regular distribution $x$ and the singular distribution $\mathrm{Pv}\{x^{-1}\}$ is well defined as the constant distribution 1:

$$x\mathrm{Pv}\{x^{-1}\} = \mathrm{Pv}\{x^{-1}\}x = 1. \tag{2.36}$$

Taken in combination with the (singular) distribution $\delta(x)$ this gives on the one hand

$$\left[\mathrm{Pv}\{x^{-1}\}x\right]\delta(x) = 1\delta(x) = \delta(x) \tag{2.37}$$

and on the other

$$\mathrm{Pv}\{x^{-1}\}[x\delta(x)] = \mathrm{Pv}\{x^{-1}\}0 = 0. \tag{2.38}$$

Thus associativity fails in a product containing two singular distributions, even though the singularities are particularly simple.

Again, consider the problem of assigning a meaning to the formal product $\delta(x)H(x)$. It seems reasonable to assume that the product $H(x)H(x)$ should be defined as

$$H^2(x) \equiv H(x)H(x) = H(x) \tag{2.39}$$

since this is consistent with the pointwise product of arbitrary Heaviside function representatives of the distribution $H(x)$. If the product is commutative then we should get

$$DH^2(x) = \delta(x)H(x) + H(x)\delta(x) = 2\delta(x)H(x)$$

while from (2.39) we have

$$DH^2(x) = DH(x) = \delta(x).$$

Hence it appears that we should define the product $\delta(x)H(x)$ to be $\frac{1}{2}\delta(x)$. On the other hand suppose we make the equally reasonable demand that the associativity property should hold for any product involving no more than one singular distribution. Then for the factors $x$, $\delta(x)$ and $H(x)$ we would get

$$x\left[\delta(x)H(x)\right] = [x\delta(x)]H(x) = 0H(x) = 0.$$

Since scalar multiples of $\delta(x)$ are the only distributions which are annihilated by (Schwartz) multiplication with $x$ it follows that $\delta(x)H(x)$ is necessarily of the form $k\delta(x)$, where $k$ is some constant. So far we have consistency with the previous result. However this means that

$$[\delta(x)H(x)]\,H(x) \;=\; k\delta(x)H(x) \;=\; k^2\delta(x) \qquad (2.40)$$

while

$$\delta(x)\,[H(x)H(x)] \;=\; \delta(x)H(x) \;=\; k\delta(x) \qquad (2.41)$$

and the associativity assumption requires that $k^2 = k$, so that $k$ can only assume the value 1 or the value 0.

The inconsistency could be removed in several ways, each of them having some undesirable implications.

(a) We could accept (2.39) and the resulting $\delta(x)H(x) = \tfrac{1}{2}\delta(x)$, but discard the assumption of associativity. This would mean that

$$[\delta(x)H(x)]\,H(x) \;=\; \tfrac{1}{2}\delta(x)H(x) \;=\; \tfrac{1}{4}\delta(x)$$

whereas

$$\delta(x)\,[H(x)H(x)] \;=\; \delta(x)H(x) \;=\; \tfrac{1}{2}\delta(x).$$

(b) We could accept (2.39) but discard commutativity. Then we would need to set $\delta(x)H(x) = k_1\delta(x)$ and $H(x)\delta(x) = k_2\delta(x)$. Since

$$DH^2(x) \;=\; k_1\delta(x) + k_2\delta(x) \;=\; \delta(x)$$

we have $k_1 + k_2 = 1$. Together with the associativity constraint this entails that we must set either $k_1 = 0$ and $k_2 = 1$ or else $k_1 = 1$ and $k_2 = 0$.

(c) Finally, of course, we could reject the assumption (2.39) itself. There is no immediately obvious alternative value for the apparently elementary product $H(x)H(x)$, and a consistent and reasonably comprehensive theory of multiplication would need to be developed before the issue could be resolved. The problem of producing such a theory has been the subject of intensive research for over half a century.

## Exercise 2.6 :

1. If $a, b$ are real constants determine the general solution in $\mathcal{D}'$ of each of the following equations

    (a) $(x - a)\mu = 0,$   (b) $(x - a)\mu = \delta_b,$
    (c) $(x - a)\mu = \delta_b',$   (d) $(x - a)(x - b)\mu = 0.$

    (Consider both the cases $a \neq b$ and $a = b$).

2. Determine the general solution in $\mathcal{D}'$ of the equation

$$(1 + x^2)(1 - x^2)^2 \mu = 0.$$

### 2.4.4 Theories of multiplicative products

The discussion of the product $\delta(x)H(x)$ in the preceding section is typical of attempts to obtain what is sometimes referred to as an *"intrinsic"* product theory for distributions; that is to say, of a definition of multiplication in which the product of two distributions is always again a distribution. Such theories are obliged to follow one or other of alternative strategies. On the one hand we can aim for consistency by limiting the domain of definition of the multiplicative operation to a subclass of distributions which are in some sense *"well-behaved"*. On the other, if we wish to be able to multiply singular distributions without constraint, then we must be prepared to give up some of the valuable properties (such as associativity) of the classical product. An alternative approach to the whole multiplication problem would be to abandon the attempt to define an intrinsic product and to allow modes of multiplication in which generalised functions other than distributions may arise. A general and comprehensive product theory was developed in this sense at an early stage by Güttinger [27] and by König [58]. This is based essentially on the Hahn-Banach extension theorem for linear functionals. For example, let $\mu$ be a continuous linear functional defined not on the whole of $\mathcal{D}$ but on the subspace $\mathcal{D}_{[0]}$ given by

$$\mathcal{D}_{[0]} = \{\varphi \in \mathcal{D} : \varphi(0) = 0\}.$$

Let $\theta$ be some particular function in $\mathcal{D}$ for which $\theta(0) = 1$. For an arbitrary $\varphi \in \mathcal{D}$ define an associated function $\check{\varphi}$ as follows:

$$\check{\varphi}(x) = \varphi(x) - \theta(x)\varphi(0).$$

Then $\check{\varphi} \in \mathcal{D}_{[0]}$ and so $\mu(\check{\varphi})$ is defined. If we can extend $\mu$ as a continuous linear functional on the whole of $\mathcal{D}$ then we should have

$$\mu(\check{\varphi}) = \mu(\varphi) - \varphi(0)\mu(\theta).$$

That is to say, given the value $\mu(\theta)$ for the particular (fixed) function $\theta$ in $\mathcal{D}$, the value of $\mu(\varphi)$ is constrained to be

$$<\mu,\varphi> = \mu_0(\breve{\varphi}) + k\varphi(0)$$

where for clarity we write $\mu_0$ for the original functional defined on $\mathcal{D}_{[0]}$, and where $k$ denotes the value assigned to $<\mu,\theta>$. The functional $\mu$ on $\mathcal{D}$ is thus defined as

$$\mu = \mu_0 + k\delta.$$

**Example 2.4.4** For any function $\varphi \in \mathcal{D}$ which vanishes at the origin the function $H_c(x)\varphi(x)$ is well defined for all $x$, is independent of the number $c$, and belongs to $\mathcal{D}_{[0]}$. We can define a functional $H(x) \bullet \delta(x)$ on the subspace $\mathcal{D}_{[0]}$ of $\mathcal{D}$ by setting

$$< H(x) \bullet \delta(x), \varphi(x) > \; = \; < \delta(x), H_c(x)\varphi(x) > \; = \; 0.$$

Then $H(x) \bullet \delta(x)$ extends on the whole of $\mathcal{D}$ to the distribution $k\delta(x)$, where $k$ is an arbitrary constant.

**Example 2.4.5** If $\varphi \in \mathcal{D}$ and $\varphi(0) = 0$ then in a neighbourhood of 0 we can write $\varphi(x) \equiv x\psi(x)$, where $\psi \in \mathcal{D}$. Hence we can define a functional $\mathrm{Pv}\{x^{-1}\} \bullet \delta(x)$ on $\mathcal{D}_{[0]}$ by

$$< \mathrm{Pv}\{x^{-1}\} \bullet \delta(x), \varphi(x) \; = \; < \delta(x), \psi(x) > \; = \; \psi(0),$$

and since

$$\psi(0) = \lim_{x \to 0} \left\{ \frac{\varphi(x)}{x} \right\} = \varphi'(0),$$

we can extend $\mathrm{Pv}\{x^{-1}\} \bullet \delta(x)$ to the whole of $\mathcal{D}$ according to

$$\mathrm{Pv}\{x^{-1}\} \bullet \delta(x) = -\delta'(x) + k\delta(x).$$

The method readily generalises. In principle given $\mu,\nu \in \mathcal{D}'$ we need to define a subspace $\mathcal{D}_{[\mu,\nu]}$ of $\mathcal{D}$ consisting of functions $\psi \in \mathcal{D}$ which vanish at singular points of $\mu$ and $\nu$. This allows the definition of an appropriate functional on $\mathcal{D}_{[\mu,\nu]}$ which can subsequently be extended to $\mathcal{D}$. However the products which result from this approach are generally neither commutative nor associative, and there is an inherent arbitrariness as the presence of the arbitrary constants in the examples given above makes clear. For these reasons the Güttinger/König approach has generally found less favour than the development of various forms of intrinsic product. This has been especially true of extended forms of distributional product based on sequential treatments of the theory of distributions such as that of Mikusinski.

These have been generally guided by the idea of developing a product which remains consistent with the Schwartz product and which further extends its domain of definition. Within the last decade however an entirely fresh attack on the multiplication problem (and on other irregular operations) by Colombeau and others has in many respects superseded all other approaches. In some respects the Colombeau theory can be said to supplement sequential theories but it requires a certain departure from the idea of an intrinsic product, and the introduction of a wide variety of generalised functions other than distributions. On the other hand it generates a product which is comprehensive and which obeys all the laws (commutativity, associativity) of a true algebra. We return to this material in detail in chapter 5.

# 3

# Generalised Functions and Fourier Analysis

## 3.1 REVIEW OF THE CLASSICAL FOURIER TRANSFORM

### 3.1.1 Fourier transforms in $L^1(\mathbb{R})$

The result
$$\delta(t) = \lim_{n\to\infty} \frac{1}{\pi} \int_0^n \cos(xt)dx \quad \text{in } \mathcal{D}',$$
was used in §2.2.2 to justify the symbolic formula
$$\delta(t) = \frac{1}{\pi} \int_0^{+\infty} \cos(xt)dx \equiv \frac{1}{2\pi} \int_{-\infty}^{+\infty} \exp(ixt)dx. \tag{3.1}$$

The following (purely formal) manipulations then give
$$\begin{aligned} f(t) &= \int_{-\infty}^{+\infty} \delta(t-\tau)f(\tau)d\tau = \int_{-\infty}^{+\infty} f(\tau) \left\{ \frac{1}{2\pi} \int_{-\infty}^{+\infty} \exp[i\omega(t-\tau)]d\omega \right\} d\tau \\ &= \frac{1}{2\pi} \int_{-\infty}^{+\infty} \exp(i\omega t) \left\{ \int_{-\infty}^{+\infty} \exp(-i\omega\tau)f(\tau)d\tau \right\} d\omega \\ &\equiv \frac{1}{2\pi} \int_{-\infty}^{+\infty} \exp(i\omega t) \hat{f}(\omega)d\omega \end{aligned} \tag{3.2}$$

where we write
$$\hat{f}(\omega) \equiv \int_{-\infty}^{+\infty} \exp(-i\omega\tau) f(\tau) d\tau. \tag{3.3}$$

Equations (3.2) and (3.3) give respectively the classical integral formulas for the direct and inverse Fourier transform of the function $f$. Moreover (3.1) might itself be interpreted as a special case of the Fourier inversion formula, with the constant function 1 playing the role of Fourier transform for the delta function $\delta(t)$.

In reality of course the definition of $\hat{f}(\omega)$ in terms of the Fourier integral of (3.3) depends on the existence of that integral, and this in turn depends on appropriate restrictions on the behaviour of the function $f(t)$. Similarly the derivation of the inversion formula, as well as its range of validity, is considerably more complicated than the simple transformation used to obtain (3.2) might suggest. In this chapter we begin by briefly reviewing some of the more important features of the classical Fourier transform and then go on to consider extension of the transforms to generalised functions.

In the classical theory it is well known that not every function has a Fourier transform which is itself a function. Similarly it turns out that not every distribution has a (generalised) Fourier transform which is itself a distribution. There is an important sub-class $\mathcal{S}'$ of $\mathcal{D}'$ whose members are called **tempered distributions**. A generalisation of the Fourier transform does exist which maps $\mathcal{S}'$ into itself. However to define Fourier transforms of arbitrary distributions it proves necessary to go beyond $\mathcal{D}'$ and to introduce a new space $\mathcal{Z}'$ of generalised functions which we call **ultradistributions**.

**Remark 3.1** *In dealing with the Fourier transform it is often convenient to follow the common engineering practice of distinguishing between the independent variable $t$ used in connection with a function $f$ and the independent variable $\omega$ used in connection with its Fourier transform $\hat{f}$. There is no real mathematical significance in this distinction but it is undeniably helpful on occasion to exploit the usual association of $t$ with "time" and of $\omega$ with "frequency". Equally there are other occasions when it is desirable to use "neutral" variables, such as $x$ and $y$, instead.*

For any function $f$ which is absolutely integrable over $\mathbb{R}$ (i.e. $f \in L^1(\mathbb{R})$) the Fourier transform is well defined everywhere on $\mathbb{R}$ by the classical Fourier integral,
$$\hat{f}(\omega) = \mathcal{F}[f](\omega) = \int_{-\infty}^{+\infty} f(t) \exp(-i\omega t) dt \ , \quad \omega \in \mathbb{R}. \tag{3.4}$$

Then $\hat{f}(\omega)$ is a function which is bounded and uniformly continuous on $\mathbb{R}$; also, from the Riemann-Lebesgue Lemma, we have $\lim_{|\omega|\to\infty} \hat{f}(\omega) = 0$. We prove two other properties of the $L^1$ transform which will be of importance in the sequel:

**Theorem 3.1 (Parseval)** *If $f$ and $g$ belong to $L^1(\mathbb{R})$ then*

$$\int_{-\infty}^{+\infty} f(x)\hat{g}(x)dx = \int_{-\infty}^{+\infty} \hat{f}(y)g(y)dy \qquad (3.5)$$

where $\hat{f} \equiv \mathcal{F}[f]$ and $\hat{g} \equiv \mathcal{F}[g]$.

**Proof:** Both sides of (3.5) do exist since $\hat{f}$ and $\hat{g}$ are bounded and continuous. By the absolute integrability of the integrands on $\mathbb{R} \times \mathbb{R}$ we have,

$$\begin{aligned}
\int_{-\infty}^{+\infty} f(x)\hat{g}(x)dx &= \int_{-\infty}^{+\infty}\int_{-\infty}^{+\infty} f(x)g(y)\exp(-ixy)dydx \\
&= \int_{-\infty}^{+\infty}\int_{-\infty}^{+\infty} f(x)g(y)\exp(-ixy)dxdy = \int_{-\infty}^{+\infty} g(y)\hat{f}(y)dy
\end{aligned}$$

as asserted. □

**Theorem 3.2 (Convolution Theorem)** *If $f, g$ are two functions in $L^1(\mathbb{R})$ then the Fourier transform of the convolution $f * g$ exists in the $L^1$-sense and we have $\mathcal{F}[f * g](x) = \hat{f}(x)\hat{g}(x)$.*

**Proof:** Since

$$\begin{aligned}
\mathcal{F}[f * g](x) &= \int_{-\infty}^{+\infty} \exp(-ixy) \int_{-\infty}^{+\infty} f(y-t)g(t)dt\,dy \\
&= \int_{-\infty}^{+\infty} g(t) \int_{-\infty}^{+\infty} \exp(-ixy)f(y-t)dy\,dt \\
&= \int_{-\infty}^{+\infty} g(t) \int_{-\infty}^{+\infty} \exp(-ix(\tau+t))f(\tau)d\tau\,dt \\
&= \int_{-\infty}^{+\infty} \exp(-ixt)g(t)dt \int_{-\infty}^{+\infty} \exp(-ix\tau)f(\tau)d\tau ,
\end{aligned}$$

where the change in order of integrations is justified by the fact that the last iterated integral is absolutely convergent, the result follows. □

So far as the inverse transform is concerned matters are less satisfactory. Not every function which is continuous on $\mathbb{R}$ and which vanishes at infinity is the transform of a function in $L^1(\mathbb{R})$. A purely formal manipulation would give

$$f(x) = \frac{1}{2\pi} \int_{-\infty}^{+\infty} \hat{f}(y)\exp(ixy)dy \qquad (3.6)$$

but there is an immediate difficulty in that $\hat{f}$ may not be an absolutely integrable function. The integral on the right-hand side of (3.6) may therefore not exist in the Lebesgue sense; the $L^1$ transform does not map $L^1(\mathbb{R})$ into $L^1(\mathbb{R})$. An inversion theorem for the $L^1$ transform can be given in the following form:

**Theorem 3.3** ($L^1$-**Inversion**) *For any $f \in L^1(\mathbb{R})$ which is of bounded variation over some neighbourhood of a point $x$, we have*

$$\frac{1}{2\pi} \lim_{R \to \infty} \int_{-R}^{+R} \hat{f}(y) \exp(ixy) dy = \frac{1}{2}[f(x^+) + f(x^-)]$$

*where $f(x^\pm) = \lim_{\tau \to 0^\pm} f(x + \tau)$.*

We seek an extended definition of the Fourier transform, which agrees with the $L^1$ transform whenever the latter exists, but which applies to a wide class of distributions and for which a symmetrical transform exists. First we review briefly some classical extensions of the transform.

**Remark 3.2** *The inversion formula is often expressed in terms of a* **summability kernel**. *A summability kernel on $\mathbb{R}$ is a family of continuous functions $\{K_\lambda(t)\}_{\lambda > 0}$ such that*

*(1)* $\forall \lambda > 0 : \int_{-\infty}^{+\infty} K_\lambda(t) dt = 1,$

*(2)* $\| K_\lambda \|_{L^1(\mathbb{R})} = O(1)$ *as $\lambda \to \infty$,*

*(3)* $\forall \varepsilon > 0 : \lim_{\lambda \to +\infty} \int_{|t| > \varepsilon} |K_\lambda(t)| \, dt = 0.$

*Then $\lim_{\lambda \to +\infty} \| f - K_\lambda * f \|_{L^1(\mathbb{R})} = 0$, for any $f \in L^1(\mathbb{R})$.*
*A common way to generate a summability kernel on $\mathbb{R}$ is to take a function $\varphi$ in $L^1(\mathbb{R})$ such that*

$$\int_{-\infty}^{+\infty} \varphi(t) dt = 1$$

*and then set $K_\lambda(t) \equiv \lambda \varphi(\lambda t)$ for $\lambda > 0$. In particular the Fejer kernel on $\mathbb{R}$ is defined by taking*

$$K(t) = \frac{1}{2\pi} \left( \frac{\sin(t/2)}{(t/2)} \right)^2$$
$$= \frac{1}{2\pi} \int_{-1}^{+1} (1 - |\omega|) \exp(i\omega t) d\omega.$$

*Then if $f$ is any function in $L^1(\mathbb{R})$ with Fourier transform $\hat{f}$, we have*

$$f(t) = \frac{1}{2\pi} \lim_{R \to +\infty} \int_{-R}^{+R} \left( 1 - \frac{|\omega|}{R} \right) \hat{f}(\omega) \exp(i\omega t) d\omega \qquad (3.7)$$

*almost everywhere on $\mathbb{R}$.*

### 3.1.2 Fourier transforms in $L^2(\mathbb{R})$

The most important classical extension of the Fourier transform is to square-integrable functions. If $f$ is any complex-valued function belonging to $L^2(\mathbb{R})$ then for each $m \in \mathbb{N}$ the function

$$\hat{F}_m(\omega) \equiv \int_{-m}^{+m} \exp(-i\omega t) f(t)\, dt$$

exists and belongs to $L^2(\mathbb{R})$. The sequence of functions $(\hat{F}_m)_{m \in \mathbb{N}}$ converges to a limit $\hat{F}(\omega)$ in the sense of the norm in $L^2(\mathbb{R})$. That is to say, there exists a function $\hat{F}$ in $L^2(\mathbb{R})$ which is such that

$$\lim_{m \to \infty} \int_{-\infty}^{+\infty} |\hat{F}(\omega) - \hat{F}_m(\omega)|^2 d\omega = 0$$

and it can be shown that

$$\int_{-\infty}^{+\infty} |\hat{F}(\omega)|^2\, d\omega = 2\pi \int_{-\infty}^{+\infty} |f(t)|^2 dt\,.$$

If, in addition, $f \in L^1(\mathbb{R})$ then $\lim_{m \to \infty} \hat{F}_m(\omega)$ exists at each point $\omega \in \mathbb{R}$ and $\hat{F}(\omega)$ coincides with the ordinary $L^1$-transform $\hat{f}(\omega)$ of $f(t)$. In general we call $\hat{F}(\omega)$ the $L^2$-transform of $f(t)$, whether or not $f(t)$ belongs to $L^1(\mathbb{R})$, and continue to use the notation $\hat{f}$ for both $L^1$ and $L^2$ transforms. In the same way for each $m > 0$ the function

$$f_m(t) \equiv \frac{1}{2\pi} \int_{-m}^{+m} \hat{f}(\omega) \exp(i\omega t) d\omega$$

exists and belongs to $L^2(\mathbb{R})$; the sequence of functions $(f_m)_{m \in \mathbb{N}}$ converges to the limit $f(t)$ in the sense of the $L^2$-norm.

With the usual notation l.i.m. for limit in the mean the theory of the Fourier transform in $L^2(\mathbb{R})$ can be expressed as follows:

**Theorem 3.4 (Plancherel)** *For any function $f \in L^2(\mathbb{R})$ there exists a function $\hat{f} \in L^2(\mathbb{R})$, called the Fourier transform of $f$, such that*

$$\hat{f}(\omega) = \underset{m \to \infty}{\text{l.i.m.}} \int_{-m}^{+m} f(t) \exp(-i\omega t) dt$$

$$f(t) = \underset{m \to \infty}{\text{l.i.m.}} \int_{-m}^{+m} \hat{f}(\omega) \exp(i\omega t) d(\frac{\omega}{2\pi})$$

*and*

$$\int_{-\infty}^{+\infty} |\hat{f}(\omega)|^2 d\omega = 2\pi \int_{-\infty}^{+\infty} |f(t)|^2 dt.$$

*Every function $f \in L^2(\mathbb{R})$ can be expressed in the form $f = \hat{g}$ for some $g \in L^2(\mathbb{R})$; the Fourier transform is thus an isometry on $L^2(\mathbb{R})$.*

For a function $f \in L^p(\mathbb{R}), 1 < p < 2$, the Fourier transform can be defined by a process similar to that for the $L^2(\mathbb{R})$ case. It is based on the Hausdorff-Young inequality which states that

$$\left\{\int_{-\infty}^{+\infty} |\hat{f}(\omega)|^q d\omega\right\}^{1/q} \leq \left\{\int_{-\infty}^{+\infty} |f(t)|^p dt\right\}^{1/p} \qquad (3.8)$$

for any $f \in L^1 \cap L^2(\mathbb{R})$, where $q = p/(p-1), 1 < p < 2$. The Fourier transform of $f \in L^p(\mathbb{R}), 1 < p < 2$, is defined as the limit in $L^q(\mathbb{R})$ of the sequence of functions

$$\hat{f}_m(\omega) = \int_{-m}^{+m} f(t) \exp(-i\omega t) dt , \quad m = 1, 2, 3, \ldots .$$

This defines a mapping from $L^p(\mathbb{R}), 1 < p < 2$, into $L^q(\mathbb{R}), q = p/(p-1)$. However this mapping is not an isometry, and the range is not even the whole of $L^q(\mathbb{R})$, in contrast to the $L^2$ case. The inversion formula has the same form as in the $L^1$ case except that convergence is in the $L^p$ sense: for $f \in L^p(\mathbb{R}), 1 < p < 2$, we have

$$f(t) = \frac{1}{2\pi} \lim_{R \to \infty} \int_{-R}^{+R} \left(1 - \frac{|\omega|}{R}\right) f(\omega) \exp(i\omega t) d\omega$$

in the $L^p(\mathbb{R})$ norm.

It is not possible to extend the Hausdorff-Young inequality for values of $p > 2$ and so a different procedure would be required if we wished to develop a Fourier transform theory for functions in $L^p(\mathbb{R})$, where $p > 2$. The key to such an extension turns out to be the duality relation for the classical $L^1(\mathbb{R})$ transform given by the Parseval theorem (theorem 3.1 above).

### 3.1.3 Functions of rapid decrease

A complex-valued function $f$ is said to be of **rapid decrease** as $|x|$ tends to infinity if and only if

$$\lim_{|x| \to \infty} |x^n f(x)| = 0$$

for all non-negative integers $n$. For example any function in $C_0^\infty$ is certainly a function of rapid decrease; so also are the functions $\exp(-x^2)$ and $1/\cosh(x)$. We denote by $\mathcal{S}(\mathbb{R})$, or simply by $\mathcal{S}$, the linear space of all complex-valued, infinitely differentiable functions $\varphi$ such that $\varphi$, together with each one of its derivatives, is a function of rapid decrease as $|x| \to \infty$. Further we say that a sequence $(\varphi_n)_{n \in \mathbb{N}}$ converges in $\mathcal{S}$ if and only if for $p, r = 0, 1, 2, \ldots$, each of the sequences $(x^p \varphi_n^{(r)}(x))_{n \in \mathbb{N}}$

is uniformly convergent. This is the mode of convergence which corresponds to the topology generated on $\mathcal{S}$ by the family of seminorms

$$\| \varphi \|_{\mathcal{S}}^{p,r} \equiv \sup_{x \in \mathbb{R}} |x^p \varphi^{(r)}(x)| \ , \quad p, r \in \mathbb{N}_0 .$$

Convergence in $\mathcal{S}$ can also be characterised by an equivalent boundedness condition. A sequence $(\varphi_n)_{n \in \mathbb{N}}$ converges in $\mathcal{S}$ if and only if

(1) for any given pair of non-negative integers $p$ and $r$,

$$|x^p \varphi_n^{(r)}(x)| \leq C_{p,r} ,$$

for all $n \in \mathbb{N}$ and all $x \in \mathbb{R}$, the constants $C_{p,r}$ being independent of $x$ and $n$,

(2) for each $r$ the sequence $(\varphi_n^{(r)}(x))_{n \in \mathbb{N}}$ converges uniformly on every compact in $\mathbb{R}$.

It is clear that the linear space $\mathcal{C}_0^\infty$ is contained in the linear space $\mathcal{S}$. Now suppose that $(\varphi_n)_{n \in \mathbb{N}} \subset \mathcal{C}_0^\infty \subset \mathcal{S}$ converges to a limit $\varphi$ in the sense of $\mathcal{D}$. Then all the $\varphi_n$ vanish outside some compact in $\mathbb{R}$, say outside $[-a, a]$. For each $p$ and $r$ we have $|x|^p \cdot |\varphi^{(r)}(x) - \varphi_n^{(r)}(x)| = 0$ identically for $|x| \geq a$, and is therefore bounded by $a^p \cdot \sup_{-a \leq x \leq +a} |\varphi^{(r)}(x) - \varphi_n^{(r)}(x)|$. Since the sequence $(\varphi_n)_{n \in \mathbb{N}}$ converges $\omega$-uniformly to $\varphi$ it follows that $a^p \cdot \sup_{-a \leq x \leq +a} |\varphi^{(r)}(x) - \varphi_n^{(r)}(x)|$ converges to zero, and therefore that

$$\lim_{n \to \infty} \left\{ \sup_{x \in \mathbb{R}} |x|^p \cdot |\varphi^{(r)}(x) - \varphi_n^{(r)}(x)| \right\} = \lim_{n \to \infty} \| (\varphi - \varphi_n) \|_{\mathcal{S}}^{p,r} = 0$$

uniformly on $\mathbb{R}$. Hence convergence in $\mathcal{D}$ implies convergence in $\mathcal{S}$. In fact we actually have the result:

**Theorem 3.5** *The space $\mathcal{D}$ is dense in the space $\mathcal{S}$ in the sense that for each $\varphi \in \mathcal{S}$ there exists $(\varphi_n)_{n \in \mathbb{N}}$ in $\mathcal{D}$ converging in $\mathcal{S}$ to $\varphi$.*

**Proof:** Let

$$\xi(x) \equiv \begin{cases} \exp[x^2/(x^2 - 1)] & \text{if } |x| < 1 \\ 0 & \text{if } |x| \geq 1 \end{cases}$$

and set $\xi_n(x) \equiv \xi(x/n)$ for $n = 1, 2, \ldots$; then for any $r \in \mathbb{N}$ we have $\xi_n^{(r)}(x) = n^{-r} \xi^{(r)}(x/n)$ and it follows that $\xi_n(x) \to 1$ as $n \to \infty$, and $\xi_n^{(r)}(x) \to 0$ as $n \to \infty$, for all $r \in \mathbb{N}$. All convergences are uniform over every bounded domain of $\mathbb{R}$. Further, for every $r \in \mathbb{N}_0$, the functions $\xi_n^{(r)}$ are uniformly bounded over $\mathbb{R}$ with respect to $n \in \mathbb{N}$.

Given any $\varphi \in \mathcal{S}$ define $\varphi_n(x) = \varphi(x) \xi_n(x)$, for $n = 1, 2, \ldots$. Then $\varphi_n \in \mathcal{D}$ for each $n \in \mathbb{N}$; also, for any $p \in \mathbb{N}$ the function $x^p \varphi(x)$ is bounded on $\mathbb{R}$, and therefore

$x^p\varphi(x)\xi_n(x)$ is uniformly bounded on $\mathbb{R}$ with respect to $n$. Moreover these functions converge uniformly over every bounded domain of $\mathbb{R}$ to $x^p\varphi(x)$ as $n \to \infty$. Now for $r \in \mathbb{N}$,

$$(\varphi\xi_n)^{(r)} = \sum_{j=0}^{r} \binom{r}{j}\varphi^{(r-j)}\xi_n^{(j)}$$

and, since $x^p\varphi^{(r-j)}(x)$ is bounded on $\mathbb{R}$ for all $j = 0, 1, \ldots, r$, it follows that the functions $x^p\{\varphi(x)\xi_n(x)\}^{(r)}$ are uniformly bounded on $\mathbb{R}$ with respect to $n \in \mathbb{N}$. Hence, as $n \to \infty$, the corresponding sequence converges uniformly over every bounded domain of $\mathbb{R}$ to $x^p\varphi(x)$. Thus $(\varphi_n)_{n\in\mathbb{N}}$ converges to $\varphi$ in $\mathcal{S}$. □

### 3.1.4 Fourier transforms in $\mathcal{S}$

If $\varphi \in \mathcal{S}$ then $\varphi$ is surely absolutely integrable over $\mathbb{R}$ and so the Fourier transform $\hat{\varphi}$ is well defined as the integral

$$\hat{\varphi}(\omega) = \mathcal{F}[\varphi](\omega) = \int_{-\infty}^{+\infty} \varphi(t)\exp(-i\omega t)\,dt \ , \ \omega \in \mathbb{R}. \tag{3.9}$$

We show that $\hat{\varphi}$ is itself a member of $\mathcal{S}$. In the first place we have

$$|h^{-1}[\hat{\varphi}(\omega + h) - \hat{\varphi}(\omega)]| \leq \int_{-\infty}^{+\infty} \left|\frac{\exp(-iht) - 1}{h}\right| |\varphi(t)|\,dt$$
$$= \int_{-\infty}^{+\infty} |\varphi(t)| \cdot |t + O(h)|\,dt$$

and since $t\varphi(t) \in L^1(\mathbb{R})$ the convergence as $|h| \to 0$ is dominated (actually uniform). Hence,

$$\hat{\varphi}'(\omega) = \int_{-\infty}^{+\infty} (-it)\varphi(t)\exp(-i\omega t)\,dt \tag{3.10}$$

and a similar argument shows that for $k = 1, 2, 3, \ldots$, we have

$$\hat{\varphi}^{(k)}(\omega) = \int_{-\infty}^{+\infty} (-it)^k \varphi(t)\exp(-i\omega t)\,dt$$

and therefore that $\hat{\varphi}$ is differentiable to any order. Again, integrating by parts and recalling that $|\varphi(t)| \to 0$ as $|t| \to \infty$, we get,

$$\hat{\varphi}(\omega) = \int_{-\infty}^{+\infty} \varphi(t)\exp(-i\omega t)\,dt$$
$$= (i\omega)^{-1} \int_{-\infty}^{+\infty} \varphi'(t)\exp(-i\omega t)\,dt$$
$$= (i\omega)^{-2} \int_{-\infty}^{+\infty} \varphi''(t)\exp(-i\omega t)\,dt$$
$$= \ldots = (i\omega)^{-m} \int_{-\infty}^{+\infty} \varphi^{(m)}(t)\exp(-i\omega t)\,dt,$$

so that, for any non-negative integer $m$,
$$|(i\omega)^m \varphi(\omega)| \le \int_{-\infty}^{+\infty} |\varphi^{(m)}(t)|\, dt < +\infty.$$
It follows that $\varphi$ must be a function of rapid decrease as $|\omega|$ tends to infinity (for if, for example, $\lim_{|\omega| \to \infty} |\hat\varphi(\omega)| \ne 0$ then $|(i\omega)\hat\varphi(\omega)|$ would not be bounded as $|\omega| \to \infty$). A similar argument shows that for arbitrary non-negative integers $m, k$, we have
$$|(i\omega)^m \hat\varphi^{(k)}(\omega)| \le \int_{-\infty}^{+\infty} |[t^k \varphi(t)]^{(m)}|\, dt < +\infty$$
and therefore that $\hat\varphi^{(k)}(\omega)$ must be a function of rapid decrease as $|\omega| \to \infty$. We have therefore established the following result:

**Theorem 3.6** *The Fourier transform $\hat\varphi$ of any function $\varphi$ which belongs to $\mathcal{S}$ is itself a member of $\mathcal{S}$.*

### 3.1.5 The Poisson formula in $\mathcal{S}$

Given $\varphi \in \mathcal{S}$ consider the doubly infinite sum of translates
$$\sum_{m=-\infty}^{+\infty} \varphi(x + m\lambda)$$
where $\lambda$ is a (fixed) positive number. For any $t \in \mathbb{R}$ and any non-negative integer $k$ we have $|(t+m\lambda)^k \varphi(t+m\lambda)| \to 0$ as $|m| \to \infty$, and so, given $\varepsilon > 0$ we can write
$$|\varphi(t + m\lambda)| < \varepsilon/|t + m\lambda|^k$$
for all sufficiently large $m$. The series therefore converges for all $t \in \mathbb{R}$; the sum function $\Phi(t)$ is clearly periodic, with period $\lambda$, and the convergence is uniform on compacts and therefore certainly on $[-\lambda/2, \lambda/2]$. In particular $\Phi(t)$ must be bounded and continuous on $[-\lambda/2, \lambda/2]$ and therefore everywhere. Direct computation of the Fourier coefficients of $\Phi(t)$ yields

$$\begin{aligned}
c_n(\Phi) &= \frac{1}{\lambda} \int_{-\lambda/2}^{+\lambda/2} \Phi(t) \exp(-2\pi i n t/\lambda)\, dt \\
&= \frac{1}{\lambda} \int_{-\lambda/2}^{+\lambda/2} \left\{ \sum_{m=-\infty}^{+\infty} \varphi(t + m\lambda) \exp(-2\pi i n t/\lambda) \right\} dt \\
&= \frac{1}{\lambda} \sum_{m=-\infty}^{+\infty} \left\{ \int_{-\lambda/2}^{+\lambda/2} \varphi(t + m\lambda) \exp(-2\pi i n t/\lambda)\, dt \right\} \\
&= \frac{1}{\lambda} \sum_{m=-\infty}^{+\infty} \int_{(2m-1)\lambda/2}^{(2m+1)\lambda/2} \varphi(t) \exp(-2\pi i n t/\lambda)\, dt \\
&= \frac{1}{\lambda} \int_{-\infty}^{+\infty} \varphi(t) \exp(-2\pi i n t/\lambda)\, dt = \frac{1}{\lambda} \hat\varphi(2\pi n/\lambda).
\end{aligned}$$

where the interchange of summation and integration signs is justified by the uniformity of the convergence.

Now since $\varphi \in \mathcal{S}$ the numbers $\hat{\varphi}(2\pi n/\lambda)$ decrease rapidly to zero as $|n| \to \infty$ (for example, $|\hat{\varphi}(2\pi n)| = o(|n|^{-2})$ as $|n| \to \infty$). Hence the Fourier series

$$\frac{1}{\lambda} \sum_{n=-\infty}^{+\infty} \hat{\varphi}(2\pi n/\lambda) \exp(2\pi int/\lambda)$$

converges uniformly on $[-\lambda/2, \lambda/2]$ (and therefore on every compact). From the classical theory of Fourier series it follows that, for all $t \in \mathbb{R}$,

$$\Phi(t) \equiv \sum_{m=-\infty}^{+\infty} \varphi(t + m\lambda) = \frac{1}{\lambda} \sum_{n=-\infty}^{+\infty} \hat{\varphi}(2\pi n/\lambda) \exp(2\pi int/\lambda). \tag{3.11}$$

Taking $t = 0$ in (3.11) we obtain

$$\sum_{m=-\infty}^{+\infty} \varphi(m\lambda) = \frac{1}{\lambda} \sum_{n=-\infty}^{+\infty} \hat{\varphi}(2\pi n/\lambda)$$

which is known as the **Poisson Formula** for $\varphi \in \mathcal{S}$. If we allow $\lambda$ to tend to infinity in (3.11) then

(a) the left-hand side tends to $\varphi(t)$ since $\varphi \in \mathcal{S}$ and so all terms in the series corresponding to non-zero values of $m$ will tend to zero,

(b) the right-hand side is plainly a Riemann-type sum which converges to an (absolutely convergent) integral, and we have

$$\varphi(t) = \frac{1}{2\pi} \int_{-\infty}^{+\infty} \hat{\varphi}(\omega) \exp(i\omega t) \, d\omega.$$

This is just the Fourier Inversion Formula for a function $\varphi$ belonging to $\mathcal{S}$, and it follows that every function $\hat{\varphi}$ in $\mathcal{S}$ is the Fourier transform of some function $\varphi$ in $\mathcal{S}$. In fact we have:

**Theorem 3.7** *The Fourier transform and its inverse are continuous linear maps from $\mathcal{S}$ onto itself.*

**Proof:** That $\mathcal{F}$ (resp. $\mathcal{F}^{-1}$) is a one-to-one map of $\mathcal{S}$ onto $\mathcal{S}$ is immediate; linearity follows from the linearity of the integral. It remains to show that $\mathcal{F}$ and $\mathcal{F}^{-1}$ are continuous. Suppose that $(\varphi_n)_{n \in \mathbb{N}}$ is a sequence of functions in $\mathcal{S}$ converging in $\mathcal{S}$ to zero. We wish to show that $(\hat{\varphi}_n)_{n \in \mathbb{N}}$ also converges in $\mathcal{S}$ to zero. For any $r, s \in \mathbb{N}_0$ we have

$$(i\omega)^r \hat{\varphi}_n^{(s)}(\omega) = \int_{-\infty}^{+\infty} [(-it)^s \varphi_n(t)]^{(r)} \exp(-i\omega t) \, dt$$

$$= (-i)^s \sum_{\nu=0}^{r} \binom{r}{\nu} \int_{-\infty}^{+\infty} (t^s)^{(\nu)} \varphi_n^{(r-\nu)}(t) \exp(-i\omega t) \, dt$$

and so,

$$|\omega^r \hat{\varphi}_n^{(s)}(\omega)| \leq \sum_{\nu=0}^{\min(r,s)} \binom{r}{\nu} \frac{s!}{(s-\nu)!} \int_{-\infty}^{+\infty} |t^{s-\nu} \varphi_n^{(r-\nu)}(t)| dt$$

$$= \sum_{\nu=0}^{\min(r,s)} \binom{r}{\nu} \frac{s!}{(s-\nu)!} \int_{-\infty}^{+\infty} \frac{|(1+t^2) t^{s-\nu} \varphi_n^{(r-\nu)}(t)|}{1+t^2} dt$$

$$\leq \pi \sum_{\nu=0}^{\min(r,s)} \binom{r}{\nu} \frac{s!}{(s-\nu)!} \left\{ \| \varphi_n \|_{\mathcal{S}}^{s-\nu, r-\nu} + \| \varphi_n \|_{\mathcal{S}}^{s-\nu+2, r-\nu} \right\}.$$

For $\nu = 0, 1, \ldots, \min(r,s)$ we have that $\| \varphi_n \|_{\mathcal{S}}^{s-\nu, r-\nu}$ and $\| \varphi_n \|_{\mathcal{S}}^{s-\nu+2, r-\nu}$ tend to zero as $n \to \infty$. Hence the above inequality shows that $\| \hat{\varphi}_n \|_{\mathcal{S}}^{r,s} \to 0$ as $n \to \infty$ uniformly on compacts for every pair $r, s \in \mathbb{N}_0$. In other words the sequence $(\hat{\varphi}_n)_{n \in \mathbb{N}}$ converges in $\mathcal{S}$ to zero.
The proof that $\mathcal{F}^{-1}$ is a continuous operator is similar. □

## Exercise 3.1 :

1. Find the (classical) Fourier transforms of each of the following:

   (a) $\exp(-|x|)$,

   (b) $\exp(-|x|)\text{sgn}(x)$,

   (c) $\exp(-x^2)$.

   Hint: for (c) note that $x^2 + iy = (x + i/2y)^2 + y^2/4$, and use the result

   $$\int_{-\infty}^{+\infty} \exp(-x^2) dx = \sqrt{\pi}.$$

2. By taking $f(x) = \exp(-x^2 - 2\pi i a x)$ and using the Poisson summation formula in $\mathcal{S}$ show that

   $$1 + 2 \sum_{m=1}^{\infty} \exp(-m^2) \cdot \cos(2\pi m a)$$
   $$= \sqrt{\pi} \exp(-\pi^2 a^2) \left\{ 1 + 2 \sum_{n=1}^{\infty} \exp(-\pi^2 n^2) \cosh(2\pi^2 n a) \right\}.$$

3. The Poisson formula is actually valid for any $f \in L^1(\mathbb{R})$ which is continuous, is of bounded variation, and is such that $|f(x)| \to 0$ as $|x| \to \infty$. Use this to show that

   $$\sum_{n=-\infty}^{+\infty} \frac{1}{a^2 + n^2} = \frac{\pi}{a} \coth(\pi/a).$$

## 3.2 GENERALISED FOURIER TRANSFORMS

### 3.2.1 Tempered distributions

If $\varphi$ is any function belonging to the space $\mathcal{S}$ then $\varphi$ can always be expressed as a limit $\varphi = \lim_{n\to\infty} \varphi_n$ where the functions $\varphi_n$ belong to $\mathcal{D}$ and the convergence is $\omega$-uniform on every compact $K \subset \mathbb{R}$.

**Definition 3.1** *A distribution $\mu \in \mathcal{D}'$ is said to be a tempered distribution (or a distribution of slow growth) if the limit*

$$\lim_{n\to\infty} <\mu, \varphi_n>$$

*exists for any sequence $(\varphi_n)_{n\in\mathbb{N}} \subset \mathcal{D}$ which converges $\omega$-uniformly on compacts to the function $\varphi \in \mathcal{S}$. We can extend $\mu$ as a continuous linear functional on $\mathcal{S}$ by setting*

$$<\mu, \varphi> = \lim_{n\to\infty} <\mu, \varphi_n>$$

*for any $\varphi \in \mathcal{S}$.*

Taking theorem 1.9 and theorem 3.5 into account it is easy to see that the subspace of $\mathcal{D}'$ consisting of all tempered distributions can be identified with the space $\mathcal{S}'$ of all continuous linear functionals on the topological vector space $\mathcal{S}$. In this sense we may write

$$\mathcal{S}' \subset \mathcal{D}'$$

and it follows that all operations that are well defined for distributions in $\mathcal{D}'$ also apply to distributions in $\mathcal{S}'$. However not all such operations applied to a tempered distribution will necessarily generate a tempered distribution. It is easy to confirm that the operations in the following list do carry $\mathcal{S}'$ into $\mathcal{S}'$:

(a) addition of distributions,

(b) multiplication of a distribution by a scalar,

(c) translation of a distribution,

(d) differentiation of a distribution.

**Example 3.2.1** Let $f(t)$ be a locally integrable function on $\mathbb{R}$ such that for some integer $m$,

$$\lim_{|t|\to\infty} |t^{-m} f(t)| = 0 .$$

Then $f$ is said to be a **function of slow growth**, and it generates a regular tempered distribution since we have

$$< \mu_{(f)}(t), \varphi(t) > \; = \; \int_{-\infty}^{+\infty} f(t)\varphi(t)dt$$
$$= \int_{-1}^{+1} f(t)\varphi(t)dt + \int_{|t|>1} f(t)\varphi(t)dt$$

and we can choose $m \in \mathbb{N}_0$ so that $t^{-m}f(t)$ is absolutely integrable on $|t| > 1$. Then we get

$$|< \mu_{(f)}, \varphi >| \; \leq \; \| \varphi \|_{\mathcal{S}}^{0,0} \int_{-1}^{+1} |f(t)|dt \; + \; \| \varphi \|_{\mathcal{S}}^{m,0} \int_{|t|>1} |t^{-m}f(t)|dt$$

which is enough to show that the functional generated by $f$ is continuous (it being obviously linear).

On the other hand it is not true that every regular tempered distribution is generated by a locally integrable function of slow growth. The function $f(x) = e^x \cos(e^x)$ is locally integrable but not of slow growth. Nevertheless it generates a (regular) tempered distribution.

**Example 3.2.2** Any distribution of compact support will be a tempered distribution. Thus the delta function $\delta_a$, and any derivative $\delta_a^{(m)}$, are tempered distributions. The sum

$$\sum_{n=-\infty}^{+\infty} \delta(t - nT)$$

is an example of a singular distribution which is not of compact support, but which is a tempered distribution.

By contrast the distribution

$$\xi = \sum_{n=1}^{+\infty} \exp(n^2)\delta(t - n)$$

is not tempered.

### 3.2.2 Fourier transforms in $\mathcal{S}'$

Let $\mu$ be any tempered distribution in $\mathcal{S}'$. Then we define the Fourier transform $\hat{\mu}$ of $\mu$ to be the linear functional on $\mathcal{S}$, itself a tempered distribution, which satisfies the formula

$$< \hat{\mu}, \varphi > \; = \; < \mu, \hat{\varphi} >, \qquad (3.12)$$

for all $\varphi$ in $\mathcal{S}$. Note that $\varphi \in \mathcal{S}$ implies $\hat{\varphi} \in \mathcal{S}$, so that (since $\mu$ is a tempered distribution) the right-hand side of (3.12) is always well defined; the definition therefore makes sense.

Equation (3.12) also serves to define the inverse Fourier transform for $\mathcal{S}'$; setting $\hat{\mu} \equiv \mathcal{F}[\mu] = \sigma$, $\mu = \mathcal{F}^{-1}[\sigma]$ and $\hat{\varphi} \equiv \mathcal{F}[\varphi] = \psi$, $\varphi = \mathcal{F}^{-1}[\psi]$, we get

$$<\mathcal{F}^{-1}[\sigma], \psi> = <\sigma, \mathcal{F}^{-1}[\psi]>, \qquad (3.13)$$

for any $\sigma \in \mathcal{S}'$ and every $\psi \in \mathcal{S}$. The inverse Fourier transform of any tempered distribution is again a tempered distribution. The usual properties of the classical Fourier transform in $L^1(\mathbb{R})$ extend to the Fourier transform in $\mathcal{S}'$. In particular we have

$$\mathcal{F}[(-it)^k \mu(t)](\omega) = D^k \hat{\mu}(\omega), \qquad (3.14)$$

$$\mathcal{F}[D^k \mu(t)](\omega) = (i\omega)^k \hat{\mu}(\omega), \qquad (3.15)$$

$$\mathcal{F}[\mu(t-a)](\omega) = \exp(-ia\omega)\hat{\mu}(\omega), \qquad (3.16)$$

for all $k \in \mathbb{N}_0$ and $a \in \mathbb{R}$.

**Remark 3.3 Convolution and the Exchange Formula:** *For any two functions $\varphi, \psi \in \mathcal{S}$ the convolution $\varphi * \psi$ is again a function belonging to $\mathcal{S}$ and a direct calculation (or theorem 3.2) shows that*

$$\mathcal{F}[\varphi * \psi](\omega) = \hat{\varphi}(\omega)\hat{\psi}(\omega). \qquad (3.17)$$

*Clearly there may be a problem in the extension of (3.17) to the Fourier transforms of (tempered) distributions. If $\hat{\mu}$ and $\hat{\nu}$ are sufficiently singular then the product $\hat{\mu} \cdot \hat{\nu}$ may not even be defined.*

The following examples of generalised Fourier transforms are particularly important.

**Example 3.2.3** Since $<\delta(t-a), \exp(-i\omega t)>$ is well defined, a natural definition of $\mathcal{F}[\delta_a](\omega)$ would be $\exp(-i\omega a)$. That this value agrees with the definition of the Fourier transform of the tempered distribution $\delta_a$ follows immediately from

$$<\delta(t-a), \hat{\varphi}(t)> = \hat{\varphi}(a)$$

and

$$<\exp(-i\omega a), \varphi(\omega)> = \int_{-\infty}^{+\infty} \varphi(\omega) \exp(-i\omega a)\, d\omega = \hat{\varphi}(a).$$

Similarly, since

$$<\exp(iat), \hat{\varphi}(t)> = \int_{-\infty}^{+\infty} \hat{\varphi}(t) \exp(iat)\, dt = 2\pi\varphi(a)$$

while
$$< 2\pi\delta(\omega - a), \varphi(\omega) > = 2\pi\varphi(a),$$
it follows that the regular tempered distribution generated by $\exp(iat)$ has the Fourier transform $2\pi\delta_a$. In particular, taking $a = 0$, we have
$$\mathcal{F}[1](\omega) = 2\pi\delta(\omega).$$

**Example 3.2.4** From the results of the example above we can immediately deduce the following results:
$$\mathcal{F}[\cos(at)](\omega) = \pi[\delta(\omega - a) + \delta(\omega + a)],$$
$$\mathcal{F}[\sin(at)](\omega) = i\pi[\delta(\omega + a) - \delta(\omega - a)],$$
$$\mathcal{F}[\delta'(t - a)](\omega) = i\omega \cdot \exp(-i\omega a),$$
$$\mathcal{F}[-it \cdot \exp(iat)](\omega) = 2\pi\delta'(\omega - a).$$

**Example 3.2.5** Since $D[\operatorname{sgn}(t)] = 2\delta(t)$ we have
$$\mathcal{F}[D\mu](\omega) = i\omega\mathcal{F}[\mu](\omega) = 2$$
where $\mu \equiv \operatorname{sgn}(t)$. The distributional solution of the equation $i\omega\hat{\mu}(\omega) = 2$ is given by
$$\hat{\mu}(\omega) = \frac{2}{i\omega} + c_0\delta(\omega)$$
where $c_0$ is an arbitrary constant. Hence
$$\mathcal{F}[\operatorname{sgn}(t)](\omega) = \frac{2}{i\omega} + c_0\delta(\omega)$$
and similarly
$$\mathcal{F}[H(t)](\omega) = \frac{1}{i\omega} + c'_0\delta(\omega)$$
where $c'_0$ is another arbitrary constant. But
$$\frac{1}{2\pi}\operatorname{Pv}\int_{-\infty}^{+\infty} \frac{\exp(ixy)}{iy}\, dy = \frac{1}{2}\operatorname{sgn}(x)$$
and therefore, for consistency, we need to set $c_0 = 0$ and $c'_0 = \pi$ to get
$$\mathcal{F}[\operatorname{sgn}(t)](\omega) = \frac{2}{i\omega} \quad ; \quad \mathcal{F}[H(t)](\omega) = \frac{1}{i\omega} + \pi\delta(\omega).$$

**Example 3.2.6** If
$$\mu(t) = \sum_{n=-\infty}^{+\infty} \delta(t - np)$$
then we can express this $p$-periodic (tempered) distribution as a generalised Fourier series and write
$$\mu(t) = \frac{1}{p} \sum_{n=-\infty}^{+\infty} \exp(in2\pi t/p)$$
which suggests the result
$$\mathcal{F}[\mu](\omega) \equiv \hat{\mu}(\omega) = \frac{2\pi}{p} \sum_{n=-\infty}^{+\infty} \delta(\omega - 2n\pi/p).$$
To confirm this let $\varphi$ be an arbitrary function in $\mathcal{S}$. Then,
$$<\hat{\mu}, \varphi> = <\frac{2\pi}{p} \sum_{n=-\infty}^{+\infty} \delta(\omega - 2n\pi/p), \varphi(\omega)> = \frac{2\pi}{p} \sum_{n=-\infty}^{+\infty} \varphi(2n\pi/p),$$
and,
$$<\mu, \hat{\varphi}> = <\sum_{n=-\infty}^{+\infty} \delta(t - np), \hat{\varphi}(t)> = \sum_{n=-\infty}^{+\infty} \hat{\varphi}(np),$$
and the Poisson formula in $\mathcal{S}$ now shows that $<\hat{\mu}, \varphi> = <\mu, \hat{\varphi}>$.

**Exercise 3.2 :**

1. Find the (distributional) Fourier transforms of each of the following:

    (a) $(1 - \exp(-at)) H_0(t)$,

    (b) $\cos^2(at)$, and $\sin^2(at)$,

    (c) $t^n H_0(t), n = 1, 2, \ldots$.

2. Show that
$$\mathcal{F}\left\{\text{Pf}\left(\frac{1}{t^n}\right)\right\} = \frac{\pi(-i)^n}{(n-1)!}\omega^{n-1}\text{sgn}(\omega) \ , \ n = 1, 2, \ldots$$

3. Let $\mu$ be a distribution with compact support. Prove that,

    (a) $\mu$ is a tempered distribution,

    (b) $\mu$ is a distribution of finite order.

## 3.3 FOURIER TRANSFORMS IN $\mathcal{E}'$

### 3.3.1 Distributions of compact support

We denote by $\mathcal{E}$ the linear space $\mathcal{C}^\infty(\mathbb{R})$ equipped with the following mode of convergence:

> A sequence $(\varphi_n)_{n \in \mathbb{N}}$ of infinitely differentiable functions is said to converge to zero in the sense of the space $\mathcal{E}$ if and only if the $\varphi_n$ converge $\omega$-uniformly to zero on every compact subset of $\mathbb{R}$ as $n \to \infty$.

Convergence in this sense corresponds to the topology generated on $\mathcal{C}^\infty$ by the family of seminorms,

$$\| \varphi \|_{[m,K]} \equiv \max_{0 \leq r \leq m} \left\{ \sup_{x \in K} |\varphi^{(r)}(x)| \right\} \tag{3.18}$$

where $m \in \mathbb{N}_0$ and $K$ is any compact subset of $\mathbb{R}$.

The dual space $\mathcal{E}'$ is the linear space of all functionals $\mu$ on $\mathcal{E}$ which are continuous in the sense that $\lim_{n \to \infty} <\mu, \varphi_n> = 0$ for every sequence $(\varphi_n)_{n \in \mathbb{N}}$ which converges to zero in $\mathcal{E}$. An argument similar to that used in theorem 1.2 shows that continuity of a linear functional in $\mathcal{E}'$ is equivalent to a certain boundedness condition:

**Theorem 3.8** *A linear functional $\mu$ on $\mathcal{E}$ belongs to $\mathcal{E}'$ if and only if there exists a compact subset $K$ of $\mathbb{R}$, a non-negative integer $m$ and a constant $C \geq 0$ such that,*

$$|<\mu, \varphi>| \leq C \max_{0 \leq r \leq m} \left\{ \sup_{x \in K} |\varphi^{(r)}(x)| \right\} \equiv C \| \varphi \|_{[m,K]}$$

*for all test functions $\varphi \in \mathcal{E}$.*

If $(\varphi_n)_{n \in \mathbb{N}}$ is a sequence of functions belonging to $\mathcal{D}$ which converges to zero in the sense of $\mathcal{D}$ then $(\varphi_n)_{n \in \mathbb{N}}$ converges to zero in $\mathcal{E}$. It can be shown that $\mathcal{D}$ is a dense subspace of $\mathcal{E}$. Hence from theorem 1.9 it follows that the restriction of any functional $\mu \in \mathcal{E}'$ to the space $\mathcal{D}$ is a distribution. What is more theorem 3.8 shows that $<\mu, \varphi> = 0$ for any $\varphi \in \mathcal{D}$ such that the support of $\varphi$ does not meet $K$; hence $\mu$ is necessarily a distribution of compact support. On the other hand, let $\mu$ be any distribution in $\mathcal{D}'$ which has compact support. Choose a function $\theta \in \mathcal{D}$ such that $\theta(x) = 1$ on the support of $\mu$ and extend $\mu$ to a functional $\mu_1$ on $\mathcal{E}$ by setting

$$\forall \varphi \in \mathcal{E} : <\mu_1, \varphi> \equiv <\mu, \theta \varphi> . \tag{3.19}$$

We leave it to the reader to confirm that this extension is uniquely defined (independently of the choice of $\theta$) and is linear and continuous on $\mathcal{E}'$. Thus we may identify $\mathcal{E}'$ with the subspace of $\mathcal{D}'$ comprising all distributions of compact support.

In the corollary 1.3.1 it was shown that every distribution in $\mathcal{E}'$ could be represented as a finite sum of finite-order derivatives of continuous functions, each vanishing identically outside some neighbourhood of the support of $\mu$. The following structure theorem for distributions of compact support is also established by Schwartz ( [97, Chapter III, Théorème XXVII])

**Theorem 3.9** *Let $\mu \in \mathcal{E}'$ be of order $m$ and have support $K$. Then $\mu$ is the sum of finitely many derivatives of order $\leq m$ of measures, each with support contained in some arbitrarily chosen neighbourhood $I$ of $K$.*

### 3.3.1.1 Bandlimited functions (classical case)

A function which is the inverse Fourier transform of a distribution of compact support is said to be **bandlimited** or to be a **function of compact spectrum**. Suppose in particular that $f(t) \in L^1(\mathbb{R})$ has a Fourier transform $\hat{f}(\omega)$ which vanishes identically outside the finite interval $[-\pi\Omega, +\pi\Omega]$. (Such a function is said to be bandlimited in the *classical* sense). The Fourier inversion integral converges for all $t$ and

$$f(t) = \frac{1}{2\pi} \int_{-\pi\Omega}^{+\pi\Omega} \hat{f}(\omega) \exp(i\omega t) d\omega \qquad (3.20)$$

everywhere in $\mathbb{R}$. The function $\hat{f}(\omega)$ is bounded and continuous, and we may differentiate under the integral sign to get

$$f^{(r)}(a) = \frac{1}{2\pi} \int_{-\infty}^{+\infty} (i\omega)^r \hat{f}(\omega) \exp(ia\omega) d\omega,$$

for all $r \geq 0$, and every $a \in \mathbb{R}$. Hence,

$$f(t) = \frac{1}{2\pi} \int_{-\pi\Omega}^{+\pi\Omega} \hat{f}(\omega) \exp[i\omega(t-a)] \exp(i\omega a) d\omega$$

$$= \frac{1}{2\pi} \int_{-\pi\Omega}^{+\pi\Omega} \hat{f}(\omega) \exp(i\omega a) \left\{ \sum_{n=0}^{\infty} \frac{[i\omega(t-a)]^n}{n!} \right\} d\omega$$

$$= \frac{1}{2\pi} \sum_{n=0}^{\infty} \frac{[i(t-a)]^n}{n!} \int_{-\pi\Omega}^{+\pi\Omega} \omega^n \hat{f}(\omega) \exp(i\omega a) d\omega \qquad (3.21)$$

which is a Taylor expansion about the point $t = a$; $f(t)$ is analytic over the entire real axis. In fact the argument remains valid if we replace throughout the real variable $t$ by a complex variable $z$. Thus a function $f(t)$ which is bandlimited in the classical sense extends to a function $f(z)$ which is defined and analytic over the entire complex plane; that is, $f$ is an entire function. Moreover, $f(z)$ satisfies a certain boundedness condition:

$$|f(z)| = \left| \frac{1}{2\pi} \int_{-\pi\Omega}^{+\pi\Omega} \exp(i\omega z) \hat{f}(\omega) \, d\omega \right| \leq \frac{V}{2\pi} \exp(\pi\Omega |y|) \qquad (3.22)$$

where

$$V \equiv \int_{-\pi\Omega}^{+\pi\Omega} |\hat{f}(\omega)| d\omega ,$$

and $y = \Im(z)$. It is clearly of interest to establish whether such properties remain true for functions which are inverse Fourier transforms of arbitrary distributions of compact support. A significant example of the difference between classically bandlimited functions and the more general case occurs in connection with the so-called **Sampling Theorem**.

### 3.3.1.2 The WKS Sampling Theorem

Let $f \in L^1(R) \cap C(\mathbb{R})$ be bandlimited to $[-\pi\Omega, \pi\Omega]$. Then, from (3.20), we can write

$$f(t) = \frac{1}{2\pi} \int_{-\pi\Omega}^{+\pi\Omega} \hat{f}(\omega) \left\{ \sum_{n=-\infty}^{+\infty} \exp(i\omega n/\Omega) \frac{\sin[\pi(\Omega t - n)]}{\pi(\Omega t - n)} \right\} d\omega$$

where for each fixed $t$ we replace $\exp(i\omega t)$, as a function of $\omega$, by its Fourier series expansion over the interval $(-\pi\Omega, \pi\Omega)$. Then

$$f(t) = \sum_{n=-\infty}^{+\infty} \left\{ \frac{1}{2\pi} \int_{-\pi\Omega}^{+\pi\Omega} \hat{f}(\omega) \exp(i\omega n/\Omega) \, d\omega \right\} \frac{\sin[\pi(\Omega t - n)]}{\pi(\Omega t - n)}$$

the interchange of summation and integration being justified by the absolute and uniform convergence of the Fourier series concerned. The Fourier inversion theorem then gives

$$f(t) = \sum_{n=-\infty}^{+\infty} f(n/\Omega) \frac{\sin[\pi(\Omega t - n)]}{\pi(\Omega t - n)} \equiv \sum_{n=-\infty}^{+\infty} f(n/\Omega) \operatorname{sinc}(\Omega t - n). \qquad (3.23)$$

The function $f(t)$ can therefore be reconstructed from the values $f(n/\Omega)$ sampled at the discrete points $\{n/\Omega\}_{n \in \mathbb{Z}}$. This is the essential content of the **Whitaker-Kotel'nikov-Shannon (WKS) Sampling Theorem** [41], which is most often stated for a bandlimited function which belongs to $L^2(\mathbb{R})$:

**Theorem 3.10 (WKS)** *If $f : \mathbb{R} \to \mathbb{R}$ has the form*

$$f(t) \equiv \frac{1}{2\pi} \int_{-\pi\Omega}^{+\pi\Omega} g(\omega) \exp(i\omega t) \, d\omega,$$

*where $g(\omega)$ is a function in $L^2(\mathbb{R})$ which vanishes almost everywhere outside $[-\pi\Omega, \pi\Omega]$ then $f(t)$ is a bounded continuous function in $L^2(\mathbb{R})$, and (3.23) holds uniformly on compacts.*

**Proof:** If $g$ is an $L^2$ function of compact support then $g$ must also belong to $L^1(\mathbb{R})$. Also $f(t) = \mathcal{F}^{-1}[g](t)$ is necessarily a bounded continuous function which belongs to $L^2$ (though not necessarily to $L^1$). The argument used to establish (3.23) above then goes through as before. □

If $f(t) = \exp(iat)$ then $f$ is the inverse Fourier transform of the singular (tempered) distribution $2\pi\delta(\omega - a)$. If $-\pi < a < +\pi$ and we take $\Omega = 1$ then the WKS expansion (3.23) becomes

$$\exp(iat) = \sum_{n=-\infty}^{+\infty} \exp(ina) \frac{\sin[\pi(t-n)]}{\pi(t-n)} \qquad (3.24)$$

For each fixed $t$ the right-hand side of (3.24) is the ordinary Fourier series expansion of $\exp(iat)$, as a function of $a$, over the interval $(-a, +a)$; thus the WKS sampling expansion remains valid. On the other hand, if $f(t) = it \cdot \exp(iat)$, so that $f$ is the inverse Fourier transform of the distribution $-2\pi\delta'(\omega - a)$, we have that $f(n) = in \cdot \exp(ian) = O(n)$ tends to infinity with $n$ and (3.23) does not hold. This shows that the fact that $f$ has a Fourier transform which is a distribution of compact support is not in itself enough to guarantee that the WKS sampling expansion will be valid.

**Exercise 3.3 :**

1. Show that $\mathcal{D}$ is a dense subspace of $\mathcal{E}$, and that $\mathcal{E}'$ is dense in $\mathcal{D}'$.

2. Prove Theorem 3.8.

3. Show that

$$\sum_{n=-\infty}^{+\infty} \frac{\sin(\pi(t-n))}{\pi(t-n)} = 1.$$

4. Suppose that $f(t)$ has a (classical) Fourier transform which belongs to $L^1(\mathbb{R})$ and which vanishes identically outside $(-(1-\varepsilon)\pi, (1-\varepsilon)\pi)$, where $0 < \varepsilon < 1$. Show that

$$f(t) = \sum_{n=-\infty}^{+\infty} f(n) \cdot \operatorname{sinc}(\varepsilon(t-n)) \cdot \operatorname{sinc}(t-n).$$

### 3.3.2 The Paley-Wiener-Schwartz Theorem

#### 3.3.2.1 Preliminaries on entire functions

We denote by $\mathcal{H}(\mathbb{C})$ the linear space of all entire functions $f(z)$, and by $\mathcal{H}_e(\mathbb{C})$ the subspace of all entire functions of exponential type; that is, $\mathcal{H}_e(\mathbb{C})$ comprises all entire functions $f(z)$ for which an inequality of the form

$$|f(z)| \leq A \cdot \exp(\tau |z|) \tag{3.25}$$

holds for some positive constants $A$ and $\tau$; in particular, such a function $f$ is said to be of exponential type $\leq \tau$. A function in $\mathcal{H}_e(\mathbb{C})$ enjoys a very important property that is not shared by all entire functions; if $f(z)$ is of exponential type and satisfies some particular growth conditions along a line in the complex plane then more can always be said about its growth over the whole complex plane.

**Theorem 3.11 (Paley-Wiener)** *The entire function $f : \mathbb{C} \to \mathbb{C}$ is of exponential type $\leq \pi\Omega$ and belongs to $L^2(\mathbb{R})$ on the real line if and only if it can be represented as*

$$f(z) = \frac{1}{2\pi} \int_{-\pi\Omega}^{+\pi\Omega} \hat{f}(\omega) \exp(i\omega z)\, d\omega \ , \quad z \in \mathbb{C}$$

*for some function $\hat{f} \in L^2(-\pi\Omega, +\pi\Omega)$.*

**Proof:** (See, for example Young [116].) □

For any real $\Omega > 0$ we denote by $\mathcal{B}^0_{\pi\Omega}$ the space of all $L^2(\mathbb{R})$ functions with Fourier transforms which vanish identically outside $(-\pi\Omega, +\pi\Omega)$, and by $\mathcal{B}^0$ the space of all $L^2(\mathbb{R})$ functions with arbitrary compact spectrum:

$$\mathcal{B}^0 = \bigcup_{\Omega > 0} \mathcal{B}^0_{\pi\Omega}.$$

From the Paley-Wiener Theorem an $L^2(\mathbb{R})$ function $f : \mathbb{R} \to \mathbb{C}$ belongs to $\mathcal{B}^0$ if and only if it may be extended onto the entire complex plane as an entire function of exponential type such that

$$|f(z)| \leq C \cdot \exp(\tau |\Im(z)|) \ , \ \forall z \in \mathbb{C}$$

for some constants $C > 0$ and $\tau \geq 0$. Hence $\mathcal{B}^0 \equiv \mathcal{B}^0(\mathbb{C})$ is a subspace of $\mathcal{H}_e(\mathbb{C})$, and is sometimes referred to as the **Paley-Wiener space**.

### 3.3.2.2 Polynomially bounded functions of compact spectrum

Let $\mathcal{B}_{\pi\Omega}(\mathbb{C})$ denote the subclass of functions in $\mathcal{H}_e(\mathbb{C})$ of exponential type $\leq \pi\Omega$ which are bounded by some polynomial on the real axis.

**Theorem 3.12 (Paley-Wiener-Schwartz)** *If $f$ is a function in $\mathcal{B}_{\pi\Omega}$ then the (generalised) Fourier transform of $f$, which we shall write as $\hat{\mu} \equiv \mathcal{F}[f]$, is a distribution of compact support in $[-\pi\Omega, +\pi\Omega]$.*

**Proof:** Given $\varepsilon > 0$, let $\hat{\varphi}_\varepsilon$ be a function in $\mathcal{D}$ with support contained in $(-\varepsilon\pi\Omega, +\varepsilon\pi\Omega)$. Its inverse Fourier transform, $\varphi_\varepsilon(t)$, will be analytically continuable into the complex plane as an entire function of exponential type $\leq \varepsilon\pi\Omega$ that decreases to zero on the real axis more rapidly than any power of $|t|^{-1}$ as $t \to \infty$. The same will be true of the product $\varphi_\varepsilon(t)f(t)$ and the Fourier transform of this product will be given by

$$\mathcal{F}[\varphi_\varepsilon f] = \hat{\varphi}_\varepsilon * \hat{\mu}.$$

By the Paley-Wiener Theorem, since $\varphi_\varepsilon f \in L^2(\mathbb{R})$, it follows that $\hat{\varphi}_\varepsilon * \hat{\mu}$ is a function of compact support contained within the interval $(-\pi\Omega(1+\varepsilon), +\pi\Omega(1+\varepsilon))$. From the family $\{\hat{\varphi}_\varepsilon\}_{\varepsilon > 0}$ we may extract a sequence converging in $\mathcal{D}'$ to the distribution $\delta(\omega)$ as $\varepsilon \to 0$. By the continuity of convolution it follows that

$$\hat{\varphi}_\varepsilon * \hat{\mu}(\omega) \to \delta(\omega) * \hat{\mu}(\omega) = \hat{\mu}(\omega)$$

as $\varepsilon \to 0$. Each of the functions $\hat{\varphi}_\varepsilon * \hat{\mu}$ has support contained in $(-\pi\Omega(1+\varepsilon), +\pi\Omega(1+\varepsilon))$; therefore, since $\varepsilon > 0$ may be taken arbitrarily small, the support of $\hat{\mu}$ must be contained in the interval $[-\pi\Omega, +\pi\Omega]$. □

**Theorem 3.13 (Converse Paley-Wiener-Schwartz)** *If $f$ is the inverse Fourier transform of a distribution $\hat{\mu}$ of compact support contained in $[-\pi\Omega, +\pi\Omega]$ then $f \in \mathcal{B}_{\pi\Omega}$.*

**Proof:** If $\hat{\mu}$ has compact support then it is a distribution of finite order and so we may write

$$\hat{\mu} = \sum_{j=0}^{r} D^j \hat{\mu}_j \tag{3.26}$$

where the $\hat{\mu}_j$ are measures of compact support contained in the interval $[-\pi\Omega, +\pi\Omega]$. For each $j = 0, 1, \ldots, r$ there exists a function $\hat{G}_j$ of bounded variation over $[-\pi\Omega, +\pi\Omega]$ such that

$$f_j(t) = \frac{1}{2\pi} <\hat{\mu}_j(\omega), \exp(i\omega t)> = \frac{1}{2\pi} \int_{-\pi\Omega}^{+\pi\Omega} \exp(i\omega t) d\hat{G}_j(\omega).$$

Further the functions $f_j$ may be extended onto the complex plane through

$$f_j(z) = \frac{1}{2\pi} \int_{-\pi\Omega}^{+\pi\Omega} \exp(i\omega z) \, d\hat{G}_j(\omega)$$

thereby showing that they are entire functions of exponential type $\leq \pi\Omega$, and are bounded on the real axis. Taking the inverse Fourier transform of both sides of (3.26) we have

$$f(t) = \sum_{j=0}^{r} (-it)^j f_j(t)$$

which extends onto the entire complex plane and is such that

$$|f(z)| \leq \sum_{j=0}^{r} |z|^j |f_j(z)| \leq \sum_{j=0}^{r} c_j |z|^j \exp(\pi\Omega |\Im(z)|).$$

This shows that $f(z)$ is an entire function of exponential type $\leq \pi\Omega$ which is polynomially bounded on the real axis. $\square$

We now set

$$\mathcal{B} \equiv \mathcal{B}(\mathbb{C}) = \bigcup_{\Omega > 0} \mathcal{B}_{\pi\Omega}(\mathbb{C}) \subset \mathcal{H}_e(\mathbb{C}) \qquad (3.27)$$

so that $\mathcal{B}(\mathbb{C})$ comprises all functions of arbitrary compact spectrum. For any $r \in \mathbb{N}$ we denote by $L^2_r(\mathbb{R})$ the space of all square integrable functions with respect to the tempered measure

$$dm_r(t) \equiv (1 + t^2)^{-r} dt \, .$$

Then to any function $f \in \mathcal{B}$ there corresponds a non-negative integer $r$ such that $f \in L^2_r(\mathbb{R})$. We can now define for each $r$ the subspace

$$\mathcal{B}^r(\mathbb{C}) \equiv \mathcal{B}^r = \mathcal{B} \cap L^2_r(\mathbb{R}) \qquad (3.28)$$

which comprises all functions of compact spectrum which grow no faster on $\mathbb{R}$ as $t \to \infty$ than $|t|^{r-\varepsilon}$, for some $\varepsilon > 1/2$. $\mathcal{B}^r$ is a Hilbert space with respect to the inner product

$$(f, g) \equiv \int_{-\infty}^{+\infty} f(t) \bar{g}(t) \, dm_r(t)$$

and moreover $\mathcal{B}^r$ is closed under convergence in the sense of the norm associated with this inner product; that is, $\mathcal{B}^r$ is a closed subspace of $L^2_r(\mathbb{R})$.

**Theorem 3.14** *If $f \in \mathcal{B}^r$ then $f(t)$ is the inverse Fourier transform of a distribution of compact support of order $\leq r$.*

**Proof:** It is immediate from the definition of $B^r$ that $\hat{\mu} \equiv \mathcal{F}[f]$ is a distribution of compact support, and therefore a distribution of finite order; it remains to show that $\hat{\mu}$ is of order $\leq r$. The function

$$g(t) \equiv \frac{f(t)}{(1-it)^r} = f(t) \frac{(1+it)^r}{(1+t^2)^r}$$

certainly belongs to $L^2(\mathbb{R})$, since $|f(t)|$ grows no faster than $|t|^{r-\varepsilon}$ for some $\varepsilon > 1/2$. Hence $\hat{g} = \mathcal{F}[g]$ exists and we have

$$\hat{\mu}(\omega) = (1+D)^r \hat{g}(\omega).$$

The function $\hat{g}$ is not, in general, a function of compact support but for any $\varphi \in C^\infty$

$$<\hat{\mu},\varphi> = <(1+D)^r \hat{g}, \lambda\varphi>$$

where $\lambda \in \mathcal{D}$ is a function such that $\lambda(\omega) = 1$ on the support of $\hat{\mu}$. If $0 < \varepsilon < 1$ we can always select $\lambda$ so that $\lambda(\omega) = 0$ for all $\omega \notin I_\varepsilon \equiv \text{supp}(\hat{\mu}) + [-\varepsilon, +\varepsilon]$. Then

$$<(1+D)^r \hat{g}, \lambda\varphi> = <\hat{g}, (1-D)^r(\lambda\varphi)> = \sum_{j=0}^r \binom{r}{j}(-1)^j <\hat{g}, D^j(\lambda\varphi)>$$

$$= \sum_{j=0}^r \binom{r}{j}(-1)^j <\hat{g}, \sum_{m=0}^j \binom{j}{m}(D^m \lambda)(D^{j-m}\varphi)>$$

$$= \sum_{j=0}^r \binom{r}{j}(-1)^j \sum_{m=0}^j \binom{j}{m} <\hat{g} D^m \lambda, D^{j-m}\varphi>$$

$$= \sum_{j=0}^r \binom{r}{j}(-1)^j \sum_{m=0}^j \binom{j}{m}(-1)^{j-m} <D^{j-m}(\hat{g} D^m \lambda), \varphi>$$

Collecting and re-distributing the terms in the double summation we obtain

$$<(1+D)^r \hat{g}, \lambda\varphi> = \sum_{m=0}^r <D^m \sum_{j=m}^r (-1)^j [\binom{r}{j}\binom{j}{j-m}) \hat{g}(D^{j-m}\lambda)], \varphi>$$

and therefore writing

$$\hat{f}_{\varepsilon m} \equiv \sum_{j=m}^r (-1)^j [\binom{r}{j}\binom{j}{j-m})\hat{g}(D^{j-m}\lambda)]$$

for $m = 0, 1, \ldots, r$, we get

$$<\hat{\mu},\varphi> = <\sum_{m=0}^r D^m \hat{f}_{\varepsilon m}, \varphi>.$$

Since $\lambda$ has support contained in $I_\varepsilon$ the products $\hat{g} D^{j-m}\lambda$ also have their supports contained in this set, and therefore so do the functions $\hat{f}_{\varepsilon m}$, for $m = 0, 1, \ldots, r$. Further, since

$\hat{g} \in L^2(\mathbb{R})$ then each $\hat{f}_{\varepsilon m} \in L^2(\mathbb{R})$ and because these functions have compact support it follows that $\hat{f}_{\varepsilon m} \in L^1(\mathbb{R})$ for $m = 0, 1, \ldots, r$. Thus we obtain

$$
\begin{aligned}
|<\hat{\mu}, \varphi>| &= \left| \sum_{m=0}^{r} (-1)^m <\hat{f}_{\varepsilon m}, \varphi^{(m)}> \right| \\
&\leq \sum_{m=0}^{r} \left| \int_{I_\varepsilon} \hat{f}_{\varepsilon m}(\omega) \varphi^{(m)}(\omega)\, d\omega \right| \\
&\leq \sum_{m=0}^{r} \left\{ \sup_{\omega \in K} |\varphi^{(m)}(\omega)| \right\} \int_K |\hat{f}_{\varepsilon m}(\omega)|\, d\omega \leq C_\varepsilon \|\varphi\|_{[r, K]}
\end{aligned}
$$

where $K$ is a compact of $\mathbb{R}$ such that $K \supset \operatorname{supp}(\hat{\mu}) + [-1, +1]$. This shows that $\hat{\mu}$ is a distribution of order $\leq r$. □

### 3.3.3 Periodic distributions

A brief introduction to periodic distributions was given in §2.2.1.2 where periodic distributions were shown to constitute a generalization of periodic functions.

#### 3.3.3.1 Definition

A distribution $\hat{\mu} \in \mathcal{D}'$ is said to be periodic in case there exists some real number $p > 0$ such that

$$<\hat{\mu}(\omega), \varphi(\omega - p)> \; = \; <\hat{\mu}(\omega), \varphi(\omega)>, \quad \forall \varphi \in \mathcal{D}.$$

Any number $p$ for which the above equality is true is called a **period** of $\hat{\mu}$; clearly, a periodic distribution $\hat{\mu}$ of period $p$ will have an infinite set of periods $\{kp\}_{k \in \mathbb{N}}$. Without loss of generality we shall take $p$ to be the smallest period. For the sake of simplicity we shall take $p = 2\pi$ throughout what follows.

Recall the definition of translate for a distribution $\hat{\mu}$ as given in §1.3.1: the translation operator $\tau_a$ is defined for any function $\varphi \in \mathcal{D}$ by $\tau_a \varphi(\omega) = \varphi(\omega - a)$, and extends to an arbitrary distribution $\hat{\mu} \in \mathcal{D}'$ according to

$$<\tau_a \hat{\mu}, \varphi> \; = \; <\hat{\mu}, \tau_{-a} \varphi>$$

for all test functions $\varphi \in \mathcal{D}$. We can therefore give a formal definition of a periodic distribution as follows:

**Definition 3.2** *A distribution $\hat{\mu} \in \mathcal{D}'$ is said to be periodic of period $2\pi$ (or, a $2\pi$-periodic distribution), if $\tau_{2\pi} \hat{\mu} = \hat{\mu}$ (where the equality sign is to be understood in the sense of $\mathcal{D}'$).*

We denote by $\mathcal{T}_{2\pi}(\mathbb{R})$, or simply by $\mathcal{T}_{2\pi}$, the set of all periodic distributions of period $2\pi$; that is, we have
$$\mathcal{T}_{2\pi} \equiv \{\hat{\mu} \in \mathcal{D}' : \tau_{2\pi}\hat{\mu} = \hat{\mu}\}$$
which is easily seen to be a linear subspace of $\mathcal{D}'$. Then we define
$$\mathcal{P}_{2\pi} \equiv \mathcal{T}_{2\pi} \cap \mathcal{E}$$
as the set of all periodic functions which are infinitely smooth. Clearly $\mathcal{P}_{2\pi}$ is a linear subspace of $\mathcal{E}$. Convergence in $\mathcal{P}_{2\pi}$ is defined as follows: a sequence $(\theta_n)_{n \in \mathbb{N}}$ is said to converge in $\mathcal{P}_{2\pi}$ to a limit function $\theta$ if and only if

**P1** $\theta_n \in \mathcal{P}_{2\pi}$ for all $n \in \mathbb{N}$,

**P2** for each non-negative integer $j$ the sequence $(\theta_n^{(j)})_{n \in \mathbb{N}}$ converges uniformly to $\theta^{(j)}$.

It follows then that the limit function $\theta$ will also be in $\mathcal{P}_{2\pi}$, that is, $\mathcal{P}_{2\pi}$ is closed under convergence.

Let $\varphi$ be a function in $\mathcal{D}$ and as in §2.2.1.2 define the periodic extension of $\varphi$ as the function $T_{2\pi}[\varphi]$ by setting
$$T_{2\pi}[\varphi](\omega) \equiv \sum_{n=-\infty}^{+\infty} \tau_{2n\pi}\varphi(\omega) = \sum_{n=-\infty}^{+\infty} \varphi(\omega - 2n\pi)$$
where, since $\varphi$ has compact support, for each $\omega \in \mathbb{R}$ the infinite sum has only a finite number of non-zero terms. It is obvious that $T_{2\pi}[\varphi]$ is a function in $\mathcal{E}$. The mapping $T_{2\pi} : \mathcal{D} \to \mathcal{E}$ is called the **periodic extension operator** of period $2\pi$. Now let $\hat{\mu}$ be a distribution in $\mathcal{E}'$. Then we define the periodic extension of period $2\pi$ of $\hat{\mu}$, denoted by $T'_{2\pi}[\hat{\mu}]$, by
$$< T'_{2\pi}[\hat{\mu}], \varphi > = < \hat{\mu}, T_{2\pi}[\varphi] >$$
for all $\varphi \in \mathcal{D}$.

**Theorem 3.15** *Let $\hat{\mu}$ be any distribution in $\mathcal{E}'$. Then $T'_{2\pi}[\hat{\mu}]$ is a periodic distribution in $\mathcal{T}_{2\pi}$.*

**Proof:** For any function $\varphi \in \mathcal{D}$ we have that $T_{2\pi}[\tau_{-2\pi}\varphi] = \tau_{-2\pi}T_{2\pi}[\varphi] = T_{2\pi}[\varphi]$ and therefore we obtain
$$\begin{aligned}< \tau_{2\pi}T'_{2\pi}[\hat{\mu}], \varphi > &= < T'_{2\pi}[\hat{\mu}], \tau_{-2\pi}\varphi > \\ &= < \hat{\mu}, T_{2\pi}[\tau_{-2\pi}\varphi] > \\ &= < \hat{\mu}, T_{2\pi}[\varphi] > = < T'_{2\pi}[\hat{\mu}], \varphi >\end{aligned}$$
which proves the theorem. $\square$

**Lemma 3.3.1** *(1) Let $\hat{\eta}$ be a distribution in $\mathcal{T}_{2\pi}$ and $\varphi$ a function in $\mathcal{D}$. Then $T'_{2\pi}[\varphi\hat{\eta}] = (T_{2\pi}[\varphi])\hat{\eta}$ ;*

*(2) If $\hat{\mu}$ is a distribution of compact support in $\mathcal{E}'$ and $\theta$ a function in $\mathcal{P}_{2\pi}$ then $T'_{2\pi}[\theta\hat{\mu}] = \theta(T'_{2\pi}[\hat{\mu}])$ .*

**Proof:** (1) If $\hat{\eta}$ is a periodic distribution in $\mathcal{T}_{2\pi}$ then $\tau_{2\pi}(\varphi\hat{\eta}) = (\tau_{2\pi}\varphi)(\tau_{2\pi}\hat{\eta}) = (\tau_{2\pi}\varphi)\hat{\eta}$ and therefore we obtain

$$T'_{2\pi}[\varphi\hat{\eta}] = \sum_{n=-\infty}^{+\infty} \tau_{2n\pi}(\varphi\hat{\eta}) = \sum_{n=-\infty}^{+\infty} (\tau_{2n\pi}\varphi)\hat{\eta}$$

$$= \left\{\sum_{n=-\infty}^{+\infty} \tau_{2n\pi}\varphi\right\}\hat{\eta} = (T_{2\pi}[\varphi])\hat{\eta}$$

as asserted.

(2) If $\theta$ is a periodic function in $\mathcal{P}_{2\pi}$ then $\tau_{2\pi}(\theta\hat{\mu}) = (\tau_{2\pi}\theta)(\tau_{2\pi}\hat{\mu}) = \theta(\tau_{2\pi}\hat{\mu})$ and so

$$T'_{2\pi}[\theta\hat{\mu}] = \sum_{n=-\infty}^{+\infty} \tau_{2n\pi}(\theta\hat{\mu}) = \sum_{n=-\infty}^{+\infty} \theta(\tau_{2n\pi}\hat{\mu})$$

$$= \theta \sum_{n=-\infty}^{+\infty} \tau_{2n\pi}\hat{\mu} = \theta\left(T'_{2\pi}[\hat{\mu}]\right)$$

as was to be proved. □

*A function $\xi \in \mathcal{D}$ is said to be a $2\pi$-**periodic partition of unity** in $\mathcal{D}$ (or, a $2\pi$-**unitary** function in $\mathcal{D}$) if $T_{2\pi}[\xi] = 1$.*

**Theorem 3.16** *(1) Every function in $\mathcal{P}_{2\pi}$ is the periodic extension of a function in $\mathcal{D}$;*

*(2) Every distribution in $\mathcal{T}_{2\pi}$ is the periodic extension of a distribution in $\mathcal{E}'$.*

**Proof:** (1) Let $\theta \in \mathcal{P}_{2\pi}$. Then for any (fixed) $2\pi$-unitary function $\xi \in \mathcal{D}$ the function $\varphi = \xi\theta$ belongs to $\mathcal{D}$. Moreover, taking the lemma 3.3.1 into account, we have that

$$T_{2\pi}[\varphi] = T_{2\pi}[\xi\theta] = (T_{2\pi}[\xi])\theta = \theta$$

which proves the first part of the theorem.

(2) Now let $\hat{\eta}$ be a distribution in $\mathcal{T}_{2\pi}$ and for any $2\pi$-unitary function $\xi \in \mathcal{D}$ define $\hat{\mu} = \xi\hat{\eta}$. Then $\hat{\mu} \in \mathcal{E}'$. Using again a result of lemma 3.3.1 we get

$$T'_{2\pi}[\hat{\mu}] = T'_{2\pi}[\xi\hat{\eta}] = (T_{2\pi}[\xi])\hat{\eta} = \hat{\eta}$$

which completes the proof. □

Denote by $\mathcal{P}'_{2\pi}$ the dual space of $\mathcal{P}_{2\pi}$, that is, the space of all continuous linear functionals on $\mathcal{P}_{2\pi}$. For any $F \in \mathcal{P}'_{2\pi}$ denote by $(F, \theta)$ the action of $F$ on the test function $\theta \in \mathcal{P}_{2\pi}$. Then we have

**Theorem 3.17** *The spaces $\mathcal{P}'_{2\pi}$ and $\mathcal{T}_{2\pi}$ are algebraically and topologically isomorphic.*

**Proof:** (a) The mapping $T_{2\pi} : \mathcal{D} \to \mathcal{P}_{2\pi}$ continuously transforms $\mathcal{D}$ into $\mathcal{P}_{2\pi}$. Its transpose, $T'_{2\pi} : \mathcal{P}'_{2\pi} \to \mathcal{D}'$, defined by $< T'_{2\pi}[F], \varphi > = (F, T_{2\pi}[\varphi])$ for any $F \in \mathcal{P}'_{2\pi}$ and every $\varphi \in \mathcal{D}$, continuously transforms $\mathcal{P}'_{2\pi}$ into $\mathcal{D}'$. Moreover

$$< \tau_{2\pi} T'_{2\pi}[F], \varphi > \;=\; < T'_{2\pi}[F], \tau_{-2\pi}\varphi >$$
$$=\; (F, T_{2\pi}[\tau_{-2\pi}\varphi]) = (F, T_{2\pi}[\varphi]) = < T'_{2\pi}[F], \varphi >$$

and therefore $T'_{2\pi}[F] \in \mathcal{T}_{2\pi} \subset \mathcal{D}'$. Hence $T'_{2\pi}$ continuously maps $\mathcal{P}'_{2\pi}$ into $\mathcal{T}_{2\pi}$.

(b) Let $\xi$ be a $2\pi$-unitary function in $\mathcal{D}$. The map $\Xi_{2\pi} : \mathcal{P}_{2\pi} \to \mathcal{D}$, defined by $\Xi_{2\pi}[\theta] = \xi\theta$, is a continuous mapping from $\mathcal{P}_{2\pi}$ into $\mathcal{D}$. Its transpose $\Xi'_{2\pi} : \mathcal{D}' \to \mathcal{P}'_{2\pi}$ is defined by

$$(\Xi'_{2\pi}[\hat{\eta}], \theta) \;=\; < \hat{\eta}, \Xi_{2\pi}[\theta] > \;=\; < \hat{\eta}, \xi\theta >$$

and therefore the restriction $\Xi'_{2\pi}|_{\mathcal{T}_{2\pi}}$ continuously maps $\mathcal{T}_{2\pi}$ into $\mathcal{P}'_{2\pi}$. In what follows we shall denote that restriction for simplicity as $\Xi'_{2\pi}$. It remains to show that one of the maps

$$T'_{2\pi} : \mathcal{P}'_{2\pi} \to \mathcal{T}_{2\pi} \;,\; \Xi'_{2\pi} : \mathcal{T}_{2\pi} \to \mathcal{P}'_{2\pi}$$

is bijective and bicontinuous; for this it is enough to prove that $T'_{2\pi}$ and $\Xi'_{2\pi}$ are mutual inverses. In fact, $T_{2\pi} \circ \Xi_{2\pi}$ is the identity on $\mathcal{P}_{2\pi}$; by transposing one deduces that $\Xi'_{2\pi} \circ T'_{2\pi}$ is the identity on $\mathcal{P}'_{2\pi}$. On the other hand for any $\hat{\eta} \in \mathcal{T}_{2\pi}$ and $\varphi \in \mathcal{D}$ we have that

$$< T'_{2\pi} \circ \Xi'_{2\pi}[\hat{\eta}], \varphi > \;=\; (\Xi'_{2\pi}[\hat{\eta}], T_{2\pi}[\varphi])$$
$$=\; < \hat{\eta}, \xi T_{2\pi}[\varphi] > \;=\; < \xi\hat{\eta}, T_{2\pi}[\varphi] >$$
$$=\; < T'_{2\pi}[\xi\hat{\eta}], \varphi > \;=\; < \hat{\eta}, \varphi >$$

which shows that $T'_{2\pi} \circ \Xi'_{2\pi}$ is the identity on $\mathcal{T}_{2\pi}$. The proof is now complete since $T'_{2\pi} = (\Xi'_{2\pi})^{-1}$ and both mappings $T'_{2\pi}$ and $\Xi'_{2\pi}$ are continuous. □

From now on we may identify the spaces $\mathcal{P}'_{2\pi}$ and $\mathcal{T}_{2\pi}$ by the isomorphism $\Xi'_{2\pi}$. Thus for any $\hat{\eta} \in \mathcal{P}'_{2\pi}$ we have

$$(\hat{\eta}, \theta) \equiv < \hat{\eta}, \xi\theta > , \; \forall \theta \in \mathcal{P}_{2\pi} \tag{3.29}$$

for any $2\pi$-unitary function $\xi \in \mathcal{D}$. If $\hat{\mu} \in \mathcal{E}'$ is such that $\hat{\eta} = T'_{2\pi}[\hat{\mu}]$ then

$$\begin{aligned}(\hat{\eta}, \theta) &= <\hat{\eta}, \xi\theta> \\ &= <T'_{2\pi}[\hat{\mu}], \xi\theta> = <\hat{\mu}, T_{2\pi}[\xi\theta]> = <\hat{\mu}, \theta> \end{aligned} \qquad (3.30)$$

and this shows that the definition (3.29) is, in fact, independent of the $2\pi$-unitary function $\xi \in \mathcal{D}$ considered.

### 3.3.3.2 Fourier series

Suppose that $\hat{\eta}$ is a periodic distribution in $\mathcal{P}'_{2\pi}$. Then, if

$$\hat{\eta}(\omega) = T_{2\pi}[\hat{\mu}](\omega) = \sum_{n \in \mathbb{Z}} \tau_{2n\pi} \hat{\mu}(\omega) \qquad (3.31)$$

where $\hat{\mu}$ is a distribution in $\mathcal{E}'$ such that $\mathrm{supp}(\hat{\mu}) \subset [-\pi, +\pi]$, we have

$$\hat{\eta}(\omega) = \sum_{m \in \mathbb{Z}} c_m(\hat{\eta}) \cdot \exp(-im\omega). \qquad (3.32)$$

in an appropriate sense to be specified in due course. Let $\xi \in \mathcal{D}$ be any unitary function. Then

$$\begin{aligned} c_m(\hat{\eta}) &= \frac{1}{2\pi} \left(\hat{\eta}(\omega), \exp(im\omega)\right) \\ &= \frac{1}{2\pi} <\hat{\eta}(\omega), \xi(\omega)\exp(im\omega)> \\ &= \frac{1}{2\pi} <\hat{\mu}(\omega), \exp(im\omega)> . \end{aligned} \qquad (3.33)$$

The series (3.32) is called the Fourier series of the periodic distribution $\hat{\eta} \in \mathcal{P}'_{2\pi}$ with Fourier coefficients $c_m(\hat{\eta})$ as defined above. This is the generalization to periodic distributions of the classical Fourier series and as will be seen in the following theorems this generalised Fourier series has a simpler structure than in the classical case. A trigonometric series whose coefficients are the Fourier coefficients of some periodic distribution $\hat{\eta}$ always converges (in the distributional sense) to $\hat{\eta}$. Every trigonometric series whose coefficients are of slow growth will always converge in $\mathcal{S}'$. Furthermore, it can be shown that if the coefficients are not of slow growth, the trigonometric series will not converge even in $\mathcal{D}'$.

**Theorem 3.18** *Let $\hat{\eta}$ be a periodic distribution in $\mathcal{P}'_{2\pi}$ and let the constants $c_m(\hat{\eta}), m = 0, \pm 1, \ldots$ be as defined in (3.33). Then $(c_m)_{m \in \mathbb{Z}}$ is a sequence of slow growth and the series (3.32) converges in $\mathcal{S}'$ to $\hat{\eta}$.*

**Proof:** The sequence $(c_m)_{m \in \mathbb{Z}}$ is of slow growth if there exist numbers $M > 0$ and $k \in \mathbb{N}_0$ such that $|c_m| < M \cdot |m|^k$ for all $m \in \mathbb{Z}$. From (3.33) and taking into account that $\hat{\mu}$ is a distribution of compact support in $[-\pi, +\pi]$ there exists a continuous function $\hat{f}$ with support in a neighbourhood of the interval $[-\pi, +\pi]$ and an integer $r \in \mathbb{N}_0$ such that

$$c_m(\hat{\eta}) = \frac{1}{2\pi} < \hat{\mu}(\omega), \exp(im\omega) >$$

$$= \frac{1}{2\pi} \int_{-\pi}^{+\pi} \hat{f}(\omega)(-1)^r [\exp(im\omega)]^{(r)} d\omega$$

for every $m \in \mathbb{Z}$. Then

$$|c_m(\hat{\eta})| \leq \left\{ \frac{1}{2\pi} \int_{-\pi}^{+\pi} |\hat{f}(\omega)| \, d\omega \right\} \cdot |m|^r$$

and taking

$$M = \frac{1}{2\pi} \int_{-\pi}^{+\pi} |\hat{f}(\omega)| \, d\omega$$

it follows that $(c_m)_{m \in \mathbb{Z}}$ is a sequence of slow growth. Finally, for any $\varphi \in \mathcal{S}$ we have

$$< \sum_{m=-\infty}^{+\infty} c_m(\hat{\eta}) \exp(-im\omega), \varphi(\omega) > = \sum_{m=-\infty}^{+\infty} c_m(\hat{\eta}) \hat{\varphi}(m)$$

the sum on the right-hand side being absolutely convergent since $\hat{\varphi} \in \mathcal{S}$ implies that $c_m(\hat{\eta})\hat{\varphi}(m) = o(|m|^{-2})$ as $|m| \to \infty$. □

From this theorem we obtain easily the following result

**Theorem 3.19 (Structure of periodic distributions)** *Every periodic distribution (of period $2\pi$) is the derivative of a certain order of a continuous $2\pi$-periodic function on $\mathbb{R}$ and conversely.*

**Proof:** Let $\hat{\eta} \in \mathcal{P}'_{2\pi}$ have the following Fourier series

$$\hat{\eta}(\omega) = \sum_{m=-\infty}^{+\infty} c_m \exp(-im\omega)$$

where $|c_m| \leq M \cdot |m|^k$ for some $M > 0$ and $k \in \mathbb{N}_0$. Then for any $r \geq k + 2$

$$\hat{\eta}(\omega) = (1 - D)^r \sum_{n=-\infty}^{+\infty} \frac{c_m}{(1 + im)^r} \exp(-im\omega)$$

and since $|c_m/(1 + im)^r| = O(|m|^{-2})$ the series

$$\sum_{m=-\infty}^{+\infty} \frac{c_m}{(1 + im)^r} \exp(-im\omega)$$

converges pointwise (and uniformly) to a continuous function $\hat{f}(\omega)$ which is periodic of period $2\pi$. Hence
$$\hat{\eta}(\omega) = (1-D)^r \hat{f}(\omega)$$
as asserted. The converse is immediate. □

**Exercise 3.4 :**

1. The function
$$F(t) = \frac{2}{\pi}\left\{1 - 2\sum_{n=1}^{\infty}\frac{\cos(2nt)}{4n^2-1}\right\}$$
   is plainly periodic with period $\pi$. Prove that it satisfies a differential equation
$$D^2 F(t) + F(t) = h(t)$$
   where $h$ is a certain distribution. By solving this equation with appropriate boundary conditions show that $F(t)$ is the Fourier series expansion of a periodic function $\sum_{n=-\infty}^{+\infty} f(t-n\pi)$, where $f(t)$ vanishes outside $(-\pi/2, +\pi/2)$, and determine $f(t)$.

2. Let $p$ be a positive integer. If $f_p(t) = (t/2\pi)^{2p}, \forall t \in (-\pi, +\pi)$ and if $f_p(t) = 0$ otherwise, find the Fourier coefficients of the periodic extension $T_{2\pi} f_p$, and hence sum the series
$$1 + \frac{1}{2^4} + \frac{1}{3^4} + \cdots + \frac{1}{n^4} + \cdots$$

3. Give an example of a $2\pi$-unitary function $\xi(t)$ in $\mathcal{D}$ and show that its Fourier transform satisfies
$$\hat{\xi}(n) = \begin{cases} 2\pi & \text{if } n=0 \\ 0 & \text{if } n = +1, -1, +2, -2, \ldots \end{cases}$$

4. Let $\mu$ be a $2\pi$-periodic distribution in $\mathcal{P}'_{2\pi}$. If $\xi$ is an arbitrary (fixed) $2\pi$-unitary function in $\mathcal{D}$ we write
$$\mu \odot \theta = <\mu, \xi\theta>$$
   for each $\theta$ in $\mathcal{P}_{2\pi}$.
   Prove that $(D\mu) \odot \theta = -\mu \odot \theta'$, and $\mu(t-\tau) \odot \theta(t) = \mu(t) \odot \theta(t+\tau)$.

5. For arbitrary $2\pi$-periodic distributions $\mu, \nu$ define their $\Delta$-convolution to be the $2\pi$-periodic distribution $\mu\Delta\nu$ given by
$$(\mu\Delta\nu) \odot \theta = \mu(t) \odot [\nu(\tau) \odot \theta(t+\tau)]$$
   where the $\odot$ operation is defined as in question 4. above. Show that $\Delta$-convolution is always associative and commutative and prove that
   
   (a) $(\mu\Delta\nu)(t-a) = \mu(t-a)\Delta\nu(t) = \mu(t)\Delta\nu(t-a)$,
   (b) $D(\mu\Delta\nu) = (D\mu)\Delta\nu = \mu\Delta(D\nu)$,
   (c) if $\mu, \nu$ have Fourier coefficients $F_n, G_n$ respectively then the Fourier coefficients of $\mu\Delta\nu$ are given by $2\pi(F_n G_n)$.

### 3.3.4 Generalised sampling theorems

#### 3.3.4.1 The Campbell Sampling Theorem

We have seen in §3.3.1.2 that the classical WKS sampling theorem does not generally hold for band-limited functions whose Fourier transforms are distributions of order greater than 0. There is, however, a general form of Sampling Theorem (due to L.L.Campbell, [7]) which is valid for band-limited functions whose Fourier transforms are arbitrary distributions of compact support. Let $\hat{\mu}(\omega)$ be a distribution with support contained in $[-\pi\Omega(1-\varepsilon), +\pi\Omega(1-\varepsilon)]$, where $0 < \varepsilon < 1$. Then the convolution

$$\hat{\mu}(\omega) * \sum_{n=-\infty}^{+\infty} \delta(\omega - 2\pi n\Omega)$$

exists and the exchange formula (3.17) is valid. The inverse Fourier transform of $\hat{\mu}(\omega)$ is a continuous function $f(t) \equiv \mathcal{F}^{-1}[\hat{\mu}(\omega)]$. Then we have

$$\begin{aligned}
\Omega \sum_{n=-\infty}^{+\infty} \hat{\mu}(\omega - 2\pi n\Omega) &= \frac{1}{2\pi} \left\{ \hat{\mu}(\omega) * 2\pi\Omega \sum_{n=-\infty}^{+\infty} \delta(\omega - 2\pi n\Omega) \right\} \\
&= \mathcal{F}\left\{ f(t) \sum_{n=-\infty}^{+\infty} \delta(t - n/\Omega) \right\} \\
&= \sum_{n=-\infty}^{+\infty} f(n/\Omega) \exp(-in\omega/\Omega) .
\end{aligned} \quad (3.34)$$

Now let $\hat{\rho}(\omega)$ be any infinitely differentiable function such that $\hat{\rho}(\omega) = 1$ on the support of $\hat{\mu}(\omega)$, and $\hat{\rho}(\omega) = 0$ for $|\omega| > \pi\Omega(1+\varepsilon)$. Using (3.34) we have

$$< \sum_{n=-\infty}^{+\infty} \hat{\mu}(\omega - 2\pi n\Omega), \hat{\rho}(\omega) \exp(i\omega t) >$$

$$= < \frac{1}{\Omega} \sum_{n=-\infty}^{+\infty} f(n/\Omega) \exp(-in\omega/\Omega), \hat{\rho}(\omega) \exp(i\omega t) > . \quad (3.35)$$

The only non-zero term on the left-hand side of (3.35) is given when $n = 0$, and evaluating this we get

$$< \hat{\mu}(\omega), \hat{\rho}(\omega) \exp(i\omega t) > = 2\pi f(t),$$

while for a typical term of the right-hand side we have

$$\begin{aligned}
< \exp(-in\omega/\Omega), \hat{\rho}(\omega) \exp(i\omega t) > &= \int_{-\infty}^{+\infty} \hat{\rho}(\omega) \exp[i\omega(t - n/\Omega)] d\omega \\
&= 2\pi \rho(t - n/\Omega) \quad (3.36)
\end{aligned}$$

where $\rho(t)$ is the inverse Fourier transform of the function $\hat{\rho}(\omega)$. Thus we have derived the following general sampling theorem:

**Theorem 3.20** *Let $f(t)$ be the inverse Fourier transform of the distribution $\hat{\mu}(\omega)$ such that $\mathrm{supp}(\hat{\mu}) \subset [-\pi\Omega(1-\varepsilon), +\pi\Omega(1-\varepsilon)]$, where $0 < \varepsilon < 1$. If the function $f$ is sampled at the instants $\{n/\Omega\}_{n\in\mathbb{Z}}$, then it can be reconstructed from its sampled values from the formula*

$$f(t) = \frac{1}{\Omega} \sum_{n=-\infty}^{+\infty} f(n/\Omega) \rho(t - n/\Omega) \qquad (3.37)$$

*where $\rho(t)$ is some function whose Fourier transform $\hat{\rho}(\omega)$ is infinitely differentiable, is equal to 1 on the support of $\hat{\mu}(\omega)$, and vanishes identically for all $\omega$ such that $|\omega| > \pi\Omega(1+\varepsilon)$.*

Some comments about this theorem are in order:

(1) The form of the Campbell sampling expansion (3.37) depends on the choice of the function $\hat{\rho}(\omega)$ (or, equivalently, on the choice of $\rho(t)$). It is not actually necessary that $\hat{\rho}(\omega)$ should be infinitely differentiable; since $\hat{\mu}(\omega)$ has compact support it is necessarily a finite-order distribution, and so $\hat{\rho}(\omega)$ need only be differentiable to some appropriate finite order.

(2) Suppose in particular that $\hat{\mu}(\omega)$ is of order 0. Then either

$$<\hat{\mu}(x), \varphi(x)> = \int_{-\infty}^{+\infty} \varphi(x) \hat{h}(x)\, dx$$

where $\hat{h}$ is a locally integrable function, or at worst

$$<\hat{\mu}(x), \varphi(x)> = \int_{-\infty}^{+\infty} \varphi(x)\, d\hat{G}(x)$$

where $\hat{G}$ is a function of bounded variation. It is then enough that $\hat{\rho}(\omega)$ should be continuous. In particular taking for $\hat{\rho}(\omega)$ a simple trapezoidal function equal to 1 for $|\omega| \leq \pi\Omega(1-\varepsilon)$ we get

$$f(t) = \sum_{n=-\infty}^{+\infty} f(n/\Omega) \mathrm{sinc}[\varepsilon(\Omega t - n)] \mathrm{sinc}(\Omega t - n) \qquad (3.38)$$

(3) The result of (3.38) can be generalised to get the so-called **Helms-Thomas self-truncating series**

$$f(t) = \sum_{n=-\infty}^{+\infty} f(n/\Omega) \mathrm{sinc}^m[\varepsilon(\Omega t - n)] \mathrm{sinc}(\Omega t - n) \qquad (3.39)$$

where $m$ is an integer greater than or equal to the order of the distribution $\hat{\mu}$.

### 3.3.4.2 Sampling expansions of mixed type

Other forms of sampling expansion can be established for functions which are inverse Fourier transforms of distributions of compact support. We first review the notation which we shall use throughout this section.

$\mathcal{E}'_{\pi\Omega}$ denotes the space of all distributions with support contained in the interval $[-\pi\Omega, +\pi\Omega]$, while $\mathcal{E}'_{\pi\Omega,r}$ denotes the subspace of all distributions in $\mathcal{E}'_{\pi\Omega}$ which are of order $\leq r$. Finally recall that $\mathcal{B}^r_{\pi\Omega}$ is the subspace of $L^2_r(\mathbb{R})$ consisting of those functions which are band-limited to $[-\pi\Omega, +\pi\Omega]$ in the distributional sense; i.e. $f$ is a member of $\mathcal{B}^r_{\pi\Omega}$ if and only if $f$ belongs to $L^2_r(\mathbb{R})$ and is such that its Fourier transform is a distribution in $\mathcal{E}'_{\pi\Omega}$. We show first that if $f \in \mathcal{B}^r_{\pi\Omega}$ then there exists a representation of the form

$$f(t) = P_r(t) \cdot \eta(t) + t^r \varphi(t) \tag{3.40}$$

where $\eta(t)$ is the inverse Fourier transform of any suitably restricted measure in $\mathcal{E}'_{\pi\Omega,0}$, $P_r(t)$ is a polynomial of degree $< r$, whose coefficients depend on $f$ and $\eta$, and $\varphi(t)$ is an $L^2(\mathbb{R})$ function which is band-limited in the classical sense to the interval $[-\pi\Omega, +\pi\Omega]$. The function $\varphi(t)$ admits a classical WKS sampling expansion which, on substitution into (3.40), allows us to derive the following theorem:

**Theorem 3.21** *Let $f \in \mathcal{B}^r_{\pi\Omega}$, and let $\hat{\eta}$ be an arbitrary distribution in $\mathcal{E}'_{\pi\Omega,0}$ such that $<\hat{\eta}(\omega),1> \neq 0$. Then the following expansion holds uniformly on compacts,*

$$f(t) = P_r(t)\eta(t) + \sum_{n=-\infty}^{+\infty} \left(\frac{\Omega t}{n}\right)^r [f(n/\Omega) - P_r(n/\Omega)\eta(n/\Omega)]\operatorname{sinc}(\Omega t - n)$$

*where $\eta = \mathcal{F}^{-1}[\hat{\eta}]$, $P_r$ is the polynomial of degree $< r$ defined by*

$$P_r(t) = \sum_{\sigma=0}^{r-1} \frac{t^\sigma}{\sigma!} \left[\frac{d^\sigma}{d\tau^\sigma} \frac{f(\tau)}{\eta(\tau)}\right]_{\tau=0}$$

*and the term corresponding to $n = 0$ in the series is to be interpreted in the sense of*

$$\lim_{\tau \to 0} \{(t/\tau)^r [f(\tau) - P_r(\tau)\eta(\tau)]\operatorname{sinc}[\Omega(t-\tau)]\} = \frac{t^r}{r!}[D^r(f/\eta)]_{\tau=0} \operatorname{sinc}(\Omega t).$$

Before proving this theorem we need some preliminary results. First note that from theorem 3.9 there exist (infinitely many) representations of the form

$$\hat{\mu} = \sum_{\sigma=0}^{r} D^\sigma \hat{\mu}_\sigma \tag{3.41}$$

where $\hat{\mu}_\sigma \in \mathcal{E}'_{\pi\Omega,0}$ for $\sigma = 0, 1, \ldots, r$. Taking inverse Fourier transforms of both sides of (3.41) we get

$$f(t) = \sum_{\sigma=0}^{r}(-it)^\sigma f_\sigma(t)$$

so that for any $\nu \in \mathbb{N}_0$

$$f^{(\nu)}(t) = \sum_{\alpha=0}^{\nu}\binom{\nu}{\alpha}(-i)^\alpha \left\{ \sum_{\sigma=\alpha}^{r} \frac{\sigma!}{(\sigma-\alpha)!}(-it)^{\sigma-\alpha} f_\sigma^{(\nu-\sigma)}(t) \right\}.$$

It follows that for $\nu < r$ we have

$$f^{(\nu)}(0) = \sum_{\sigma=0}^{\nu}\binom{\nu}{\sigma}\sigma!(-i)^\sigma f_\sigma^{(\nu-\sigma)}(0), \qquad (3.42)$$

while for $\nu \geq r$,

$$f^{(\nu)}(0) = \sum_{\sigma=0}^{r}\binom{\nu}{\sigma}\sigma!(-i)^\sigma f_\sigma^{(\nu-\sigma)}(0). \qquad (3.43)$$

Hence, given $\hat{\mu}$, we can choose measures $\hat{\mu}_\sigma, \sigma = 0, 1, 2, \ldots, r-1$, arbitrarily provided they satisfy the $r$ equations

$$\sum_{\sigma=0}^{\nu}\binom{\nu}{\sigma}\sigma!\frac{(-i)^\sigma}{2\pi}<\hat{\mu}_\sigma(\omega),(i\omega)^{\nu-\sigma}> = f^{(\nu)}(0), \quad 0 \leq \nu \leq r-1. \qquad (3.44)$$

We can therefore establish the following result:

**Lemma 3.3.2** *Let $f \in \mathcal{B}^r_{\pi\Omega}, 0 \leq r < \infty$. Given any set of measures $\{\hat{\mu}_\sigma\}_{\sigma=0,1,\ldots,r-1}$, in $\mathcal{E}'_{\pi\Omega,0}$ which are such that the equations (3.44) are satisfied the Fourier transform $\hat{\mu}$ of $f$ admits the representation*

$$\hat{\mu} = \sum_{\sigma=0}^{r-1} D^\sigma \hat{\mu}_\sigma + D^r \hat{\varphi}$$

*where $\hat{\varphi}$ is necessarily a regular distribution in $\mathcal{E}'_{\pi\Omega,0}$ defined by a function in $L^2(-\pi\Omega, +\pi\Omega)$.*

**Proof:** It is only necessary to confirm the asserted properties of $\hat{\mu}_r = \hat{\varphi}$. First we have

$$f_r(t) = \frac{1}{(-it)^r}\left\{f(t) - \sum_{\sigma=0}^{r-1}(-it)^\sigma f_\sigma(t)\right\}$$

where $f$, and each of the $f_\sigma$, extend to the complex plane as entire functions of first order of growth and of exponential type $\leq \pi\Omega$, $f$ being $O(|t|^{r-\varepsilon})$ with $\varepsilon > 1/2$, and $f_\sigma$ being $O(1)$ as $|t| \to \infty$. But since $\lim_{t\to 0} f_r(t) = [(-i)^{-r}/r!]\{f^{(r)}(0)-\sum_{\sigma=0}^{r-1}\binom{r}{\sigma}(-i)^\sigma\sigma!f_\sigma^{(r-\sigma)}(0)\}$ then $f_r(t)$ is bounded at $t=0$. Hence $f_r$ also extends as an entire function of the same order of growth and exponential type. Also $f(t)/(-it)^r$ is $O(|t|^{-\varepsilon}), \varepsilon > 1/2$, as $|t| \to \infty$ and so $f_r \in L^2(\mathbb{R})$. The result therefore follows from the Paley-Wiener theorem. □

Provided they satisfy (3.44) the measures $\hat{\mu}_\sigma, \sigma = 0, 1, \ldots, r-1$ are otherwise arbitrary. The following corollary to lemma 3.3.2 provides a particularly simple form for these measures.

**Corollary 3.3.1** *Let $\hat{\eta}$ be a measure in $\mathcal{E}'_{\pi\Omega,0}$ such that $<\hat{\eta}(\omega), 1>$ is not zero. Then if $Q_r(D)$ is the differential operator of order $r$ defined by $Q_r(D) = \sum_{\sigma=0}^{r-1} a_\sigma D^\sigma$ we may write $\hat{\mu} = Q_r(D)\hat{\eta} + D^r \hat{\varphi}$ provided the coefficients $a_\sigma, \sigma = 0, 1, \ldots, r-1$, satisfy the equations*

$$\sum_{\sigma=0}^{\nu} \binom{\nu}{\sigma} \sigma! \frac{(-i)^\sigma}{2\pi} < \hat{\eta}(\omega), (i\omega)^{\nu-\sigma} > a_\nu = f^{(\nu)}(0),$$

*for $\nu = 0, 1, \ldots, r-1$.*

**Proof:** (Theorem 3.21) Let $f \in \mathcal{B}^r_{\pi\Omega}$; then $\hat{\mu} = Q_r(D)\hat{\eta} + D^r \hat{\varphi}$, and so

$$f(t) = P_r(t)\eta(t) + (-it)^r \varphi(t), \qquad (3.45)$$

where

$$P_r(t) = \mathcal{F}^{-1}[Q_r(D)](t) = Q_r(-it) = \sum_{\sigma=0}^{r-1} b_\sigma t^\sigma,$$

with $b_\sigma = (-i)^\sigma a_\sigma$. Since $\varphi$ is a classically band-limited function in $L^2(\mathbb{R})$ we have

$$\varphi(t) = \sum_{n=-\infty}^{+\infty} \varphi(n/\Omega) \operatorname{sinc}(\Omega t - n)$$

where the convergence is uniform on compacts. Using the corollary 3.3.1 we can now express $\varphi(n/\Omega)$ in terms of $f(n/\Omega)$ etc. to obtain the required form for the sampling expansion given in the theorem. It only remains to derive the stated form for the coefficients $b_\sigma$ and for $\varphi(0)$. First we have that

$$\sum_{\sigma=0}^{\nu} \binom{\nu}{\sigma} \sigma! \eta^{(\nu-\sigma)}(0) \cdot b_\sigma = f^{(\nu)}(0), \qquad (3.46)$$

for $\nu = 0, 1, \ldots, r-1$. The remainder of the proof proceeds by induction: the given form holds for $\nu = 0$ since then (3.46) is $\eta(0) b_0 = f(0)$, which implies that

$$b_0 = \frac{1}{0!} \left[ \frac{f(\tau)}{\eta(\tau)} \right]_{\tau=0}.$$

Now suppose that

$$b_\sigma = \frac{1}{\sigma!} \left[ \frac{d^\sigma}{d\tau^\sigma} \frac{f(\tau)}{\eta(\tau)} \right]_{\tau=0}$$

for all $\sigma$ such that $\sigma = 0, 1, \ldots, \nu - 1$. Then from (3.46),

$$\sum_{\sigma=0}^{\nu-1} \binom{\nu}{\sigma} \sigma! \eta^{(\nu-\sigma)}(0) b_\sigma + \nu! \eta(0) b_\nu = f^{(\nu)}(0),$$

or

$$b_\nu = \frac{1}{\nu! \eta(0)} \left\{ f^{(\nu)}(0) - \sum_{\sigma=0}^{\nu-1} \binom{\nu}{\sigma} \sigma! \eta^{(\nu-\sigma)}(0) \frac{1}{\sigma!} \left[ \frac{d^\sigma}{d\tau^\sigma} \frac{f(\tau)}{\eta(\tau)} \right]_{\tau=0} \right\}$$

which, after some manipulation, reduces to

$$b_\nu = \frac{1}{\nu! \eta(0)} \left\{ \frac{d^\nu}{d\tau^\nu} \left[ \frac{f(\tau)}{\eta(\tau)} \right]_{\tau=0} \right\} \cdot \eta(0) = \frac{1}{\nu!} \left[ \frac{d^\nu}{d\tau^\nu} \frac{f(\tau)}{\eta(\tau)} \right]_{\tau=0}.$$

(This result agrees with the fact that since $f$ and $\varphi$ are both analytic in $t$ and the last term in (3.45) is $O(t^r)$, then $P_r(t)$ is the Taylor expansion about the origin of $f(t)/\eta(t)$ up to order $r - 1$). Finally from (3.43) we get

$$f^{(r)}(0) = \sum_{\sigma=0}^{r-1} \binom{r}{\sigma} \sigma! \eta^{(r-\sigma)}(0) b_\sigma + r! (-i)^r \varphi(0)$$

and so proceeding as in the above inductive argument,

$$\varphi(0) = \frac{1}{(-i)^r r!} \left\{ f^{(r)}(0) - \sum_{\sigma=0}^{r-1} \binom{r}{\sigma} \eta^{(r-\sigma)}(0) \left[ \frac{d^\sigma}{d\tau^\sigma} \frac{f(\tau)}{\eta(\tau)} \right]_{\tau=0} \right\}$$

$$= \frac{1}{(-i)^r r!} \left[ \frac{d^r}{d\tau^r} \frac{f(\tau)}{\eta(\tau)} \right]_{\tau=0}.$$

Hence,

$$\lim_{\tau \to 0} \left( \frac{t}{\tau} \right)^r [f(\tau) - P_r(\tau) \eta(\tau)] \text{sinc} \Omega(t - \tau) = \lim_{\tau \to 0} \{ t^r \cdot \text{sinc}[\Omega(t - \tau)] \cdot (-i)^r \varphi(\tau) \}$$

$$= \frac{t^r}{r!} \text{sinc}(\Omega t) \cdot \left[ \frac{d^r}{d\tau^r} \frac{f(\tau)}{\eta(\tau)} \right]_{\tau=0},$$

which completes the proof. □

**Remark 3.4** *Particular choices of the measure $\hat{\eta}$, and of its corresponding inverse transform $\eta$, yield sampling expansions derived elsewhere in the literature. For example, taking*

$$\hat{\eta}(\omega) = \pi[\delta(\omega + \pi\Omega) + \delta(\omega - \pi\Omega)]$$

*gives a sampling expansion suggested by Pfaffelhuber [85], while the choice*

$$\hat{\eta}(\omega) = 2\pi \delta(\omega - a) , \quad -\pi\Omega \leq a \leq +\pi\Omega,$$

*gives that due to Lee [62].*

**Exercise 3.5 :**

1. If $f = \mathcal{F}^{-1}[\hat{\mu}]$ where $\hat{\mu}$ is a distribution of order 0 and has compact support contained in $[-\pi\Omega, +\pi\Omega]$, prove that $f \in \mathcal{B}_{\pi\Omega}^0$.

   Give an example of a function $f$ belonging to the class $\mathcal{B}_{\pi\Omega}^0$ which is the inverse Fourier transform of a distribution of order 1.

2. If $\hat{\mu}(\omega)$ is a distribution of order 0 then, as indicated above in the text, we can choose $\hat{\rho}(\omega)$ to obtain a form of Campbell's Sampling Expansion as follows:

$$f(t) = \sum_{n=-\infty}^{+\infty} f(n/\Omega)\text{sinc}(\Omega t - n)\text{sinc}[\varepsilon(\Omega t - n)].$$

   Use this result to show that we may allow $\varepsilon \to 0$ and so derive the classical WKS expansion when $f \in L^p(\mathbb{R})$ for some $p$ such that $1 \leq p < +\infty$.

## 3.4 FOURIER TRANSFORMS AND ULTRADISTRIBUTIONS

### 3.4.1 Fourier transforms of functions in $\mathcal{D}$

The definition of Fourier transform based on the Parseval relation which we used for tempered distributions cannot be extended to arbitrary distributions. When $\mu$ is a tempered distribution ($\mu \in \mathcal{S}'$) we can define $\hat{\mu}$ by

$$<\hat{\mu}, \varphi> = <\mu, \hat{\varphi}> \qquad (3.47)$$

for any $\varphi \in \mathcal{S}$ because then we also have $\hat{\varphi} \in \mathcal{S}$ and both sides of (3.47) have meaning. If $\mu$ is a distribution which is not tempered (so that $\mu \in \mathcal{D}'\backslash\mathcal{S}'$) then the right-hand side of (3.47) will have a sense for any $\hat{\varphi} \in \mathcal{D}$ but the left-hand side needs a new interpretation. This is because $\varphi = \mathcal{F}^{-1}[\hat{\varphi}]$ will not be in $\mathcal{D}$ but will belong to a new class of test functions. The Fourier transform $\hat{\mu}$ of $\mu$ will therefore not be a distribution but a functional defined on this new class. Hence we need to establish those properties which characterise functions which are inverse Fourier transforms of functions in $\mathcal{D}$.

Let $\hat{\varphi}(t)$ be an arbitrary function in $\mathcal{D}$ with support contained in the interval $[-a, a]$ for some given $a \geq 0$. Since $\hat{\varphi}(t) \in \mathcal{D} \subset \mathcal{S}$ it must be the inverse Fourier transform of a certain function $\varphi(\omega)$ belonging to $\mathcal{S}$ given by

$$\varphi(\omega) \equiv \frac{1}{2\pi} \int_{-a}^{+a} \hat{\varphi}(t) \exp(i\omega t) \, dt . \qquad (3.48)$$

Then,

**(1)** the integral (3.48) converges uniformly over every bounded domain in the complex plane, and defines a function $\varphi(z)$ of a complex variable,

**(2)** the integrand $\hat{\varphi}(t)\exp(izt)$ is a continuous function of the variables $(z,t) \in \mathbb{C} \times \mathbb{R}$,

**(3)** the integrand is an analytic function of $z$ for every real $t$.

It follows that $\varphi(z)$ is an entire function. Further, integrating by parts $k$ times we get

$$(-iz)^k \varphi(z) = \frac{1}{2\pi} \int_{-a}^{+a} \hat{\varphi}^{(k)}(t) \exp(izt)\, dt$$

whence

$$|z^k \varphi(z)| \leq C_k \cdot \exp(a|\Im(z)|), \tag{3.49}$$

where

$$C_k \equiv \frac{1}{2\pi} \int_{-a}^{+a} |\hat{\varphi}^{(k)}(t)|\, dt\ .$$

Conversely let

$$\hat{\varphi}(t) = \int_{-\infty}^{+\infty} \varphi(\omega) \exp(-i\omega t)\, d\omega$$

where $\varphi(\omega)$ extends to an entire function $\varphi(z)$ satisfying inequalities of the form (3.49) for $k = 0, 1, 2, \ldots$. This means that, for all $y = \Im(z)$ in any fixed finite interval, $\varphi(\omega + iy)$ goes to zero faster than any power of $|\omega|^{-1}$ as $|\omega| \to \infty$; hence by Cauchy's Theorem we may shift the path of integration onto any line parallel to the $\omega$-axis. Then, for every $y$, we have

$$\hat{\varphi}(t) = \int_{-\infty}^{+\infty} \varphi(\omega + iy) \exp\left(-i(\omega + iy)t\right) d\omega$$
$$= \exp(yt) \int_{-\infty}^{+\infty} \varphi(\omega + iy) \exp(-i\omega t) d\omega$$

where the last integral converges uniformly; differentiating we get

$$\hat{\varphi}'(t) = \int_{-\infty}^{+\infty} (-iz)\varphi(z) \exp(-izt) d\omega$$
$$= \exp(yt) \int_{-\infty}^{+\infty} (-iz)\varphi(z) \exp(-i\omega t) d\omega\ . \tag{3.50}$$

Since the last integral is again uniformly convergent for all $t$ in $\mathbb{R}$, the formal differentiation under the integral sign is justified and so $\hat{\varphi}'(t)$ does exist as given by (3.50). Repeating the argument shows that $\hat{\varphi}(t)$ is infinitely smooth, with

$$\hat{\varphi}^{(k)}(t) = \int_{-\infty}^{+\infty} (-iz)^k \varphi(z) \exp(-izt) d\omega. \tag{3.51}$$

Finally we show that $\hat{\varphi}(t)$ must have compact support contained in the interval $[-a, a]$, where $a$ is as given in the inequalities (3.49). Using these inequalities in the cases $k = 0$ and $k = 2$ we get

$$|\varphi(z)| \leq \exp(a|y|) \cdot \inf\left\{C_0, \frac{C_2}{|z|^2}\right\} \leq C \, \frac{\exp(a|y|)}{1+\omega^2}.$$

where $C = \max\{C_0, C_2\}$. Hence, for every $y$,

$$\begin{aligned}|\hat{\varphi}(t)| &\leq \left|\exp(yt)\int_{-\infty}^{+\infty}\varphi(\omega+iy)\exp(-i\omega t)d\omega\right|\\ &\leq \exp(ty+a|y|)\int_{-\infty}^{+\infty}C(1+\omega^2)^{-1}d\omega = C\pi \cdot \exp(ty+a|y|).\end{aligned}$$

Now let $u < 0$ and write $y = u|t|/t$; then

$$|\hat{\varphi}(t)| \leq C\pi \cdot \exp(u|t|+a|u|) = C\pi \cdot \exp[u(|t|-a)]$$

and thus for $|t| > a$, the right-hand side tends to zero as $u \to -\infty$; this implies that $|\hat{\varphi}(t)| = 0$ for $|t| > a$.

### 3.4.2 The space $\mathcal{Z}$ of test functions

In the preceding section we have shown that a necessary and sufficient condition for a function $\hat{\varphi}(t)$ to belong to the space $\mathcal{D}$ is that its inverse Fourier transform can be extended onto the complex plane as an entire function satisfying the inequalities (3.49). If $\mathcal{Z}$ denotes the linear space of all functions $\varphi$ whose Fourier transforms $\hat{\varphi}$ are functions in $\mathcal{D}$ then we can express this result as follows:

> **Theorem 3.22** *A necessary and sufficient condition for a function $\varphi$ to belong to the space $\mathcal{Z}$ is that $\varphi(z)$ is an entire function which satisfies a set of inequalities of the form*
> 
> $$|z^k \varphi(z)| \leq C_k \cdot \exp(a|y|),$$
> 
> *where $a > 0$ and $k = 0, 1, 2, \ldots$.*

A sequence $(\varphi_n)_{n \in \mathbb{N}}$ is said to converge to a limit $\varphi$ in the space $\mathcal{Z}$ if the following conditions are fulfilled:

**(1)** each $\varphi_n$ belongs to $\mathcal{Z}$,

**(2)** there exist constants $a > 0$ and $C_j > 0$, where $j = 0, 1, 2, \ldots$ (not depending on $n$) such that for all $z \in \mathbb{C}$,

$$|z^j \varphi_n(z)| \leq C_j \cdot \exp(a|\Im(z)|), \quad j = 0, 1, 2, \ldots,$$

**(3)** the sequence $(\varphi_n)_{n \in \mathbb{N}}$ converges uniformly on every bounded domain of the complex plane.

Under this definition the limit function $\varphi$ necessarily belongs to $\mathcal{Z}$; the linear space $\mathcal{Z}$ is closed under convergence. However the real motivation for the definition is shown by the next theorem.

**Theorem 3.23** *A sequence $(\varphi_n)_{n \in \mathbb{N}}$ converges in $\mathcal{Z}$ to the limit function $\varphi \in \mathcal{Z}$ if and only if the sequence of Fourier transforms $(\hat{\varphi}_n)_{n \in \mathbb{N}}$ converges in $\mathcal{D}$ to the limit function $\hat{\varphi} \in \mathcal{D}$.*

**Proof:** (i) Let $(\hat{\varphi}_n)_{n \in \mathbb{N}}$ converge in $\mathcal{D}$ to $\hat{\varphi}$; suppose that the supports of all the $\hat{\varphi}_n$ and of $\hat{\varphi}$ are contained in $[-a, a]$. Then $\varphi_n$ and $\varphi$ are all members of $\mathcal{Z}$ and for any $j \in \mathbb{N}_o$,

$$|z^j \varphi_n(z)| = \left| \frac{1}{2\pi} \int_{-a}^{+a} \hat{\varphi}_n^{(j)}(\omega) \exp(i\omega z) d\omega \right|$$

$$\leq \frac{1}{2\pi} \left\{ \int_{-a}^{+a} |\hat{\varphi}_n^{(j)}(\omega)| d\omega \right\} \exp(a|\Im(z)|)$$

$$\leq \left\{ \frac{a}{\pi} \sup_{-a \leq \omega \leq +a} |\hat{\varphi}_n^{(j)}(\omega)| \right\} \exp(a|\Im(z)|).$$

Since $(\hat{\varphi}_n)_{n \in \mathbb{N}}$ converges to $\hat{\varphi} \in \mathcal{D}$ it follows that, for each $j \in \mathbb{N}_0$, $\sup_{|\omega| \leq +a} |\hat{\varphi}^{(j)}(\omega)|$ is uniformly bounded for all $n \in \mathbb{N}$. Thus,

$$|z^j \varphi_n(z)| \leq C_j \cdot \exp(a|\Im(z)|) \ , \ j = 0, 1, 2, \ldots,$$

where

$$C_j = \frac{a}{\pi} \sup_{n \in \mathbb{N}} \left\{ \sup_{-a \leq \omega \leq +a} |\hat{\varphi}_n^{(j)}(\omega)| \right\}.$$

Moreover

$$|\varphi_n(z) - \varphi(z)| \leq \frac{a}{\pi} \left\{ \sup_{-a \leq \omega \leq +a} |\hat{\varphi}_n(\omega) - \hat{\varphi}(\omega)| \right\} \cdot \exp(a|\Im(z)|)$$

and, since $\hat{\varphi}_n \to \hat{\varphi}$ in $\mathcal{D}$ and $\exp(a|\Im(z)|)$ is bounded on each bounded domain of the complex plane, it follows that

$$|\varphi_n(z) - \varphi(z)| \to 0$$

uniformly on compacts in $\mathbb{C}$. Thus $(\varphi_n)_{n \in \mathbb{N}}$ converges to $\varphi$ in $\mathcal{Z}$.

(ii) Conversely, let $(\varphi_n)_{n \in \mathbb{N}}$ converge in $\mathcal{Z}$ to $\varphi$. Then all the $\hat{\varphi}_n$ and $\hat{\varphi}$ belong to $\mathcal{D}$ and have their supports contained in the same compact, say $-a \leq \omega \leq +a$. For each $j \in \mathbb{N}_o$,

$$|\hat{\varphi}_n^{(j)}(\omega) - \hat{\varphi}^{(j)}(\omega)| = \left| \int_{-\infty}^{+\infty} (-it)^j [\varphi_n(t) - \varphi(t)] \exp(-i\omega t) dt \right|$$

$$\leq \int_{-\infty}^{+\infty} \left| t^j (1 + t^2) \frac{\varphi_n(t) - \varphi(t)}{1 + t^2} \right| dt$$

$$\leq \pi \cdot \sup_{t \in \mathbb{R}} \left| (t^j + t^{j+2})(\varphi_n(t) - \varphi(t)) \right|$$

From the definition of convergence in $\mathcal{Z}$ the right-hand side of this inequality converges to zero and therefore $(\hat{\varphi}_n)_{n\in\mathbb{N}}$ must converge in $\mathcal{D}$ to $\hat{\varphi}$. □

Since every function in $\mathcal{Z}$ is an entire function it cannot be zero on any non-degenerate interval of the real line unless it vanishes everywhere; thus $\mathcal{D} \cap \mathcal{Z} = \emptyset$. On the other hand we have:

**Theorem 3.24** $\mathcal{Z}$ *is a dense subspace of* $\mathcal{S}$.

**Proof:** That $\mathcal{Z}$ is a proper linear subspace of $\mathcal{S}$ is clear. It remains to show that for each $\varphi \in \mathcal{S}$ there exists a sequence $(\varphi_n)_{n\in\mathbb{N}}$ contained in $\mathcal{Z}$ which converges in $\mathcal{S}$ to $\varphi$. The Fourier transform maps $\mathcal{S}$ onto $\mathcal{S}$ and so $\hat{\varphi}$ belongs to $\mathcal{S}$. The space $\mathcal{D}$ is dense in $\mathcal{S}$ so that we can choose a sequence $(\hat{\varphi}_n)_{n\in\mathbb{N}}$ of functions in $\mathcal{D}$ which converges in $\mathcal{S}$ to $\hat{\varphi}$. By the continuity of the inverse Fourier transform as a mapping of $\mathcal{S}$ onto itself the sequence $(\varphi_n)_{n\in\mathbb{N}} \subset \mathcal{Z}$ converges in $\mathcal{S}$ to $\varphi$. □

### 3.4.3 Definition of ultradistributions

We denote by $\mathcal{Z}'$ the dual of the space $\mathcal{Z}$; that is $\mathcal{Z}'$ is the space of all functionals defined on $\mathcal{Z}$ which are linear and continuous with respect to the mode of convergence which we have defined on $\mathcal{Z}$. Thus $\hat{\mu}$ belongs to $\mathcal{Z}'$ if and only if

**(1)** for all $\varphi, \psi$ in $\mathcal{Z}$ and for all $a, b$ in $\mathbb{C}$ we have

$$<\hat{\mu}, a\varphi + b\psi> = a<\hat{\mu},\varphi> + b<\hat{\mu},\psi>;$$

**(2)** if $(\varphi_n)_{n\in\mathbb{N}} \subset \mathcal{Z}$ converges in $\mathcal{Z}$ to zero then the sequence of numbers $(<\hat{\mu},\varphi_n>)_{n\in\mathbb{N}}$ converges in $\mathbb{C}$ to zero.

The members of $\mathcal{Z}'$ are called **ultradistributions** and are in general distinct from distributions. Nevertheless some distributions are also ultradistributions. In particular we have:

**Theorem 3.25** *The space* $\mathcal{S}'$ *is contained in* $\mathcal{Z}'$; *that is, every tempered distribution is also an ultradistribution.*

**Proof:** $\mathcal{Z}$ is a linear subspace of $\mathcal{S}$ and so every tempered distribution is certainly a linear functional on $\mathcal{Z}$. Continuity follows from the fact that if a sequence $(\varphi_n)_{n\in\mathbb{N}}$ converges in $\mathcal{Z}$ to $\varphi$ then it also converges in $\mathcal{S}$ to $\varphi$. To see this note first that the hypothesis implies that $(\hat{\varphi}_n)_{n\in\mathbb{N}}$ converges in $\mathcal{D}$ to $\hat{\varphi}$. But then, for $j = 0, 1, 2, \ldots$, the sequence $(t^j \hat{\varphi}_n(t))_{n\in\mathbb{N}}$ converges in $\mathcal{D}$ to $t^j \hat{\varphi}(t)$. Hence $(\varphi_n^{(j)})_{n\in\mathbb{N}}$ converges in $\mathcal{Z}$ to $\varphi^{(j)}$. From the definition of convergence in $\mathcal{Z}$ it follows that, for each pair of integers $m, j \in \mathbb{N}_0$, the sequence $(\omega^m \varphi_n^{(j)}(\omega))_{n\in\mathbb{N}}$ converges to $\omega^m \varphi^{(j)}(\omega)$ uniformly on $\mathbb{R}$. □

In particular the delta function and all its derivatives are ultradistributions. Operations on ultradistributions can be defined in the same kind of way as for distributions. Thus we have:

(1) Addition of ultradistributions,

$$\forall \varphi \in \mathcal{Z} \; : \; <\hat{\mu}+\hat{\sigma},\varphi> \; \equiv \; <\hat{\mu},\varphi> \; + \; <\hat{\sigma},\varphi>,$$

(2) Multiplication by a constant,

$$\forall a \in \mathbb{C}, \forall \varphi \in \mathcal{Z} \; : \; <a\hat{\mu},\varphi> \; \equiv \; a<\hat{\mu},\varphi>,$$

(3) Translation of an ultradistribution, $\tau_a : \hat{\mu} \to \tau_a\hat{\mu}$, where $a$ may be any real or complex number,

$$\forall \varphi \in \mathcal{Z} \; : \; <\tau_a\hat{\mu}(\omega),\varphi(\omega)> \; \equiv \; <\hat{\mu}(\omega),\varphi(\omega+a)>,$$

(4) Transposition of an ultradistribution,

$$\forall \varphi \in \mathcal{Z} \; : \; <\hat{\mu}(-\omega),\varphi(\omega)> \; \equiv \; <\hat{\mu}(\omega),\varphi(-\omega)>,$$

(5) Differentiation of an ultradistribution,

$$\forall \varphi \in \mathcal{Z} \; : \; <D\hat{\mu},\varphi> \; \equiv \; <\hat{\mu},-\varphi'>.$$

The definition of translate allows a meaning to be assigned to the symbol $\hat{\mu}(z)$ where $z$ is a complex variable, $z = \omega + iy$ say. For every function $\varphi(\omega)$ belonging to $\mathcal{Z}$ we write,

$$<\hat{\mu}(z),\varphi(\omega)> \; \equiv \; <\hat{\mu}(\omega+iy),\varphi(\omega)> \; = \; <\hat{\mu}(\omega),\varphi(\omega-iy)> \quad (3.52)$$

In (3.52) the parameter $y$ is understood to be held constant during evaluation of the expression, so that $\hat{\mu}(z)$ is defined as an ultradistribution on any line in the $z$-plane parallel to the $\omega$ axis. Thus for the ultradistribution $\delta(z)$ we have

$$<\delta(z),\varphi(\omega)> \; = \; <\delta(\omega+iy),\varphi(\omega)> \; = \; <\delta(\omega),\varphi(\omega-iy)> \; = \; \varphi(-iy).$$

Except when $y = 0$, $\delta(\omega + iy)$ represents an ultradistribution which is not a distribution since $\varphi(-iy)$ will not generally be defined for $\varphi \in \mathcal{D}$. On the other hand the function $\exp(t^2)$ defines a (regular) distribution which is not an ultradistribution; $\mathcal{Z}'$ and $\mathcal{D}'$ meet but do not coincide.

### 3.4.4 Convergence of ultradistributions

A sequence $(\hat{\mu}_n)_{n \in \mathbb{N}}$ is said to converge in $\mathcal{Z}'$ if, for each $\varphi \in \mathcal{Z}$, the sequence $(<\hat{\mu}_n, \varphi>)_{n \in \mathbb{N}}$ converges in $\mathbb{C}$. As $\varphi$ traverses $\mathcal{Z}$ the limits of the sequences $(<\hat{\mu}_n, \varphi>)_{n \in \mathbb{N}}$ define a functional $\hat{\mu}$ on $\mathcal{Z}$ which is itself an ultradistribution so that $\mathcal{Z}'$ is closed under convergence. An infinite series of ultradistributions is said to converge in $\mathcal{Z}'$ if the sequence of its partial sums converges in $\mathcal{Z}'$.

Ultradistributions are not merely infinitely differentiable, like distributions, but also have the property that they can be expanded in Taylor-type series. Thus we have the result:

**Theorem 3.26** *Let $\hat{\mu}$ be an ultradistribution and $a$ any complex number. Then $\hat{\mu}$ admits the expansion,*

$$\hat{\mu}(\omega + a) = \sum_{n=0}^{\infty} \frac{a^n}{n!} D^n \hat{\mu}(\omega).$$

**Proof:** Let $\varphi$ be any member of $\mathcal{Z}$. Then $\varphi$ has a Taylor series expansion

$$\varphi(\omega - a) = \sum_{n=0}^{\infty} \frac{(-a)^n}{n!} \varphi^{(n)}(\omega)$$

which converges for all $\omega$. Now consider the Fourier transforms of the partial sums of this series,

$$\sum_{n=0}^{m} \frac{(-a)^n}{n!} (it)^n \hat{\varphi}(t). \tag{3.53}$$

As $m \to \infty$ these converge in $\mathcal{D}$ to $\exp(-iat)\hat{\varphi}(t)$, and it follows that the infinite series (3.53) converges in $\mathcal{Z}$. Hence we obtain

$$
\begin{aligned}
<\hat{\mu}(\omega + a), \varphi(\omega)> &= <\hat{\mu}(\omega), \varphi(\omega - a)> \\
&= \lim_{m \to \infty} <\hat{\mu}(\omega), \sum_{n=0}^{m} \frac{(-a)^n}{n!} \varphi^{(n)}(\omega)> \\
&= \lim_{m \to \infty} <\sum_{n=0}^{m} \frac{a^n}{n!} D^n \hat{\mu}(\omega), \varphi(\omega)> = <\sum_{n=0}^{\infty} \frac{a^n}{n!} D^n \hat{\mu}(\omega), \varphi(\omega)>
\end{aligned}
$$

which proves the theorem. □

In particular we have the expansion

$$\delta(\omega + a) = \sum_{n=0}^{\infty} \frac{a^n}{n!} \delta^{(n)}(\omega),$$

where $a$ may be any given complex number and the infinite series of delta functions converges in the sense of $\mathcal{Z}'$.

### Exercise 3.6 :

1. Show that $\mathcal{Z}$, $(\mathcal{Z}')$, is closed under the operations:

   (a) addition and multiplication by scalars;
   
   (b) (complex) translation;
   
   (c) dilatation: $\varphi(a\omega) \in \mathcal{Z}, a > 0$;
   
   (d) differentiation;
   
   (e) transposition: $\varphi(-\omega) \in \mathcal{Z}$;
   
   (f) multiplication by an entire function $\psi(z)$ satisfying
   $$|\psi(z)| \leq C(1+|z|^m)\exp(b|y|).$$

2. Prove that an infinite series of ultradistributions may be differentiated term by term.

### 3.4.5 Fourier transforms of arbitrary distributions

The Fourier transform of an arbitrary distribution in $\mathcal{D}'$ is defined in terms of a Parseval-type equation:
$$<\hat{\mu}, \varphi> = <\mu, \hat{\varphi}> \tag{3.54}$$
in which as $\hat{\varphi}$ traverses $\mathcal{D}$ so $\varphi$ traverses $\mathcal{Z}$. Thus $\hat{\mu}$ is well defined as that functional which assigns to each $\varphi$ in $\mathcal{Z}$ the same number that $\mu \in \mathcal{D}'$ assigns to $\hat{\varphi} \in \mathcal{D}$. As is easily confirmed this functional $\hat{\mu} \equiv \mathcal{F}[\mu]$ is linear and continuous on $\mathcal{Z}$ and is therefore an ultradistribution. Conversely, to each ultradistribution in $\mathcal{Z}'$ there corresponds, through (3.54), a distribution in $\mathcal{D}'$; equation (3.54) therefore serves to define the inverse Fourier transform also.

**Theorem 3.27** *The Fourier transform is a continuous linear mapping of $\mathcal{D}'$ onto $\mathcal{Z}'$. Similarly, the inverse Fourier transform is a continuous linear mapping of $\mathcal{Z}'$ onto $\mathcal{D}'$.*

**Proof:** The Fourier transform is clearly linear. Now suppose that the sequence $(\mu_n)_{n \in \mathbb{N}}$ converges in $\mathcal{D}'$ to $\mu$. Then the distribution $\mu$ has a Fourier transform $\hat{\mu}$, and we have
$$<\hat{\mu}_n, \varphi> = <\mu_n, \hat{\varphi}>,$$
for all $n \in \mathbb{N}$ and
$$<\mu_n, \hat{\varphi}> \to <\mu, \hat{\varphi}> = <\hat{\mu}, \varphi>$$
so that
$$<\hat{\mu}_n, \varphi> \to <\hat{\mu}, \varphi>$$

for each $\varphi \in \mathcal{Z}$. This shows that the Fourier transform is continuous. The proof for the inverse Fourier transform is similar. □

We can also prove an extended form of the Exchange Formula:

**Theorem 3.28** *Let $\mu, \nu$ be distributions in $\mathcal{D}$ such that $\nu$ has compact support. Then the convolution $\mu * \nu$ and the product $\hat{\mu} \cdot \hat{\nu}$ are both well defined and $\mathcal{F}[\mu * \nu] = \hat{\mu} \cdot \hat{\nu}$.*

**Proof:** Since $\nu$ has compact support the convolution $\mu * \nu$ certainly exists as a distribution and therefore does have a Fourier transform $\mathcal{F}[\mu * \nu]$. Further the Fourier transform $\hat{\nu}$ of $\nu$ is a tempered distribution which may be extended into the entire complex plane as an entire function bounded by $p(z)\exp(b|\Im(z)|)$, where $p(z)$ is a polynomial (see theorem 3.13). Hence if $\varphi$ is any function in $\mathcal{Z}$ the product $\hat{\nu}(\omega)\varphi(\omega)$ makes sense in $\mathcal{Z}$. Writing $\check{\nu}(t) \equiv \nu(-t)$ we have

$$\begin{aligned}
<\mathcal{F}[\mu * \nu], \varphi> &= <\mu * \nu, \hat{\varphi}> = <\mu(t), <\nu(\tau), \hat{\varphi}(t+\tau)>> \\
&= <\mu(t), \check{\nu} * \hat{\varphi}(t)> = <\hat{\mu}(\omega), \mathcal{F}^{-1}[\check{\nu} * \hat{\varphi}](\omega)> \\
&= <\hat{\mu}(\omega), \hat{\nu}(\omega)\varphi(\omega)> = <\hat{\mu}(\omega)\hat{\nu}(\omega), \varphi(\omega)>
\end{aligned}$$

and the result follows. □

**Exercise 3.7 :**

**1.** Prove the following properties for the Fourier transforms of arbitrary distributions:

(a) $\mathcal{F}[(-it)^n \mu(t)](\omega) = D^n \hat{\mu}(\omega);$

(b) $\mathcal{F}[D^n \mu(t)](\omega) = (i\omega)^n \hat{\mu}(\omega);$

(c) $\mathcal{F}[\mu(t-\tau)](\omega) = \exp(-i\omega\tau)\hat{\mu}(\omega);$

(d) $\mathcal{F}[\exp(iat)\mu(t)](\omega) = \hat{\mu}(\omega - a);$

where $n \in \mathbb{N}_0$, $\tau \in \mathbb{R}$ and $a \in \mathbb{C}$.

**2.** Find $\mathcal{F}[\sinh(at)], \mathcal{F}[\cosh(at)]$, and $\mathcal{F}[t^k \exp(at)]$.

**3.** If $\mu \in \mathcal{E}'$ prove that for $\alpha \in \mathbb{C}$,

$$\mathcal{F}\left\{\exp(\alpha t) \sum_{n=-\infty}^{+\infty} \mu(t-n)\right\} = 2\pi \sum_{n=-\infty}^{+\infty} \hat{\mu}(2\pi n)\delta(\omega + i\alpha - 2\pi n)$$

### 3.4.6 Structure of ultradistributions

Let $\hat{\mu}$ be an ultradistribution in $\mathcal{Z}'$. Then there exists $\mu \in \mathcal{D}'$ such that

$$<\hat{\mu}, \varphi> = <\mu, \hat{\varphi}>$$

for all $\varphi \in \mathcal{Z}$ (and therefore for all $\hat{\varphi} \in \mathcal{D}$). Since every distribution is locally a finite-order derivative of a continuous function then, for each $a > 0$, the restriction of $\mu$ to $\mathcal{D}_{[-a,+a]}$ is a finite-order derivative (say of order $r \in \mathbb{N}_0$) of a continuous function whose support may be required to be contained in a neighbourhood of $[-a, +a]$. That is we have

$$<\hat{\mu}, \varphi> = <f_a, (-1)^r \hat{\varphi}^{(r)}>$$

where $f_a$ is a continuous function with support contained in a neighbourhood of $[-a, +a]$ and $\hat{\varphi} \in \mathcal{D}_{[-a,+a]}$. Substituting

$$\hat{\varphi}^{(r)}(t) = \int_{-\infty}^{+\infty} \varphi(\omega)(-i\omega)^r \exp(-i\omega t) d\omega$$

we get

$$<\hat{\mu}, \varphi> = \int_{-a}^{+a} f_a(t) \int_{-\infty}^{+\infty} (i\omega)^r \varphi(\omega) \exp(-i\omega t) d\omega dt$$
$$= \int_{-\infty}^{+\infty} (i\omega)^r \varphi(\omega) \left\{ \int_{-a}^{+a} f_a(t) \exp(-i\omega t) dt \right\} d\omega$$

where the interchange between integrations is easily justified by the Fubini theorem. The function

$$\hat{f}_a(\omega) = \int_{-a}^{+a} f_a(t) \exp(-i\omega t) dt$$

can be continued into the complex plane as an entire function of order of growth $\leq 1$ and exponential type $\leq a$ and is bounded on the real axis. Hence, defining $\hat{F}_a(z) = (iz)^r \hat{f}_a(z)$, we have

> The restriction of an ultradistribution $\mu \in \mathcal{Z}'$ to $\mathcal{Z}_a \equiv \mathcal{F}\{\mathcal{D}_{[-a,+a]}\}$ is given by the formula
>
> $$<\hat{\mu}, \varphi> = \int_{-\infty}^{+\infty} \hat{F}_a(\omega) \varphi(\omega) d\omega \qquad (3.55)$$
>
> where $\hat{F}_a(z)$ is an entire function of exponential type $\leq a$ which increases no more rapidly than some power of $|\Re(z)|$ on the real axis.

Note that the function $\hat{F}_a(z)$ is not determined uniquely. In fact for any $b \in \mathbb{R}$ such that $|b| \geq a$ we have that $<\exp(-ib\omega), \varphi(\omega)> = \hat{\varphi}(b) \equiv 0$ for all $\varphi \in \mathcal{Z}_a$. Hence any function of the form $\hat{F}_a(\omega) + \exp(-ib\omega)$ with $|b| \geq a$ may be used in the right-hand side of (3.55).

Let $\hat{\mu} \in \mathcal{Z}'$ be such that $\mu \equiv \mathcal{F}^{-1}[\hat{\mu}]$ is a distribution of finite order in $\mathcal{D}'_{\text{fin}}$. Then $\hat{\mu} \in \mathcal{Z}'$ is itself called a **finite-order ultradistribution**; denote by $\mathcal{Z}'_{\text{fin}}$ the space of all finite-order ultradistributions.

Finite-order ultradistributions can be characterised in terms of infinite-order differential operators. In order to obtain such a characterisation we need some preliminary results.

### 3.4.6.1 Locally bounded functions on $\mathbb{R}$ and entire functions

Let
$$A(z) = \sum_{n=0}^{\infty} a_n z^n$$
be an entire function, and denote by $A(-iD)$ the infinite-order differential operator defined on the space $\mathcal{Z}$ by
$$A(-iD)[\varphi](\omega) = \sum_{n=0}^{\infty} a_n (-iD)^n \varphi(\omega).$$
For any $\varphi \in \mathcal{Z}$ there exists a unique function $\hat{\varphi} \in \mathcal{D}$ such that
$$\varphi(\omega) = \frac{1}{2\pi} \int_{-\infty}^{+\infty} \hat{\varphi}(t) \exp(i\omega t)\, dt$$
where the integral in fact extends only over some compact subset of $\mathbb{R}$. Then, for any $n \in \mathbb{N}_0$,
$$(-iD)^n \varphi(\omega) = \frac{1}{2\pi} \int_{-\infty}^{+\infty} t^n \hat{\varphi}(t) \exp(i\omega t)\, dt$$
and therefore,
$$\sum_{n=0}^{\infty} a_n (-iD)^n \varphi(\omega) = \frac{1}{2\pi} \sum_{n=0}^{\infty} a_n \int_{-\infty}^{+\infty} t^n \hat{\varphi}(t) \exp(i\omega t)\, dt$$
$$= \frac{1}{2\pi} \int_{-\infty}^{+\infty} \left\{ \sum_{n=0}^{\infty} a_n t^n \right\} \hat{\varphi}(t) \exp(i\omega t)\, dt$$
$$= \frac{1}{2\pi} \int_{-\infty}^{+\infty} A(t) \hat{\varphi}(t) \exp(i\omega t)\, dt,$$
where the interchange of integration and summation is justified by local integrability of all functions involved. Thus we have shown that
$$A(-iD)[\varphi](\omega) = \mathcal{F}^{-1}[A\varphi](\omega)$$

and, since $\hat{\varphi} \to A\hat{\varphi}$ is a continuous mapping from $\mathcal{D}$ into $\mathcal{D}$ for every $A \in \mathcal{H}|_{\mathbb{R}}$, the space of restrictions to $\mathbb{R}$ of functions in $\mathcal{H}$, it follows that

$$A(-iD) : \mathcal{Z} \to \mathcal{Z}$$

is a continuous mapping of $\mathcal{Z}$ into $\mathcal{Z}$.

**Lemma 3.4.1** *Let $f : \mathbb{R}_0^+ \to \mathbb{C}$ be a locally bounded function. Then there exists an entire function $\Phi : \mathbb{C} \to \mathbb{C}$ such that $\Phi(t) \geq |f(t)|$ for all $t \in \mathbb{R}_0^+$. Moreover we may choose $\Phi$ to have no zeros.*

**Proof:** (a) Define the function $f^* : \mathbb{R}_0^+ \to \mathbb{R}$ by setting

$$f^*(x) = \sup_{0 \leq t \leq x} |f(t)|,$$

for every $x \geq 0$. Then $f^*$ is a non-negative, increasing function in $\mathbb{R}_0^+$ such that $f^*(t) \geq |f(t)|$ for all $t \in \mathbb{R}_0^+$. Now let $(a_n)_{n \in \mathbb{N}}$ and $(b_n)_{n \in \mathbb{N}}$ be two sequences of real numbers such that

(1) $0 < a_1 \leq a_2 \leq \ldots \leq a_n \leq \ldots$ and
$0 < b_1 \leq b_2 \leq b_2 \leq \ldots \leq b_n \leq \ldots,$

(2) $\forall n \in \mathbb{N}, a_n/b_n = k > 1,$

(3) $\lim_{n \to \infty} b_n = +\infty.$

Then for each $n$ there exists $m_n \in \mathbb{N}$ such that

$$\left(\frac{a_n}{b_n}\right)^{m_n} \geq f^*(a_{n-1})$$

where, without loss of generality, we may always suppose that $m_1 < m_2 < \ldots < m_n < \ldots$. The power series

$$\sum_{n=1}^{\infty} \left(\frac{z}{b_n}\right)^{m_n}$$

has infinite radius of convergence (since by hypothesis $b_n \to +\infty$ with $n$). The entire function

$$\Phi(z) = f^*(a_1) + \sum_{n=1}^{\infty} \left(\frac{z}{b_n}\right)^{m_n}$$

is non-negative and increasing on $\mathbb{R}_0^+$. For any $t$ in the interval $[a_n, a_{n+1}]$ we get

$$\Phi(t) \geq \Phi(a_n) \geq \left(\frac{a_n}{b_n}\right)^{m_n} \geq f^*(a_{n+1}) \geq f^*(t) \geq |f(t)|$$

and, since $a_n \to +\infty$,

$$\Phi(t) \geq |f(t)|, \ \forall t \in [a_1, +\infty).$$

Further, for all $t \in [0, a_1]$ we have

$$\Phi(t) \geq f^*(a_1) \geq f^*(t) \geq |f(t)|,$$

and so the first part of the result follows.

(b) Let $f$ be a function satisfying the conditions of the hypothesis, and consider the (non-negative) auxiliary function defined on $\mathbb{R}_0^+$ by

$$f_1^*(x) = \log\left\{1 + \sup_{0 \leq t \leq x} |f(t)|\right\}$$

From (a) there exists an entire function $\Phi_1 : \mathbb{C} \to \mathbb{C}$ such that $\Phi_1(t) \geq f_1^*(t)$ for all $t \in \mathbb{R}_0^+$. Then the function defined for all $t \in \mathbb{R}_0^+$ by

$$\Phi(t) = \exp[\Phi_1(t)] \geq \exp[f_1^*(t)]$$
$$= 1 + \sup_{0 \leq x \leq t} |f(x)| \geq |f(t)|$$

completely satisfies the lemma. □

**Corollary 3.4.1** *If $f : \mathbb{R} \to \mathbb{C}$ is locally bounded then there exists an entire function $\Phi : \mathbb{C} \to \mathbb{C}$ without zeros, such that $\Phi(t) \geq |f(t)|$ for all $t \in \mathbb{R}$.*

**Proof:** It is enough to define

$$f^*(x) = \begin{cases} \sup_{-x \leq t \leq +x} |f(t)| & \text{, if } x \geq 0, \\ \sup_{+x \leq t \leq -x} |f(t)| & \text{, if } x < 0. \end{cases}$$

□

### 3.4.6.2 A structure theorem in $\mathcal{Z}'_{\text{fin}}$

Let $\hat{\mu}$ be a finite-order ultradistribution in $\mathcal{Z}'_{\text{fin}}$. Then there exists $\mu \in \mathcal{D}'_{\text{fin}}$ such that $<\hat{\mu}, \varphi> = <\mu, \hat{\varphi}>$ for all $\varphi \in \mathcal{Z}$ and therefore such that

$$<\hat{\mu}, \varphi> = (-1)^r \int_{-\infty}^{+\infty} f(t)\hat{\varphi}^{(r)}(t)dt$$

for some given $f \in \mathcal{C}(\mathbb{R})$ and $r \in \mathbb{N}_0$ (depending on $\hat{\mu}$). It follows that there exists an entire function $\Phi : \mathbb{C} \to \mathbb{C}$, without zeros, which is such that $f(t)/\Phi(t)$ is bounded on $\mathbb{R}$ and therefore such that $A(z) \equiv (1 + z^2)\Phi(z)$ is an entire function, without real zeros such that $f(t)/A(t)$ belongs to $L^1(\mathbb{R})$. Hence

$$<\hat{\mu}, \varphi> = <\frac{f(t)}{A(t)}, (-1)^r A(t)\hat{\varphi}^{(r)}(t)> \qquad (3.56)$$

and so, defining
$$g_A(t) = \frac{f(t)}{A(t)} \in L^1(\mathbb{R})$$
we obtain
$$\begin{aligned}
<\hat{\mu}, \varphi> &= <g_A(t), (-1)^r A(t) \hat{\varphi}^{(r)}> \\
&= <\hat{g}_A(\omega), A(-iD)[(-i\omega)^r \varphi(\omega)]> \\
&= <\hat{g}_A(\omega), \sum_{k=0}^{r} \binom{r}{k}(-i\omega)^k A^{(r-k)}(-iD)\varphi(\omega)> \\
&= <\sum_{k=0}^{r} A^{(r-k)}(iD)\left[\binom{r}{k}(-i\omega)^k \hat{g}_A(\omega)\right], \varphi(\omega)>
\end{aligned}$$
where
$$\hat{g}_A(\omega) = \mathcal{F}[g_A](\omega)$$
is a uniformly continuous and bounded function. Thus we have obtained
$$\hat{\mu}(\omega) = \sum_{k=0}^{r} A^{(r-k)}(iD)\left[\binom{r}{k}(-i\omega)^k \hat{g}_A(\omega)\right] ; \qquad (3.57)$$
that is, $\hat{\mu} \in \mathcal{Z}'_{\text{fin}}$ is given as a finite sum of what might be described as derivatives of infinite order of continuous functions of power growth defined on $\mathbb{R}$.

## 3.5  PERIODIC ULTRADISTRIBUTIONS

The space $\mathcal{B} = \mathcal{F}\{\mathcal{E}'\}$, defined in §3.3.2.2, comprises all functions (ultradistributions) of compact spectrum, and by the Paley-Wiener Theorem we have $\mathcal{B} \subset \mathcal{H}$; that is to say, each function $f \in \mathcal{B}$ may be extended into the complex plane as an entire function which is polynomially bounded on the real axis. Now we denote by $\mathcal{V}_{2\pi} \equiv \mathcal{V}_{2\pi}(\mathbb{R})$ the space of all ultradistributions which are periodic, of period $2\pi$:
$$\mathcal{V}_{2\pi} = \{\hat{\sigma} \in \mathcal{Z}' : \tau_{2\pi}\hat{\sigma} = \hat{\sigma}\} . \qquad (3.58)$$
$\mathcal{V}_{2\pi}$ is given the induced topology of $\mathcal{Z}'$.

The space of all $2\pi$-periodic functions of compact spectrum is given by the intersection
$$\mathcal{Q}_{2\pi} = \mathcal{V}_{2\pi} \cap \mathcal{B}$$
and consists of all $2\pi$-periodic functions which are analytically continuable into the complex plane as entire functions. Convergence in $\mathcal{Q}_{2\pi}$ is defined as follows: a sequence $(\theta_n)_{n \in \mathbb{N}}$ is said to converge in $\mathcal{Q}_{2\pi}$ to $\theta$ if

**(1)** for every $n = 1, 2, \ldots$ the function $\theta_n$ is in $\mathcal{Q}_{2\pi}$ and,

**(2)** for each non-negative integer $j = 0, 1, 2, \ldots$, the sequence $(\theta_n^{(j)})_{n \in \mathbb{N}}$ converges uniformly to $\theta^{(j)}$.

The limit function $\theta$ necessarily belongs to $\mathcal{Q}_{2\pi}$, and so the space is closed under convergence.

For any function $\psi \in \mathcal{Z}$ we define the function

$$T_{2\pi}[\psi](\omega) = \sum_{n=-\infty}^{+\infty} \tau_{2n\pi}\psi(\omega) \equiv \sum_{n=-\infty}^{+\infty} \psi(\omega - 2n\pi) \qquad (3.59)$$

where the series converges uniformly and absolutely. In fact, since

$$\mathcal{F}^{-1}\left[T_{2\pi}[\psi]\right](t) = \sum_{n=-\infty}^{+\infty} \varphi(t) \cdot \exp(i 2\pi n t)$$

where $\varphi \in \mathcal{D}$ and $\psi = \mathcal{F}[\varphi]$, we have

$$\mathcal{F}^{-1}\left[T_{2\pi}[\psi]\right](t) = \varphi(t) \cdot \sum_{n=-\infty}^{+\infty} \exp(i 2\pi n t)$$
$$= \varphi(t) \cdot \sum_{n=-\infty}^{+\infty} \tau_{2n\pi}\delta(t) = \sum_{n=-\infty}^{+\infty} \varphi(2n\pi)\delta(t - 2n\pi).$$

Since $\varphi(2n\pi) = 0$ for all sufficiently large $|n|$ it follows that $\mathcal{F}^{-1}[T_{2\pi}[\psi]]$ belongs to $\mathcal{E}'$ and so that $T_{2\pi}[\psi] \in \mathcal{B}$. The operator $T_{2\pi}$ maps $\mathcal{Z}$ continuously into $\mathcal{B}$; $T_{2\pi}[\psi]$ is a $2\pi$-periodic function, called the $2\pi$-*periodic transform of* $\psi \in \mathcal{Z}$, and therefore $T_{2\pi}[\psi]$ belongs to $\mathcal{Q}_{2\pi} \equiv \mathcal{V}_{2\pi} \cap \mathcal{B}$ for every $\psi \in \mathcal{Z}$.

Now let $\hat{\mu}$ be an ultradistribution belonging to the dual $\mathcal{B}'$ of $\mathcal{B}$. We can define the $2\pi$-periodic extension $T'_{2\pi}[\hat{\mu}]$ of $\hat{\mu}$ by

$$< T'_{2\pi}[\hat{\mu}], \psi > \equiv < \hat{\mu}, T_{2\pi}[\psi] >$$

for all $\psi \in \mathcal{Z}$. Since $T_{2\pi}[\tau_{-2\pi}\psi] = \tau_{-2\pi}T_{2\pi}[\psi] = T_{2\pi}[\psi]$ we have

$$< \tau_{2\pi} T'_{2\pi}[\hat{\mu}], \psi > \;=\; < T'_{2\pi}[\hat{\mu}], \tau_{-2\pi}\psi >$$
$$=\; < \hat{\mu}, T_{2\pi}[\tau_{-2\pi}\psi] > \;=\; < \hat{\mu}, T_{2\pi}[\psi] > \;=\; < T'_{2\pi}[\hat{\mu}], \psi >$$

and it follows that $T'_{2\pi}[\hat{\mu}]$ belongs to $\mathcal{V}_{2\pi}$.

The formal parallels with the results of §3.3.3 on periodic distributions are clear. A function $\xi \in \mathcal{Z}$ is said to be a $2\pi$-periodic partition of unity in $\mathcal{Z}$ if $T_{2\pi}[\xi] = 1$. In particular if $\varphi \in \mathcal{D}$ is such that $\varphi(0) = 1$ and if $\psi = \hat{\varphi}$ then the function

$\xi = \psi * \chi_{[-\pi,+\pi]}$, where $\chi_{[-\pi,+\pi]}$ is the characteristic function of $[-\pi,+\pi]$, is a $2\pi$-periodic partition of unity in $\mathcal{Z}$. By using arguments of the same general nature as those for the proofs of lemma 3.3.1 and theorems 3.16, 3.17 we can obtain the following results:

**Lemma 3.5.1** *(1) If $\hat{\sigma} \in \mathcal{V}_{2\pi}$ and $\psi \in \mathcal{Z}$ then $T_{2\pi}[\psi\hat{\sigma}] = (T_{2\pi}[\psi])\hat{\sigma}$.*
*(2) If $\hat{\mu} \in \mathcal{B}'$ and $\theta \in \mathcal{Q}_{2\pi}$ then $T_{2\pi}[\theta\hat{\mu}] = (T_{2\pi}[\hat{\mu}])\theta$.*

**Theorem 3.29** *(1) Every function in $\mathcal{Q}_{2\pi}$ is the $2\pi$-periodic extension of a function in $\mathcal{Z}$. (2) Every ultradistribution in $\mathcal{V}_{2\pi}$ is the $2\pi$-periodic extension of an ultradistribution in $\mathcal{B}'$.*

**Theorem 3.30** *$\mathcal{V}_{2\pi}$ may be identified with the dual space $\mathcal{Q}'_{2\pi}$.*

Thus, for any $\hat{\sigma}$ in $\mathcal{Q}'_{2\pi} \equiv \mathcal{V}_{2\pi}$ we may write

$$(\hat{\sigma}, \theta) \equiv <\hat{\sigma}, \xi\theta> \tag{3.60}$$

for all $\theta \in \mathcal{Q}_{2\pi}$ and all $2\pi$-periodic partitions of unity $\xi \in \mathcal{Z}$. Moreover $\hat{\mu} = \xi\hat{\sigma}$ is an ultradistribution in $\mathcal{B}'$ such that $\hat{\sigma} = T_{2\pi}[\hat{\mu}]$. Finally, the trigonometric series

$$\sum_{n=-\infty}^{+\infty} c_n(\hat{\sigma}) \cdot \exp(-in\omega) \tag{3.61}$$

where

$$\begin{aligned} c_n(\hat{\sigma}) &= \frac{1}{2\pi}(\hat{\sigma}(\omega), \exp(in\omega)) \\ &= \frac{1}{2\pi} <\hat{\mu}(\omega), \exp(in\omega)> \end{aligned} \tag{3.62}$$

is called the Fourier series of the ultradistribution $\hat{\sigma}$.

**Exercise 3.8** :

1. Prove Lemma 3.5.1.

2. Prove Theorem 3.29 and Theorem 3.30.

3. Prove that every $2\pi$-periodic test function in $\mathcal{Q}_{2\pi}$ is a finite linear combination of functions in the set $\{\exp(in\omega)\}_{n\in\mathbb{Z}}$:

$$\theta(\omega) = \sum_{n \in \mathbb{Z}_k} c_n(\theta) \exp(in\omega)$$

where $\mathbb{Z}_k$ is a finite subset of $\mathbb{Z}$.

## 3.6 ULTRADISTRIBUTIONS OF EXPONENTIAL TYPE

Ultradistributions of exponential type were introduced into the mathematical literature by J. S. Silva [103] in terms of certain equivalence classes of analytic functions. Such equivalence classes are sometimes called Cauchy-Stieltjes transforms of the ultradistributions in question, and their study will be developed in the next chapter. At present we study ultradistributions of exponential type as elements of a dual space denoted by $\mathcal{Z}'_{\exp}$. The space $\mathcal{Z}_{\exp}$ is a subspace of $\mathcal{S}$, the Schwartz space of functions of rapid decrease. The Fourier transform on $\mathcal{Z}_{\exp}$ is a topological isomorphism, just as it is on $\mathcal{S}$.

The term *exponential ultradistribution* is perhaps slightly misleading since, on the one hand, $\mathcal{D}' \cap \mathcal{Z}'_{\exp}$ contains $\mathcal{S}'$ as a proper subspace whereas on the other hand $\mathcal{Z}'_{\exp}$ is not entirely contained in $\mathcal{Z}'$. However the name is well established in the literature and we shall continue to use it here. Moreover we approach the study of $\mathcal{Z}'_{\exp}$ via a particular subspace $\mathcal{K}'$ of $\mathcal{D}'$ which consists of distributions which can legitimately be described as of exponential type. This has been studied by several authors, notably J. S. Silva [103] and Zielezny [119] together with the space $\mathcal{U}' \subset \mathcal{Z}'$ of its Fourier transforms.

### 3.6.1 Distributions of exponential type

Denote by $\mathcal{K}$ the space of all infinitely differentiable functions $\hat{\varphi}$ such that, for all $p \in \mathbb{N}$, the products $\exp(p|\omega|)\hat{\varphi}^{(q)}(\omega)$ are bounded and continuous on $\mathbb{R}$ for all $q$ such that $0 \leq q \leq p$. It follows that for each $\hat{\varphi} \in \mathcal{K}$ the expressions

$$\| \varphi \|_{\mathcal{K},p} = \max_{0 \leq q \leq p} \left\{ \sup_{\omega \in \mathbb{R}} \left[ \exp(p|\omega|)|\hat{\varphi}^{(q)}(\omega)| \right] \right\} \tag{3.63}$$

are finite for every $p \in \mathbb{N}_0$, and define a family of norms on $\mathcal{K}$. If $\mathcal{K}_p$ denotes the completion of $\mathcal{K}$ with respect to the norm $\| \cdot \|_{\mathcal{K},p}$ defined in (3.63) then for $0 \leq p \leq p'$ we have $\mathcal{K}_{p'} \subset \mathcal{K}_p$ and it follows that

$$\mathcal{K} = \bigcap_{p=0}^{\infty} \mathcal{K}_p \tag{3.64}$$

and therefore that $\mathcal{K}$ is a countably normed space (and is actually complete). A sequence $(\hat{\varphi}_n)_{n \in \mathbb{N}} \subset \mathcal{K}$ converges to zero in $\mathcal{K}$ if it is bounded in each of the norms $\| \cdot \|_{\mathcal{K},p}$ and for each $j = 0, 1, 2, \ldots$, the sequence $\left( \hat{\varphi}_n^{(j)} \right)_{n \in \mathbb{N}}$ converges to zero uniformly on compacts of $\mathbb{R}$.

**Lemma 3.6.1** *$\mathcal{D}$ is a dense linear subspace of $\mathcal{K}$.*

**Proof:** Let $\hat{\theta} \in \mathcal{D}$ be such that $\hat{\theta}(\omega) = 1$ for $|\omega| \leq 1$, and $\hat{\theta}(\omega) = 0$ for $|\omega| \geq 2$, and for each $p \in \mathbb{N}_0$ define $t_p = \max_{q \leq p} \left\{ \sup_{\omega \in \mathbb{R}} |\hat{\theta}^{(q)}(\omega)| \right\} < \infty$. Then for each $n = 1, 2, \ldots$, we have

$$\max_{q \leq p} \left\{ \sup_{\omega \in \mathbb{R}} |\hat{\theta}^{(q)}(\omega/n)| \right\} \leq t_p,$$

for all $p \in \mathbb{N}_0$. For any $\hat{\varphi} \in \mathcal{K}$ the sequence $\hat{\varphi}_n = \hat{\varphi}(\omega)\hat{\theta}(\omega/n)$, $n = 1, 2, \ldots$, belongs to $\mathcal{D}$ and converges uniformly to $\hat{\varphi}$ on compacts (since, starting with some appropriate number $n \geq n_0$, the function $\hat{\varphi}_n$ coincides with $\hat{\varphi}$ on that compact). Moreover, for any $q$ such that $0 \leq q \leq p$, we have

$$\begin{aligned}
\exp(p|\omega|)|\hat{\varphi}_n^{(q)}(\omega)| &= \exp(p|\omega|) \cdot |\sum_{j=0}^{q} \binom{q}{j} \hat{\theta}^{(j)}(\omega/k) \hat{\varphi}^{(q-j)}(\omega)| \\
&\leq \sum_{j=0}^{q} \binom{q}{j} t_q \exp(p|\omega|) |\hat{\varphi}^{(j)}(\omega)| \leq C_q
\end{aligned}$$

where for each $q \geq p$, $p \in \mathbb{N}_0$, $C_q$ is a constant independent of $n$. Hence the sequence $(\hat{\varphi}_n)_{n \in \mathbb{N}} \subset \mathcal{D}$ converges in $\mathcal{K}$ to $\hat{\varphi} \in \mathcal{K}$. □

From the above lemma and taking theorem 1.9 into account it follows that

$$\mathcal{K}' \subset \mathcal{D}'$$

where $\mathcal{K}'$ is the space of all **distributions of exponential type**.

As referred to in §1.2.4 the balls $\{\hat{\varphi} \in \mathcal{K} : \|\hat{\varphi}\|_{\mathcal{K},p} < \varepsilon\}$ constitute a basis of the neighbourhoods of zero in the countably normed space $\mathcal{K}$. The boundedness of $\mu \in \mathcal{K}'$ on such a neighbourhood of zero is equivalent to its boundedness relative to the norm $\|\cdot\|_{\mathcal{K},p}$, i.e., to an inequality of the form

$$|<\mu, \hat{\varphi}>| \leq C \|\hat{\varphi}\|_{\mathcal{K},p} \tag{3.65}$$

where $C$ is a constant. The least $p \in \mathbb{N}_0$ for which (3.65) holds is the order of the distribution $\mu \in \mathcal{K}'$. Hence every distribution of exponential type is a finite order distribution. The set of all distributions of exponential type of order $\leq p$, that is to say the set of all linear functionals which are continuous with respect to the norm of the space $\mathcal{K}_p$, is a subspace of $\mathcal{K}'$ which coincides with the dual $\mathcal{K}'_p$ of $\mathcal{K}_p$. Hence

$$\mathcal{K}' = \bigcup_{p=0}^{\infty} \mathcal{K}'_p. \tag{3.66}$$

The elements of $\mathcal{K}'$ can be characterised as follows:

**Theorem 3.31** *A distribution $\mu \in \mathcal{D}'$ is in $\mathcal{K}'$ if and only if $\mu$ can be represented in the form*

$$\mu(t) = D^r \left[\exp(a|t|) \cdot f(t)\right] \tag{3.67}$$

*for some numbers $r \in \mathbb{N}_0$, $a \in \mathbb{R}$, and a bounded continuous function $f$ on $\mathbb{R}$.*

From general results on duals of countably normed spaces (see Gel'fand and Shilov [26, vol. 2]) it is easy to construct a proof for this theorem which is therefore left as an exercise.

For each $\mu \in \mathcal{K}'$ the smallest number $r \in \mathbb{N}_0$ for which (3.67) holds is uniquely determined; in the sequel we assume that $r$ always denotes that smallest value. On the other hand if (3.67) holds for some fixed $a \in \mathbb{R}$ then we also have

$$\mu(t) = D^r \left\{\exp(a'|t|)\left[\exp\left((a-a')|t|\right) f(t)\right]\right\} = D^r \left[\exp(a'|t|) \cdot F(t)\right]$$

where, for all $a' \geq a$, $F(t)$ is a bounded continuous function on $\mathbb{R}$. Accordingly we define the exponential type $b \geq 0$ of a distribution $\mu \in \mathcal{K}'$ by

$$b = \inf\left\{a \geq 0 : \left[\forall \varepsilon > 0, \exists f_\varepsilon \in C^0 \cap L^\infty(\mathbb{R}) : \mu = D^r \left[\exp\left((a+\varepsilon)|t|\right) \cdot f_\varepsilon(t)\right]\right]\right\}$$

where, for any $\varepsilon > 0$, $f_\varepsilon$ is a bounded continuous function on $\mathbb{R}$. Thus we have the following representation for a distribution $\mu \in \mathcal{K}'$,

$$\mu(t) = D^r \left[\exp\left((b+\varepsilon)|t|\right) \cdot f_\varepsilon(t)\right].$$

### 3.6.2 Silva tempered ultradistributions

The inverse Fourier transform of a function $\hat{\varphi} \in \mathcal{K}_p$ defined by

$$\varphi(\omega) = \frac{1}{2\pi} \int_{-\infty}^{+\infty} \exp(i\omega t)\hat{\varphi}(t)\, dt \;, \quad \omega \in \mathbb{R}$$

is an infinitely differentiable function which may be continued into the complex plane through

$$\begin{aligned}\varphi(z) &= \frac{1}{2\pi} \int_{-\infty}^{+\infty} \exp(izt)\hat{\varphi}(t)\, dt \\ &= \frac{1}{2\pi} \int_{-\infty}^{+\infty} [\exp(-ty)\hat{\varphi}(t)] \exp(itx)\, dt \end{aligned} \tag{3.68}$$

Since this integral converges uniformly on compact subsets of the open strip of the complex plane $\Lambda_p = \{z \in \mathbb{C} : |\Im(z)| < p\}$ it follows that $\varphi(z)$ is a function analytic on that strip and for which we have

$$\| \varphi \|_{\mathcal{U},p} = \sup_{z \in \Lambda_p} \{(1+|z|)^p |\varphi(z)|\} < +\infty.$$

For each $p \in \mathbb{N}_0$ let $\mathcal{U}_p$ denote the space of all such functions $\varphi$ which are the inverse Fourier transforms of functions in $\mathcal{K}_p$. Then the space $\mathcal{U}$ defined by

$$\mathcal{U} \equiv \bigcap_{p=0}^{\infty} \mathcal{U}_p \tag{3.69}$$

with the topology defined by the system of norms $(\| \cdot \|_{\mathcal{U},p})_{p \in \mathbb{N}_0}$ is a complete countably normed space. The functions in $\mathcal{U}$ may be extended into the complex plane as entire functions of rapid descent on horizontal strips (and of arbitrary growth on verticals). Recalling that $\mathcal{H} \equiv \mathcal{H}(\mathbb{C})$ is the space of all entire functions equipped with the topology of uniform convergence on compacts in the $z$-plane we have

$$\mathcal{U} \subset \mathcal{S} \cap \mathcal{H},$$

that is, the members of $\mathcal{U}$ are those functions in $\mathcal{S}$ which may be extended into the complex plane as entire functions of rapid descent on strips. The Fourier transform is a topological isomorphism of $\mathcal{K}$ onto $\mathcal{U}$. A sequence $(\varphi_n)_{n \in \mathbb{N}} \subset \mathcal{U}$ converges to zero in $\mathcal{U}$ if and only if the sequence of its Fourier transforms $(\hat{\varphi}_n)_{n \in \mathbb{N}} \subset \mathcal{K}$ converges in $\mathcal{K}$ to zero. Since $\mathcal{D}$ is a dense subspace of $\mathcal{K}$ it follows that $\mathcal{Z}$ is a dense subspace of $\mathcal{U}$ which, in turn, is a dense subspace of $\mathcal{S}$. Denoting by arrows the (continuous) embeddings of the function spaces discussed so far we have the diagram shown below.

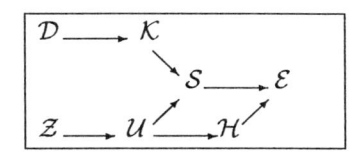

$$\begin{array}{ccc}
\mathcal{D} \longrightarrow \mathcal{K} & & \\
& \searrow \mathcal{S} \longrightarrow \mathcal{E} & \\
\mathcal{Z} \longrightarrow \mathcal{U} \nearrow \mathcal{H} \nearrow &
\end{array}$$

$\mathcal{U}'$, the dual of the space $\mathcal{U}$, is the space of all **Silva tempered ultradistributions**. Since the Fourier transform is a topological isomorphism of $\mathcal{K}$ onto $\mathcal{U}$ it follows that $\mathcal{U}'$ can also be characterised as the space of all Fourier transforms of distributions of exponential type. For any Silva tempered ultradistribution $\hat{\mu}$ there exists $\mu \in \mathcal{K}'$ such that

$$<\hat{\mu}, \varphi> = <\mu, \hat{\varphi}>$$

for all $\varphi \in \mathcal{U}$ (and therefore for all $\hat{\varphi} \in \mathcal{K}$). Hence there exists $b \geq 0$ and $r \in \mathbb{N}_0$ such that

$$\begin{aligned}<\hat{\mu}, \varphi> &= <\exp[(b+\varepsilon)|t|]f_\varepsilon(t), (-1)^r \hat{\varphi}^{(r)}(t)> \\ &= <g_\varepsilon(t), (-1)^r \cosh(b't)\hat{\varphi}^{(r)}(t)> \end{aligned} \quad (3.70)$$

where

$$g_\varepsilon(t) = \frac{\exp[(b+\varepsilon)|t|]}{\cosh(b't)} f_\varepsilon(t)$$

is a bounded continuous $L^1(\mathbb{R})$ function for every $b' > b$ and $0 < \varepsilon \leq b' - b$. Therefore we have that $\hat{g}_\varepsilon$ is in $\mathcal{C}(\mathbb{R}) \cap L^\infty(\mathbb{R})$ and so

$$\begin{aligned}<\hat{\mu}, \varphi> &= <\hat{g}_\varepsilon(\omega), (-1)^r \cosh(-ib'D)[(-i\omega)^r \varphi(\omega)]> \\ &= <(i\omega)^r \cosh(ib'D)\hat{g}_\varepsilon(\omega), \varphi(\omega)>;\end{aligned}$$

that is,

$$\hat{\mu}(\omega) = (i\omega)^r \cosh(ib'D)[\hat{g}_\varepsilon](\omega) \quad (3.71)$$

in the sense of the Silva tempered ultradistributions.

### 3.6.3 The space $\mathcal{Z}_{\exp}$

For any $j \in \mathbb{N}$ let $\mathcal{Z}_{\exp,j}$ be the subspace of $\mathcal{K}_j \cap \mathcal{U}_j$ constituted by all functions $\varphi$ for which

$$\|\varphi\|_{\exp,j} = \max_{k \leq j} \left\{ \sup_{z \in \Lambda_j} \left[\exp(j|\Re(z)|)|\varphi^{(k)}(z)|\right] \right\}$$

is finite. Functions in $\mathcal{Z}_{exp,j}$ are analytic functions in $\Lambda_j$ which belong to $\mathcal{K}_j$ on every horizontal line contained in $\Lambda_j$. If we identify functions by restriction then, for any $j \in \mathbb{N}$, we have that

$$\mathcal{Z}_{\exp,j+1} \subset \mathcal{Z}_{\exp,j}$$

and therefore the space

$$\mathcal{Z}_{\exp} = \bigcap_{j=1}^{\infty} \mathcal{Z}_{\exp,j} \quad (3.72)$$

with the topology generated by the system of semi-norms $(\|\cdot\|_{\exp,j})_{j \in \mathbb{N}}$ is a complete countably normed space. The functions in $\mathcal{Z}_{\exp}$ are entire functions over the finite complex plane which belong to $\mathcal{K}$ on every horizontal line (and have arbitrary growth on verticals). The space $\mathcal{Z}_{\exp}$ enjoys a very important property, which is stated in the following theorem:

**Theorem 3.32** *The Fourier transform (and similarly the inverse Fourier transform) is a topological automorphism on $\mathcal{Z}_{\exp}$.*

**Proof:** The proof of this theorem depends on the following lemma, which is a straightforward application of Cauchy's Theorem.

**Lemma 3.6.2** *If $\varphi \in \mathcal{Z}_{\exp}$ then for any (fixed) $\zeta \in \mathbb{C}$ the equality*
$$\int_{-\infty}^{+\infty} \exp(-i\zeta x)\varphi(x)\,dx = \int_{-\infty+i\tau}^{+\infty+i\tau} \exp[-i\zeta(x+i\tau)]\varphi(x+i\tau)d(x+i\tau)$$
*holds for any given $\tau \in \mathbb{R}$.*

(a) From the lemma it follows that for any $\tau \in \mathbb{R}$
$$\begin{aligned}\hat{\varphi}(\zeta) &= \int_{-\infty}^{+\infty} \exp(-i\zeta x)\varphi(x)dx \\ &= \int_{-\infty+i\tau}^{+\infty+i\tau} \exp[-i\zeta(x+i\tau)]\varphi(x+i\tau)d(x+i\tau) \\ &= \int_{-\infty+i\tau}^{+\infty+i\tau} \exp[-i(x\xi-\tau\eta)]\{\exp(x\eta+\tau\xi)\varphi(x+i\tau)\}d(x+i\tau)\end{aligned}$$

and therefore
$$\exp(-\tau\xi)\hat{\varphi}(\xi+i\eta) = \int_{-\infty}^{+\infty} \exp[-i(x\xi-\tau\eta)]\{\exp(x\eta)\varphi(x+i\tau)\}dx$$

where the right-hand side is well defined since from the fact that $\varphi$ belongs to $\mathcal{Z}_{\exp}$ it follows that $\exp(x\eta)\varphi(x+i\tau)$ belongs to $L^1(\mathbb{R})$ for arbitrarily fixed $\eta, \tau \in \mathbb{R}$.

Now for any $j \in \mathbb{N}$ let $k \in \mathbb{N}$ be such that $k \le j$. Since for any fixed $\eta, \tau \in \mathbb{R}$ the function $\exp(x\eta)[i(x+i\tau)]^k \varphi(x+i\tau)$ also belongs to $L^1(\mathbb{R})$ we obtain
$$\exp(-\tau\xi)\hat{\varphi}^{(k)}(\xi+i\eta) = \int_{-\infty}^{+\infty} \exp[-i(x\xi-\tau\eta)]\{\exp(x\eta)[i(x+i\tau)]^k \varphi(x+i\tau)\}dx \quad (3.73)$$

from which it follows immediately that $\|\hat{\varphi}\|_{\exp,j}$ is finite. Since this is valid for any $j \in \mathbb{N}$ then $\hat{\varphi}$ belongs to $\mathcal{Z}_{\exp}$. We have therefore proved that
$$\varphi \in \mathcal{Z}_{\exp} \Rightarrow \hat{\varphi} \in \mathcal{Z}_{\exp}.$$

By similar arguments, using the Fourier inversion formula, we can prove that
$$\hat{\varphi} \in \mathcal{Z}_{\exp} \Rightarrow \varphi \in \mathcal{Z}_{\exp}.$$

Moreover the uniqueness of the Fourier transform establishes that $\mathcal{F}$ and $\mathcal{F}^{-1}$ are one-to-one correspondences of $\mathcal{Z}_{\exp}$ onto itself.

(b) Taking (3.73) into account it is easy to see that the Fourier transform (and its inverse) is a continuous linear operation on $\mathcal{Z}_{\exp}$.[1] □

---

[1] If we had the equality $\mathcal{Z}_{\exp} = \mathcal{K} \cap \mathcal{U}$ then this theorem would follow trivially from the fact that $\mathcal{F}\{\mathcal{K}\} = \mathcal{U}$ and $\mathcal{F}^{-1}\{\mathcal{U}\} = \mathcal{K}$. However we only know that $\mathcal{Z}_{\exp}$ is a subspace of $\mathcal{K} \cap \mathcal{U}$ and therefore the above proof is necessary.

**Theorem 3.33** $\mathcal{Z}_{\exp}$ *is continuously and densely embedded into* $\mathcal{U}$.

**Proof:** (a) $\mathcal{Z}_{\exp}$ is contained in $\mathcal{U}$, and for any $\varphi \in \mathcal{Z}_{\exp}$ and $j \in \mathbb{N}$ we have

$$\begin{aligned}
\|\hat{\varphi}\|_{\mathcal{U},j} &= \sup_{z \in \Lambda_j}[(1+|z|^j)|\varphi(z)|] \\
&= \sup_{z \in \Lambda_j}\left\{\frac{(1+|z|)^j}{\exp(j|\Re(z)|)} \cdot [\exp(j|\Re(z)|)|\varphi(z)|]\right\} \\
&\leq C_j \left\{\max_{k \leq j}\sup_{z \in \Lambda_j}[\exp(j|\Re(z)|)|\varphi^{(k)}(z)|]\right\} = C_j \cdot \|\hat{\varphi}\|_{\exp,j}
\end{aligned}$$

where $C_j$ is a positive constant (depending only on $j \in \mathbb{N}$). Hence $\mathcal{Z}_{\exp}$ is continuously embedded in $\mathcal{U}$.

(b) It remains to show that every function in $\mathcal{U}$ may be approximated (with respect to the family of semi-norms $\|\cdot\|_{\mathcal{U},j}, j \in \mathbb{N}$) by a sequence of $\mathcal{Z}_{\exp}$-functions.
Denote by $\theta$ a function in $\mathcal{Z}_{\exp}$ such that $\theta(0) = 1$ and consider the sequence $(\theta_n)_{n \in \mathbb{N}}$ defined by

$$\theta_n(z) = \theta(z/n) \quad , \quad n \in \mathbb{N}\,;\, z \in \mathbb{C}.$$

Since, for each fixed $n \in \mathbb{N}$ and $j \in \mathbb{N}$ we have

$$\begin{aligned}
\|\theta_n\|_{\exp,j} &= \max_{k \leq j}\left\{\sup_{z \in \Lambda_j}[\exp(j|\Re(z)|)|\theta_n^{(k)}(z)|]\right\} \\
&= \max_{k \leq j}\left\{\frac{1}{n^k}\sup_{z \in \Lambda_j}[\exp(j|\Re(z)|)|\theta^{(k)}(z/n)|]\right\} \\
&\leq \max_{k \leq j}\left\{\sup_{z \in \Lambda_j}[\exp(j|\Re(z)|)|\theta^{(k)}(z/n)|]\right\} \leq \|\theta\|_{\exp,j},
\end{aligned}$$

then $\theta_n \in \mathcal{Z}_{\exp}$ for all $n \in \mathbb{N}$. Now consider an arbitrary function $\varphi \in \mathcal{U}$ and define the sequence $(\varphi_n)_{n \in \mathbb{N}}$ by setting

$$\varphi_n(z) = \varphi(z)\theta_n(z) \quad , \quad n \in \mathbb{N}\,;\, z \in \mathbb{C}.$$

Since, for any $n \in \mathbb{N}$ and $j \in \mathbb{N}$, we have

$$\begin{aligned}
\|\varphi_n\|_{\exp,j} &= \max_{k \leq j}\left\{\sup_{z \in \Lambda_j}\left[\exp(j|\Re(z)|)|\varphi_n^{(k)}(z)|\right]\right\} \\
&= \max_{k \leq j}\left\{\sup_{z \in \Lambda_j}\left[\exp(j|\Re(z)|)\left|\sum_{m=0}^{k}\binom{k}{m}\varphi^{(m)}(z)\theta_n^{(k-m)}(z)\right|\right]\right\} \\
&\leq \max_{k \leq j}\sum_{m=0}^{k}\binom{k}{m}\|\theta\|_{\exp,j}\cdot\|\varphi^{(m)}\|_{\mathcal{U},j} = C_j^*
\end{aligned}$$

where $C_j^*$ is a positive constant (depending only on $j$ and $\varphi$ but not on $n$) then $\varphi_n \in \mathcal{Z}_{\exp}$ for all $n \in \mathbb{N}$. Moreover if $j \in \mathbb{N}$ we have that

$$\begin{aligned}
\| \varphi_n - \varphi \|_{\mathcal{U},j} &= \sup_{z \in \Lambda_j} \left\{ (1+|z|)^j |\varphi_n(z) - \varphi(z)| \right\} \\
&= \sup_{z \in \Lambda_j} \left\{ (1+|z|)^j |\varphi(z)| \cdot |\theta_n(z) - 1| \right\} \\
&\leq \| \varphi \|_{\mathcal{U},j} \cdot \left\{ \sup_{z \in \Lambda_j} |\theta_n(z) - 1| \right\}
\end{aligned}$$

and since $\theta_n(z) \to 1$ as $n \to \infty$ uniformly on compacts then $\lim_{n \to \infty} \| \varphi_n - \varphi \|_{\mathcal{U},j} = 0$; that is $\varphi_n \to \varphi$ in $\mathcal{U}$ (as $n \to \infty$). □

Taking into account that $\mathcal{K} = \mathcal{F}\{\mathcal{U}\}, \mathcal{Z}_{\exp} = \mathcal{F}\{\mathcal{Z}_{\exp}\}$ and that $\mathcal{U}$ is a dense subspace of $\mathcal{S}$ we get

**Corollary 3.6.1** $\mathcal{Z}_{\exp}$ *is continuously and densely embedded into the spaces $\mathcal{K}$ and $\mathcal{S}$.*

Denoting by an arrow the continuous embeddings of dense subspaces we have

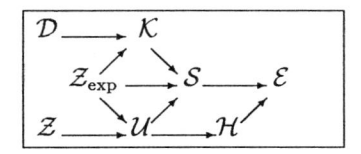

$$\begin{array}{c}
\mathcal{D} \longrightarrow \mathcal{K} \\
\mathcal{Z}_{\exp} \longrightarrow \mathcal{S} \longrightarrow \mathcal{E} \\
\mathcal{Z} \longrightarrow \mathcal{U} \longrightarrow \mathcal{H}
\end{array}$$

$\mathcal{Z}'_{\exp}$, the topological dual of $\mathcal{Z}_{\exp}$, is called the **space of ultradistributions of exponential type**. In view of the theorem 3.33 and its corollary we have the following dual inclusion relations between the various spaces of generalised functions:

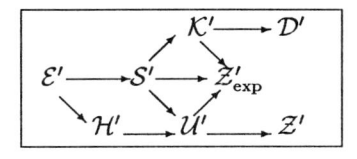

$$\begin{array}{c}
\mathcal{K}' \longrightarrow \mathcal{D}' \\
\mathcal{E}' \longrightarrow \mathcal{S}' \longrightarrow \mathcal{Z}'_{\exp} \\
\mathcal{H}' \longrightarrow \mathcal{U}' \longrightarrow \mathcal{Z}'
\end{array}$$

That is $\mathcal{Z}'_{\exp}$ contains $\mathcal{U}'$ and $\mathcal{K}'$ as proper subspaces.

Taking (3.72) into account we may write

$$\mathcal{Z}'_{\exp} = \bigcup_{j=1}^{\infty} \mathcal{Z}'_{\exp,j} \tag{3.74}$$

which means that any ultradistribution of exponential type may extend to some $\mathcal{Z}_{\exp,j}$ for a suitable $j \in \mathbb{N}$ (depending only on the particular ultradistribution concerned).

# 4

# Analytic Representation of Generalised Functions

## 4.1 THE CAUCHY TRANSFORM

### 4.1.1 Introduction

The delta function $\delta(x)$ can be expressed in terms of ordinary functions in many different ways. Consider in particular its representation as the distributional limit of the delta sequence $(n/\pi(n^2x^2+1))_{n\in\mathbb{N}}$, or equivalently as the limit of the delta net $\{\varepsilon/\pi(x^2+\varepsilon^2)\}_{0<\varepsilon<1}$. For any function $\varphi \in \mathcal{D}$ we have

$$
\begin{aligned}
<\delta(x),\varphi(x)> &= \lim_{\varepsilon\downarrow 0}\int_{-\infty}^{+\infty}\varphi(x)\left\{\frac{\varepsilon}{\pi(x^2+\varepsilon^2)}\right\}dx \\
&= \lim_{\varepsilon\downarrow 0}\int_{-\infty}^{+\infty}\varphi(x)\left\{-\frac{1}{2\pi i}\right\}\left\{\frac{1}{x+i\varepsilon}-\frac{1}{x-i\varepsilon}\right\}dx.
\end{aligned}
$$

We can write this in the form

$$<\delta(x),\varphi(x)> = \lim_{\varepsilon\downarrow 0}\int_{-\infty}^{+\infty}\varphi(x)\left\{\delta^{\circ}(x+i\varepsilon)-\delta^{\circ}(x-i\varepsilon)\right\}dx$$

where for all $z \in \mathbb{C}\backslash\mathbb{R}$ the function $\delta^\circ(z)$ can be defined as

$$\delta^\circ(z) = \frac{1}{2\pi i} <\delta(x), \frac{1}{x-z}> \equiv -\frac{1}{2\pi i z}$$

Thus the generalised function $\delta(x)$ admits a representation in terms of an ordinary function $\delta^\circ(z)$ of the complex variable $z$; this function $\delta^\circ(z)$ is analytic in each of the half-planes $\{z \in \mathbb{C} : \Im(z) > 0\}$ and $\{z \in \mathbb{C} : \Im(z) < 0\}$, and in fact extends to a function defined and analytic for all $z$ in the complement of the support of $\delta(x)$. Now let $\varphi$ be any function analytic on a neighbourhood $V$ of the origin. By the Cauchy Residue Theorem we have

$$\varphi(0) = \frac{1}{2\pi i} \oint_\Gamma \frac{\varphi(z)}{z} dz$$

where $\Gamma$ is any simple closed counterclockwise contour containing the origin and lying wholly inside $V$. Then we can write

$$\varphi(0) = -\oint_\Gamma \varphi(z) \frac{(-1)}{2\pi i z} dz = -\oint_\Gamma \varphi(z)\delta^\circ(z)dz = \oint_{\Gamma'} \varphi(z)\delta^\circ(z)dz$$

where $\Gamma'$ is the contour $\Gamma$ re-oriented in the clockwise sense. This gives a rigorous interpretation of the symbolic integral representing the sampling operation of the delta function, at least for functions $\varphi$ which are analytic in some neighbourhood of the origin.

It is not difficult to see how this result might be generalised, at least in so far as distributions of compact support are concerned. In the first place the above representation for $\delta(x)$ allows us to express any test-function $\varphi \in \mathcal{D}$ in the following form,

$$\varphi(x) = <\delta(y-x), \varphi(y)> = \lim_{\varepsilon \downarrow 0} \int_{-\infty}^{+\infty} \varphi(y) \{\delta^\circ(y-x+i\varepsilon) - \delta^\circ(y-x-i\varepsilon)\} dy$$

$$\equiv \lim_{\varepsilon \downarrow 0} \int_{-\infty}^{+\infty} \varphi(y) \frac{(-1)}{2\pi i} \left\{\frac{1}{y-x+i\varepsilon} - \frac{1}{y-x-i\varepsilon}\right\} dy.$$

Let $\mu$ be any distribution with compact support; then it is not difficult to make the following argument rigorous:

$$<\mu(x), \varphi(x)> = \lim_{\varepsilon \downarrow 0} <\mu(x), \int_{-\infty}^{+\infty} \varphi(y) \frac{(-1)}{2\pi i} \left\{\frac{1}{y-x+i\varepsilon} - \frac{1}{y-x-i\varepsilon}\right\} dy>$$

$$= \lim_{\varepsilon \downarrow 0} \int_{-\infty}^{+\infty} \varphi(y) \left\{\frac{1}{2\pi i} <\mu(x), \frac{1}{x-y-i\varepsilon}> \right.$$

$$\left. -\frac{1}{2\pi i} <\mu(x), \frac{1}{x-y+i\varepsilon}>\right\} dy$$

$$\equiv \lim_{\varepsilon \downarrow 0} \int_{-\infty}^{+\infty} \varphi(y) \{\mu^\circ(y+i\varepsilon) - \mu^\circ(y-i\varepsilon)\} dy \qquad (4.1)$$

where we define the function $\mu^\circ(z)$ by

$$\mu^\circ(z) = \frac{1}{2\pi i} <\mu(x), \frac{1}{x-z}>, \quad \Im(z) \neq 0. \tag{4.2}$$

$\mu^\circ$ is a function of the complex variable $z$ which is defined and analytic on each of the half-planes $\{z \in \mathbb{C} : \Im(z) > 0\}$ and $\{z \in \mathbb{C} : \Im(z) < 0\}$.

This suggests the possibility of a general representation theory for distributions in terms of functions of a complex variable which are analytic off the real axis. Some modification of the approach sketched above will be necessary since for arbitrary $\mu$ in $\mathcal{D}'$ we cannot expect convergence of limits of the form (4.1). We cannot even define such a representation for the Heaviside distribution $H(x)$ in this simple and straightforward way. To develop a general theory of analytic representation we need to recall some basic facts about analytic continuation and the analytic representation of ordinary functions (regular distributions). As will be seen the subject is intimately linked with the theory of the (generalised) Fourier transform. It turns out that analytic representation is possible not only for all distributions in $\mathcal{D}'$ but also for ultradistributions. In fact we are led to consider further generalisations of the function concept, in the form of the so-called *hyperfunctions* of Sato.

A function $f(x)$ defined on $\mathbb{R}$ is said to extend analytically to the whole (finite) complex plane if there exists a function $F(z)$, defined and analytic for all $z$, which coincides with $f(x)$ on $\mathbb{R}$. Analytic extensions are not possible in general. The requirement that $F(z)$ be analytic is very restrictive. For example, an analytic function (which is not identically zero) can have, at worst, only isolated zeros. This means that no function $f(x)$ on $\mathbb{R}$ which vanishes on an interval, but which is not identically zero, can be extended analytically in the above sense. It is of little use to drop the requirement that $F(z)$ should be analytic on the real axis itself since we have the classical result,

**Theorem 4.1 (Painlevé)** *If $F(z)$ is analytic on $\mathbb{C}\backslash\mathbb{R}$ and continuous on $\mathbb{R}$, then $F(z)$ must be analytic on $\mathbb{C}$.*

On the other hand, it is possible to establish results of the following type:

**Theorem 4.2** *Let $f : \mathbb{R} \to \mathbb{C}$ be a continuous function which belongs to $L^p(\mathbb{R})$ for some finite $p \geq 1$. Then there exists a function $f^\circ(z)$, defined and analytic on $\mathbb{C}\backslash\mathbb{R}$, such that*

$$\lim_{\varepsilon \downarrow 0}[f^\circ(x + i\varepsilon) - f^\circ(x - i\varepsilon)] = f(x)$$

*the convergence being locally uniform on $\mathbb{R}$.*

The difference
$$f^\circ(x + i\varepsilon) - f^\circ(x - i\varepsilon)$$
is the jump made by $f^\circ(z)$ as we pass from just below to just above the real axis. Thus while it is not possible to represent an arbitrary continuous $L^p(\mathbb{R})$-function as the restriction to $\mathbb{R}$ of a function analytic on a complex neighbourhood of $\mathbb{R}$, every such function may be represented in terms of the jump of a function analytic on $\mathbb{C}\backslash\mathbb{R}$.

### 4.1.2 Analytic functions and generalised functions

The representation of a function $f(x), x \in \mathbb{R}$, by an associated analytic function $f^\circ(z), z \in \mathbb{C}\backslash\mathbb{R}$, arises naturally in connection with the problem of defining (classically) Fourier transforms for classes of functions larger than $L^1(\mathbb{R})$ or $L^2(\mathbb{R})$. The generalisation to functions of polynomial growth, due to T. Carlemann [8], develops from the following simple idea: Let $\hat{f}(\omega), \omega \in \mathbb{R}$, belong to $L^1(\mathbb{R})$; then its inverse Fourier transform $f = \mathcal{F}^{-1}[\hat{f}]$ defined by

$$f(z) = \frac{1}{2\pi} \int_{-\infty}^{+\infty} \exp(i\omega z) \hat{f}(\omega) d\omega \;,\; z \in \mathbb{R} \tag{4.3}$$

can be written as

$$f(z) = f_+(z) - f_-(z) \tag{4.4}$$

where

$$f_+(z) = +\frac{1}{2\pi} \int_0^{+\infty} \exp(i\omega z) \hat{f}(\omega) d\omega$$
$$f_-(z) = -\frac{1}{2\pi} \int_{-\infty}^0 \exp(i\omega z) \hat{f}(\omega) d\omega. \tag{4.5}$$

For complex values of $z$ the function $f_+(z)$ is analytic on the open half-plane

$$\mathbb{C}^+ \equiv \{z \in \mathbb{C} : \Im(z) > 0\}$$

while $f_-(z)$ is analytic on the open half-plane

$$\mathbb{C}^- \equiv \{z \in \mathbb{C} : \Im(z) < 0\}$$

and both are continuous functions on the common boundary, $\Im(z) = 0$.

The inverse Fourier transform therefore appears as the difference between the functions $f_+, f_-$, which are analytic on the upper and lower half-planes $\mathbb{C}^+$ and $\mathbb{C}^-$ respectively. Defining $f^\circ(z)$ to be equal to $f_+(z)$ for $\Im(z) > 0$ and equal to $f_-(z)$

for $\Im(z) < 0$ we have that, for any $\varepsilon > 0$, $f^\circ(x + i\varepsilon) - f^\circ(x - i\varepsilon)$ is the inverse Fourier transform of $\exp(-\varepsilon|\omega|)\hat{f}(\omega), \omega \in \mathbb{R}$. Hence

$$\begin{aligned} f(x) &= \lim_{\varepsilon \downarrow 0} \frac{1}{2\pi} \int_{-\infty}^{+\infty} \exp(-\varepsilon|\omega|)\hat{f}(\omega) \exp(i\omega x) d\omega \\ &= \lim_{\varepsilon \downarrow 0} \{f^\circ(x + i\varepsilon) - f^\circ(x - i\varepsilon)\} \end{aligned} \qquad (4.6)$$

the convergence being uniform on $\mathbb{R}$.

In the case when $f$ is also a function in $L^1(\mathbb{R})$ we would have

$$\begin{aligned} f_+(z) &= \frac{1}{2\pi} \int_0^{+\infty} \exp(i\omega z)\hat{f}(\omega) d\omega \\ &= \frac{1}{2\pi} \int_0^{+\infty} \exp(i\omega z) \int_{-\infty}^{+\infty} \exp(-i\omega t) f(t) dt d\omega \\ &= \frac{1}{2\pi} \int_{-\infty}^{+\infty} f(t) \int_0^{+\infty} \exp(i\omega(z-t)) d\omega dt \\ &= \frac{1}{2\pi i} \int_{-\infty}^{+\infty} \frac{f(t)}{t - z} dt, \quad \Im(z) > 0 \end{aligned}$$

where the change in order of integrations is easily justified by the Fubini theorem. Similarly, we would also have

$$f_-(z) = \frac{1}{2\pi i} \int_{-\infty}^{+\infty} \frac{f(t)}{t - z} dt, \quad \Im(z) < 0.$$

Hence $f_+$ and $f_-$ are together represented by the function

$$f^\circ(z) = \frac{1}{2\pi i} \int_{-\infty}^{+\infty} \frac{f(t)}{t - z} dt, \quad \Im(z) \neq 0 \qquad (4.7)$$

which is sometimes referred to as the *Cauchy representation* of the function $f(t), t \in \mathbb{R}$. Considerable confusion exists with regard to terminology here; some authorities refer to (4.7) as the Cauchy transform of $f(t)$, others as the Stieltjes transform, and so on. We adopt a specific meaning for the term Cauchy transform in a restricted context (see Definition 4.1 below), and refer to the more general form as the Cauchy-Stieltjes transform of $f$. If we denote the transform defined in (4.7) by $\mathbf{S}_r[\cdot]$ then the inverse transformation is given by

$$\mathbf{S}_r^{-1}[f^\circ](x) = \lim_{\varepsilon \downarrow 0}\{f^\circ(x + i\varepsilon) - f^\circ(x - i\varepsilon)\} \qquad (4.8)$$

and (4.6) takes the form

$$f(x) = \mathbf{S}_r^{-1}[f^\circ](x), \quad x \in \mathbb{R}, \qquad (4.9)$$

the convergence being uniform on $\mathbb{R}$.

**Remark 4.1** *Note that $f^\circ$ is not the only possible analytic representative for the function $f$; if $\alpha : \mathbb{C} \to \mathbb{C}$ is any entire function then $f^\circ + \alpha$ also represents $f$ in the same sense.*

If $\hat{f}$ does not belong to $L^1(\mathbb{R})$ but merely satisfies a polynomial growth condition then the functions $f_+(z), f_-(z)$, defined in (4.5), will still be analytic in the open half-planes $\mathbb{C}^+, \mathbb{C}^-$ respectively, although they do not generally extend to the real axis. The inverse transformation cannot have the meaning expressed by (4.8) above. Accordingly Carleman was led to consider the pair $(f_+, f_-)$ as the generalised inverse Fourier transform of the function $\hat{f}(\omega), \omega \in \mathbb{R}$, satisfying those growth conditions. This generalisation of classical Fourier theory is unsatisfactory in that the inverse transform of the function $\hat{f}$ is no longer a function but an ordered pair of analytic functions (of suitably restricted growth). Carleman re-established symmetry by treating such ordered pairs as *"generalised functions"* and extending to them his new definition of Fourier transform. This gave an elegant Fourier theory for pairs of analytic functions of polynomial growth which coincides with the classical Fourier transform for those pairs representing functions in $L^1(\mathbb{R})$.

The Carleman theory of generalised functions outlined above again suggests the possibility of obtaining analytic representations of distributions. This ought to facilitate the description of operations on distributions such as addition, scalar multiplication, differentiation, translation and so on, since these would be performed directly on the analytic representatives, that is on functions in the ordinary sense of the word. The space of Sato Hyperfunctions on the line effectively contains the required analytic representatives of distributions together with other mathematical objects which as yet have not been interpreted as elements of dual spaces.

### 4.1.3 The Cauchy transform and its inverse

#### 4.1.3.1 The Cauchy representation

Given an arbitrary complex valued function $f$ defined on $\mathbb{R}$ consider the integral

$$\frac{1}{2\pi i} \int_{-\infty}^{+\infty} \frac{f(t)}{t - z} \, dt \qquad (4.10)$$

where $z$ is a complex parameter. Then we have the result

**Theorem 4.3** *If the integral (4.10) converges for some complex number $z_0 \notin \mathbb{R}$ then*

$$f^\circ(z) = \frac{1}{2\pi i} \int_{-\infty}^{+\infty} \frac{f(t)}{t - z} \, dt$$

### Sec. 4.1]  The Cauchy Transform

*defines a function analytic outside the real axis. Moreover, if $f(t) \equiv 0$ on some open interval $(a, b)$, then $f^\circ(z)$ extends analytically into the strip $\{z \in \mathbf{C} : a < \Re(z) < b\}$.*

**Proof:** (a) For any (fixed) $z \in \mathbf{C}\backslash\mathbb{R}$ we have

$$\frac{1}{t-z} = \frac{1}{t-z_0}\frac{t-z_0}{t-z} = \frac{1}{t-z_0}\left\{1 + \frac{z-z_0}{t-z}\right\}$$

and it follows that $f(t)/(t-z)$ is integrable over $\mathbb{R}$ if and only if $f(t)/[(t-z_0)(t-z)]$ is integrable over $\mathbb{R}$. Since $f(t)/(t-z_0)$ is integrable over $\mathbb{R}$ it has a bounded continuous antiderivative $F(t)$, and so the integral

$$\int_{-\infty}^{+\infty} \frac{F(t)}{(t-z)^2}\,dt$$

converges. Hence, for every $z \in \mathbf{C}\backslash\mathbb{R}$, the integral

$$\int_{-\infty}^{+\infty} \frac{f(t)}{(t-z_0)(t-z)}\,dt = \lim_{a\to-\infty, b\to+\infty}\left[\frac{F(t)}{t-z}\right]_a^b + \int_{-\infty}^{+\infty}\frac{F(t)}{(t-z)^2}\,dt = \int_{-\infty}^{+\infty}\frac{F(t)}{(t-z)^2}\,dt$$

also converges. The function

$$f^\circ(z) = \frac{1}{2\pi i}\int_{-\infty}^{+\infty}\frac{f(t)}{t-z}\,dt$$
$$= \frac{1}{2\pi i}\int_{-\infty}^{+\infty}\frac{f(t)}{t-z_0}\,dt + \frac{z-z_0}{2\pi i}\int_{-\infty}^{+\infty}\frac{f(t)}{(t-z_0)(t-z)}\,dt$$

is therefore well defined for all $z \in \mathbf{C}\backslash\mathbb{R}$.

Next we show that $f^\circ(z)$ is analytic on $\mathbf{C}\backslash\mathbb{R}$. If $z_1 \in \mathbf{C}\backslash\mathbb{R}$ and $F_1(t)$ is a primitive of $f(t)/(t-z_1), t \in \mathbb{R}$, then $F_1(t), t \in \mathbb{R}$, is bounded on $\mathbb{R}$ and thus

$$\frac{f^\circ(z)-f^\circ(z_1)}{z-z_1} = \frac{1}{2\pi i}\int_{-\infty}^{+\infty}\frac{f(t)}{(t-z)(t-z_1)}\,dt = \frac{1}{2\pi i}\int_{-\infty}^{+\infty}\frac{F_1(t)}{(t-z)^2}\,dt$$

where the last integral is absolutely convergent and therefore defines a continuous function of $z \in \mathbf{C}\backslash\mathbb{R}$. Hence we get

$$\frac{d}{dz}f^\circ(z_1) = \lim_{z\to z_1}\frac{f^\circ(z)-f^\circ(z_1)}{z-z_1} = \frac{1}{2\pi i}\int_{-\infty}^{+\infty}\frac{F_1(t)}{(t-z_1)^2}\,dt, \quad \Im(z_1) \neq 0$$

which shows that $f^\circ$ is analytic on $\mathbf{C}\backslash\mathbb{R}$.

(b) Suppose that $f(t) \equiv 0$ a.e. for $a < t < b$, (where $a, b$ may take finite or infinite values). Then for any $z \in \mathbf{C}\backslash\mathbb{R}$, we have

$$f^\circ(z) = \frac{1}{2\pi i}\left\{\int_{-\infty}^{a} + \int_{b}^{+\infty}\right\}\frac{f(t)}{t-z}\,dt.$$

A similar argument to that given in part (a) of the proof shows that the integral

$$\frac{1}{2\pi i} \int_{-\infty}^{+\infty} \frac{f(t)}{t-\xi} \, dt$$

converges for any $\xi \in (a,b)$. It follows that $f^\circ(\xi)$ is well defined. On the other hand, since

$$f^\circ(z) - f^\circ(\xi) = \frac{z-\xi}{2\pi i} \left\{ \int_{-\infty}^{a} + \int_{b}^{+\infty} \right\} \frac{f(t)}{(t-z)(t-\xi)} \, dt$$

we can conclude that $f^\circ$ is continuous at $\xi$. It follows from the Painlevé Theorem 4.1 that $f^\circ$ is an analytic function on the strip $\{z \in \mathbf{C} : a < \Re(z) < b\}$. □

The convergence of the integral in (4.10) for some $z_0 \in \mathbf{C}\backslash\mathbb{R}$ implies its convergence for every other $z \in \mathbf{C}\backslash\mathbb{R}$. Hence we can take, in particular, $z_0 = i$ in the above theorem and define $L^{i,1}(\mathbb{R})$ to be the linear space of all (equivalence classes of) functions $f : \mathbb{R} \to \mathbf{C}$ for which $f(t)/(t-i)$ is in $L^1(\mathbb{R})$. $L^{i,1}(\mathbb{R})$ can be equipped with the norm defined by

$$\| f \|_{i,1} = \int_{-\infty}^{+\infty} \frac{|f(t)|}{\sqrt{1+t^2}} \, dt$$

for any $f \in L^{i,1}(\mathbb{R})$.

**Theorem 4.4** *Let $f : \mathbb{R} \to \mathbf{C}$ be a function in $L^p(\mathbb{R})$ for some finite $p \geq 1$. Then $f$ belongs to $L^{i,1}(\mathbb{R})$ and there exists $C > 0$ such that*

$$|f^\circ(z)| \leq C/|\Im(z)|^{1/p}$$

*for all $z \in \mathbf{C}\backslash\mathbb{R}$.*

**Proof:** For any $z = x + iy$ such that $y \neq 0$ we get

$$|f^\circ(z)| \leq \frac{1}{2\pi} \int_{-\infty}^{+\infty} \frac{|f(t)|}{[(t-x)^2 + y^2]^{1/2}} \, dt$$

$$\leq \frac{1}{2\pi} \|f\|_p \cdot \left\{ \int_{-\infty}^{+\infty} [(t-x)^2 + y^2]^{-q/2} \, dt \right\}^{1/q} \leq \frac{C_q \|f\|_p}{2\pi |y|^{1/p}}$$

where $p^{-1} + q^{-1} = 1$, and it can be shown that

$$C_q = \left\{ 1 - q\frac{\Gamma(1/2)\Gamma((q-1)/2)}{\Gamma(q/2)} \right\}^{1/q} , \quad 1 \leq q < +\infty$$

while $C_q = 1$ if $q = +\infty$. The final result follows taking any constant $C > 0$ so that $C \geq C_q \| f \|_p / 2\pi$. □

**Definition 4.1** For any $f \in L^{i,1}(\mathbb{R})$ the function $f^\circ : \mathbb{C}\backslash\mathbb{R} \to \mathbb{C}$ defined by
$$f^\circ(z) = \frac{1}{2\pi i} \int_{-\infty}^{+\infty} \frac{f(t)}{t-z} dt, \quad \Im(z) \neq 0$$
is called the **Cauchy transform** of $f$ and is denoted by $\mathbf{S}_r[f]$.

The Cauchy operator $\mathbf{S}_r$ is a linear map of $L^{i,1}(\mathbb{R})$ into the space $\mathcal{H}(\mathbb{C}\backslash\mathbb{R})$ of all functions from $\mathbb{C}\backslash\mathbb{R}$ into $\mathbb{C}$ which are analytic outside the real axis. It has the following properties:

1. If $f \in L^{i,1}(\mathbb{R})$ and $a \in \mathbb{R}$, then
$$\mathbf{S}_r[\tau_a f](z) = \tau_a \mathbf{S}_r[f](z), \quad \Im(z) \neq 0.$$

2. If $f$ is an (almost everywhere) differentiable function such that both $f$ and $f'$ belong to $L^{i,1}(\mathbb{R})$ then
$$\mathbf{S}_r[D_t f](z) = D_z \mathbf{S}_r[f](z), \quad \Im(z) \neq 0,$$
where we write $D_t \equiv d/dt, D_z \equiv d/dz$.

More generally, for any $k \in \mathbb{N}_0$ we have
$$\mathbf{S}_r[D_t^k f](z) = D_z^k \mathbf{S}_r[f](z), \quad \Im(z) \neq 0,$$
provided $f, f', \ldots, f^{(k)}$ all belong to $L^{i,1}(\mathbb{R})$.

3. If $f$ and an indefinite integral $\Im_a f$ of $f$ both belong to $L^{i,1}(\mathbb{R})$ then
$$\mathbf{S}_r[f](z) = D_z \mathbf{S}_r[\Im_a f](z), \quad \Im(z) \neq 0.$$

More generally, for any $k \in \mathbb{N}_0$ we have
$$D_z^k \mathbf{S}_r[\Im_a^k f](z) = \mathbf{S}_r[f](z), \quad \Im(z) \neq 0.$$
provided $f, \Im_a f, \ldots, \Im_a^k f$ all belong to $L^{i,1}(\mathbb{R})$.

**Exercise 4.1 :**

1. Find $f^\circ(z)$ and $g^\circ(z)$ if $f(t) = (t+i)^{-1}$ and $g(t) = (t^2+1)^{-1}$.

2. Prove the properties of the operator $\mathbf{S}_r : L^{i,1}(\mathbb{R}) \to \mathcal{H}(\mathbb{C}\backslash\mathbb{R})$ listed above.

### 4.1.3.2 The inverse Cauchy transform

The following theorem, quoted from Meneses [71], shows how functions in $L^{i,1}(\mathbb{R})$ may be represented by the Cauchy transform.

**Theorem 4.5** *If $f \in L^{i,1}(\mathbb{R})$ and $f^\circ = \mathbf{S}_r[f]$, then we have*

$$\mathbf{S}_r^{-1}[f^\circ](x) \equiv \lim_{\varepsilon \downarrow 0}[f^\circ(x+i\varepsilon) - f^\circ(x-i\varepsilon)] = \frac{1}{2}[f(x^+) + f(x^-)]$$

*for all $x \in \mathbb{R}$ for which both $f(x^+)$ and $f(x^-)$ exist.*

**Proof:** Let $x \in \mathbb{R}$. For $z \notin \mathbb{R}$ define $f^\circ(z) = f_+^\circ(z) + f_-^\circ(z)$ where

$$f_+^\circ(z) = \frac{1}{2\pi i}\int_x^{+\infty} \frac{f(t)}{t-z}\,dt \;,\quad f_-^\circ(z) = \frac{1}{2\pi i}\int_{-\infty}^x \frac{f(t)}{t-z}\,dt.$$

If $g_+(t) = f(t) - f(x^+)$ then $g_+(t) \to 0$ as $t \to x^+$. Hence given $\delta > 0$ there exists $\alpha > 0$ such that $|g_+(t)| < \delta$ for $0 < t - x < \alpha$. Write

$$f_+^\circ(z) = \varphi_+(z) + \psi_+(z) \equiv \frac{1}{2\pi i}\int_x^{x+\alpha} \frac{f(t)}{t-z}\,dt + \frac{1}{2\pi i}\int_{x+\alpha}^{+\infty} \frac{f(t)}{t-z}\,dt$$

so that

$$f_+^\circ(x+iy) - f_+^\circ(x-iy) - \frac{1}{2}f(x^+) = \varphi_+(x+iy) - \varphi_+(x-iy)$$
$$+ \psi_+(x+iy) - \psi_+(x-iy) - \frac{1}{2}f(x^+)$$
$$= \frac{y}{\pi}\int_x^{x+\alpha}\frac{f(t)}{(t-x)^2+y^2}\,dt - \frac{1}{2}f(x^+) + \psi_+(x+iy) - \psi_+(x-iy)$$
$$= \frac{y}{\pi}\int_x^{x+\alpha}\frac{g_+(t)+f(x^+)}{(t-x)^2+y^2}\,dt - \frac{1}{2}f(x^+) + \psi_+(x+iy) - \psi_+(x-iy)$$
$$= \frac{1}{\pi}f(x^+)\arctan\left(\frac{\alpha}{y}\right) - \frac{1}{2}f(x^+)$$
$$+ \frac{y}{\pi}\int_x^{x+\alpha}\frac{g_+(t)}{(t-x)^2+y^2}\,dt + \psi_+(x+iy) - \psi_+(x-iy).$$

Therefore we have $|f_+^\circ(x+iy) - f_+^\circ(x-iy) - \frac{1}{2}f(x^+)| \leq I_1 + I_2 + I_3$ where

$$I_1 = \left|\frac{1}{\pi}f(x^+)\arctan\left(\frac{\alpha}{y}\right) - \frac{1}{2}f(x^+)\right|,$$

$$I_2 = \left|\frac{y}{\pi}\int_x^{x+\alpha}\frac{g_+(t)}{(t-x)^2+y^2}\,dt\right|,$$

$$I_3 = |\psi_+(x+iy) - \psi_+(x-iy)|.$$

Then,

(i) since $\lim_{y\to 0^+} \frac{1}{\pi} f(x^+) \arctan\left(\frac{\alpha}{y}\right) = \frac{1}{2} f(x^+)$ we can find $\varepsilon_1 = \varepsilon_1(\delta) > 0$ such that $I_1 < \delta/4$ for $0 < y < \varepsilon_1$;

(ii) for every $y > 0$ the bound given for $|g_+(t)|$ gives

$$I_2 \leq \frac{\delta}{\pi} \int_x^{x+\alpha} \frac{y}{(t-x)^2 + y^2}\, dt = \frac{\delta}{\pi} \arctan\left(\frac{\alpha}{y}\right) < \frac{\delta}{2};$$

(iii) by theorem 4.3 the function $\psi_+$ extends analytically into the complement in $\mathbf{C}$ of the real interval $[x + \alpha, +\infty)$; therefore there exists a real number $\varepsilon_2 = \varepsilon_2(\delta) > 0$ such that $I_3 < \delta/4$ for $0 < y < \varepsilon_2$.

Hence, for $\varepsilon = \min\{\varepsilon_1, \varepsilon_2\}$, we have

$$0 < y < \varepsilon \;\Rightarrow\; |\{f_+^\circ(x + iy) - f_+^\circ(x - iy)\} - \frac{1}{2} f(x^+)| < \delta$$

and so

$$\lim_{y\downarrow 0}\{f_+^\circ(x + iy) - f_+^\circ(x - iy)\} = \frac{1}{2} f(x^+).$$

A similar argument shows that

$$\lim_{y\downarrow 0}\{f_-^\circ(x + iy) - f_-^\circ(x - iy)\} = \frac{1}{2} f(x^-)$$

from which the general result follows. □

**Corollary 4.1.1** *If $f \in L^{i,1}(\mathbb{R})$ and $f^\circ = \mathbf{S}_r[f]$ then $\mathbf{S}_r^{-1}[f^\circ](x) = f(x)$ at all points of continuity.*

If $f \in L^{i,1}(\mathbb{R})$ is a continuously differentiable function up to order $m$ the above result may be sharpened to the following theorem (see, for example, H. Bremermann [4]).

**Theorem 4.6** *Let $f$ be a $C^m(\mathbb{R})$-function such that $f, f', \ldots f^{(m)}$ all belong to $L^{i,1}(\mathbb{R})$. If $f^\circ = \mathbf{S}_r[f]$ then for all $k = 0, 1, \ldots, m$ and all $x \in \mathbb{R}$ we have*

$$f^{(k)}(x) \equiv \mathbf{S}_r^{-1}[D_z^k f^\circ](x) = \lim_{\varepsilon \downarrow 0}\left\{\frac{d^k f^\circ}{dz^k}(x + i\varepsilon) - \frac{d^k f^\circ}{dz^k}(x - i\varepsilon)\right\}$$

*the convergence being locally uniform on $\mathbb{R}$.*

However the continuity of $f$ is not enough to guarantee that the one-sided limits

$$\lim_{\varepsilon\downarrow 0} f^\circ(x + i\varepsilon) \;;\; \lim_{\varepsilon\downarrow 0} f^\circ(x - i\varepsilon)$$

do always converge locally uniformly (as continuous functions). Nevertheless it is possible to establish the following result

**Theorem 4.7** *Let $f : \mathbb{R} \to \mathbb{C}$ be a function in $L^{i,1}(\mathbb{R})$ and $\Omega$ an open interval of $\mathbb{R}$ such that*

*(a) $f$ is absolutely continuous on $\Omega$,*

*(b) $f'$ is locally bounded on $\Omega$.*

*Then, if $f^\circ = \mathbf{S}_r[f]$, the one-sided limits*

$$\lim_{\varepsilon \downarrow 0} f^\circ(x + i\varepsilon) = f^+(x) \quad ; \quad \lim_{\varepsilon \downarrow 0} f^\circ(x - i\varepsilon) = f^-(x)$$

*exist locally uniformly in $\Omega$; $f^+$ and $f^-$ are continuous functions on $\Omega$ and*

$$f(x) = f^+(x) - f^-(x)$$

*for all $x \in \Omega$.*

**Proof:** Let $[\alpha, \beta]$ be any compact sub-interval of $\Omega$ and let $I = [a, b]$ be any interval such that $[\alpha, \beta] \subset (a, b) \subset \Omega$. If we write

$$\varphi(z) = \frac{1}{2\pi i} \int_I \frac{f(t)}{t - z} dt \quad ; \quad \theta(z) = \frac{1}{2\pi i} \int_{\mathbb{R} \setminus I} \frac{f(t)}{t - z} dt$$

then $f^\circ(z) = \varphi(z) + \theta(z), \Im(z) \neq 0$ and, by Theorem 4.3, $\theta$ extends analytically into the complement in $\mathbb{C}$ of $\mathbb{R} \setminus I$. For any $x \in (a, b)$,

$$\lim_{y \downarrow 0} f^\circ(x \pm iy) = \lim_{y \downarrow 0} \varphi(x \pm iy) + \theta(x)$$

provided the limits on the right-hand side do exist. To prove the theorem it is enough to show that these limits exist uniformly on every interval of the form $[\alpha, \beta]$ as $y \downarrow 0$. Let $z = x + iy$ where $x \in [\alpha, \beta]$ and $y > 0$; then

$$\varphi''(z) = \frac{2!}{2\pi i} \int_a^b \frac{f(t)}{(t - z)^3} dt$$

$$= \frac{1}{2\pi i} \left[ \frac{f(a)}{(a - z)^2} - \frac{f(b)}{(b - z)^2} \right] + \frac{1}{2\pi i} \int_a^b \frac{f'(t)}{(t - z)^2} dt.$$

Hence we may write,

$$\varphi'(z) = \varphi'(z_0) + \int_{z_0}^z \varphi''(\zeta) \, d\zeta$$

$$= \varphi'(z_0) + \frac{1}{2\pi i} \left[ \frac{f(a)}{(a - \zeta)} - \frac{f(b)}{(b - \zeta)} \right]_{z_0}^z + \frac{1}{2\pi i} \int_{z_0}^z \int_a^b \frac{f'(t)}{(t - \zeta)^2} dt d\zeta,$$

where $z_0 = x + y_0, x \in [\alpha, \beta]$ and $y_0 > 0$.

By hypothesis there exists $C > 0$ such that $|f'(t)| \le 2C$ for all $t \in [a,b]$, so that

$$\left| \frac{1}{2\pi i} \int_a^b \frac{f'(t)}{(t-z)^2} \, dt \right| \le \frac{C}{\pi} \int_a^b \frac{dt}{(t-x)^2 + y^2} = \frac{C}{\pi y} \left[ \arctan \frac{t-x}{y} \right]_a^b \le \frac{C}{y}.$$

Therefore, writing $A = |\varphi'(z_0)| + (1/\pi)|f(a)/(a-\alpha)| + (1/\pi)|f(b)/(b-\beta)|$, we get

$$|\varphi'(z)| \le A + \left| \int_{y_0}^y \frac{C}{y} \, dy \right| \le B + C|\log(y)|,$$

where $B = A + C|\log(y_0)|$. Now let $z_1 = x + iy_1$, $z_2 = x + iy_2$ be any two complex numbers such that $\alpha \le x \le \beta$ and $0 < y_1, y_2 < 1$. Then

$$\varphi(x + iy_2) - \varphi(x + iy_1) = i \int_{y_1}^{y_2} \varphi'(x + iy) \, dy$$

and therefore

$$\begin{aligned}
|\varphi(x + iy_2) - \varphi(x + iy_1)| &\le \int_{y_1}^{y_2} |\varphi'(x + iy)| \, dy \\
&\le \int_{y_1}^{y_2} [B + C|\log(y)|] \, dy = \int_{y_1}^{y_2} [B - C\log(y)] \, dy \\
&\le (B + C)|y_1 - y_2| + C \left( |y_2 \log(y_2)| + |y_1 \log(y_1)| \right).
\end{aligned}$$

If $y_1, y_2 \downarrow 0$ then $y_1 - y_2 \to 0$, $y_1 \log(y_1) \to 0$ and $y_2 \log(y_2) \to 0$ and therefore $\varphi(x + iy)$ converges uniformly on $[\alpha, \beta]$. Representing the limit function by $f^+(x) - \theta(x)$ we therefore obtain

$$\lim_{y \downarrow 0} f^\circ(x + iy) = f^+(x)$$

uniformly on $[\alpha, \beta]$. Similarly we may show that

$$\lim_{y \downarrow 0} f^\circ(x - iy) = f^-(x)$$

uniformly on $[\alpha, \beta]$. Since $[\alpha, \beta]$ is arbitrary in $\Omega$ the last part of the theorem follows from corollary 4.1.1. □

By induction this result generalises to the following:

**Theorem 4.8** *Let $f : \mathbb{R} \to \mathbb{C}$ be a function in $L^{i,1}(\mathbb{R})$ such that if $\Omega$ is an open interval of $\mathbb{R}$ then the restriction of $f$ to $\Omega$ belongs to $\mathcal{C}^p(\Omega)$ for some $p \ge 1$. Then the functions $f^+, f^- : \Omega \to \mathbb{C}$ as defined in theorem 4.7 belong to $\mathcal{C}^{p-1}(\mathbb{R})$. The result also holds for $p = +\infty$, provided that (for consistency) we then take $p - 1 = +\infty$.*

### 4.1.3.3 The Bremermann Space $\mathcal{O}'_{-1}$

Let $f : \mathbb{R} \to \mathbb{C}$ be a continuous function which is not infinitely differentiable. Then, for some positive integer $k$, the derivative $D^k f$ does not exist as a function, although it is certainly well defined as a distribution. Suppose moreover that $f \in L^{i,1}(\mathbb{R})$. Then $f^\circ = \mathbf{S}_r[f]$ is a function defined and analytic on $\mathbb{C}\backslash\mathbb{R}$ and so also is $D_z^k f^\circ$. However, although the inverse Cauchy transform $\mathbf{S}_r^{-1}[f^\circ]$ is well defined in the sense given by the preceding theorems, this is no longer the case for $\mathbf{S}_r^{-1}[D_z^k f^\circ]$. For any $\varphi \in \mathcal{D}$ and any given $\varepsilon > 0$, consider the integral

$$\begin{aligned} I(\varepsilon) &= \int_{-\infty}^{+\infty} [D_z^k f^\circ(x + i\varepsilon) - D_z^k f^\circ(x - i\varepsilon)]\varphi(x) dx \\ &= \int_{-\infty}^{+\infty} [f^\circ(x + i\varepsilon) - f^\circ(x - i\varepsilon)]^{(k)} \varphi(x) dx \\ &= \int_{-\infty}^{+\infty} [f^\circ(x + i\varepsilon) - f^\circ(x - i\varepsilon)](-1)^k \varphi^{(k)}(x) dx. \end{aligned}$$

Since by Theorem 4.6 $f^\circ(x + i\varepsilon) - f^\circ(x - i\varepsilon)$ tends to $f(x)$ as $\varepsilon \downarrow 0$ uniformly on compacts and since $\varphi$ is a function of compact support we can interchange limits and integration to obtain

$$\lim_{\varepsilon \downarrow 0} I(\varepsilon) = \int_{-\infty}^{+\infty} f(x)(-1)^k \varphi^{(k)}(x)\, dx \;=\; <D^k f, \varphi>.$$

This shows that $D_z^k f^\circ(x + i\varepsilon) - D_z^k f^\circ(x - i\varepsilon)$ converges to $\mu \equiv D^k f$, as $\varepsilon \downarrow 0$, in the sense of $\mathcal{D}'$. The inverse Cauchy transform, $\mathbf{S}_r^{-1}$, has therefore a well defined meaning provided we interpret the limits in the sense of distributions, that is,

$$<\mathbf{S}_r^{-1}[D_z^k f^\circ], \varphi> = \lim_{\varepsilon \downarrow 0} <D_z^k f^\circ(x + i\varepsilon) - D_z^k f^\circ(x - i\varepsilon), \varphi(x)>$$

for any $\varphi \in \mathcal{D}$. Therefore we define

$$\mu^\circ(z) = \mathbf{S}_r[\mu](z) = D_z^k f^\circ(z), \quad \Im(z) \neq 0 \qquad (4.11)$$

to be the Cauchy transform of the distribution $\mu \equiv D^k f \in \mathcal{D}'$. Moreover we have

$$\begin{aligned} \mu^\circ(z) &= D_z^k \left\{ \frac{1}{2\pi i} \int_{-\infty}^{+\infty} \frac{f(t)}{t - z} dt \right\} = \frac{k!}{2\pi i} \int_{-\infty}^{+\infty} \frac{f(t)}{(t - z)^{k+1}} dt \\ &= \frac{k!}{2\pi i} <f(t), \frac{1}{(t - z)^{k+1}}> = \frac{1}{2\pi i} <\mu(t), \frac{1}{t - z}>, \quad \Im(z) \neq 0 \end{aligned}$$

which shows that the distribution $\mu$ under the stated conditions extends, as a functional, to the family of $\mathcal{C}^\infty$-functions $\theta_z(t) = (1/2\pi i)(t - z)^{-1}, t \in \mathbb{R}$ for all complex numbers $z \notin \mathbb{R}$.

A distribution $\mu \in \mathcal{D}'$ has a Cauchy transform (in the sense described above) if it extends to this family $\{\theta_z\}_{\Im(z)\neq 0}$ of $C^\infty$-functions. The largest space of distributions which admit such a Cauchy representation has been characterized in Bremermann [4] as follows:

**Definition 4.2** *Let $\mathcal{O}_{-1} \equiv \mathcal{O}_{-1}(\mathbb{R})$ be the space of all $C^\infty$-functions $\varphi : \mathbb{R} \to \mathbb{C}$ for which there exist constants $C_j > 0, j = 0, 1, 2, \ldots$, such that the inequalities*

$$|\varphi^{(j)}(t)| \leq C_j \cdot |t|^{-1}$$

*hold for all $t \in \mathbb{R}$ and all $j \in \mathbb{N}_0$.*

A sequence $(\varphi_n)_{n \in \mathbb{N}}$ is said to converge in $\mathcal{O}_{-1}$ if and only if

(a) $\forall n \in \mathbb{N} : \varphi_n \in \mathcal{O}_{-1}$,

(b) for every $j \in \mathbb{N}_0$ the sequence $\left(\varphi_n^{(j)}\right)_{n \in \mathbb{N}}$ converges uniformly on compacts of $\mathbb{R}$,

(c) there exist positive constants $C_j, j = 0, 1, 2, \ldots$ (independent of $n \in \mathbb{N}$) such that $|\varphi_n^{(j)}(t)| \leq C_j \cdot |t|^{-1}$, for all $n \in \mathbb{N}$, all $t \in \mathbb{R}$ and all $j \in \mathbb{N}_0$.

Note that if $\varphi \in \mathcal{O}_{-1}$ then the derivatives $\varphi^{(j)}$ for all $j \in \mathbb{N}$ also belong to $\mathcal{O}_{-1}$. $\mathcal{O}_{-1}$ is a test-function space and $\mathcal{D}$ is a dense subspace of $\mathcal{O}_{-1}$.

**Definition 4.3** $\mathcal{O}'_{-1} \equiv \mathcal{O}'_{-1}(\mathbb{R})$, *the space of all continuous linear functionals on $\mathcal{O}_{-1}$, is a subspace of $\mathcal{D}'$ which will be called the* **Bremermann space** *of distributions.*

Any function $f \in L^{i,1}(\mathbb{R})$ defines a regular distribution belonging to $\mathcal{O}'_{-1}$. Also any distribution of compact support is a member of $\mathcal{O}'_{-1}$; $\mathcal{O}'_{-1}$ is a space of distributions which is intermediate between $\mathcal{E}'$ and $\mathcal{D}'$.

**Theorem 4.9** *If $\mu \in \mathcal{O}'_{-1}$ then*

$$\mu^\circ(z) = \frac{1}{2\pi i} < \mu(t), \frac{1}{t-z} >$$

*is an analytic function in $\mathbb{C} \backslash \mathrm{supp}(\mu)$. The function $\mu^\circ(z)$ will be called the Cauchy transform of $\mu$ and denoted by $\mathbf{S}_r[\mu](z)$. It represents the distribution $\mu$ in the sense that*

$$\lim_{\varepsilon \downarrow 0} \int_{-\infty}^{+\infty} [\mu^\circ(x+i\varepsilon) - \mu^\circ(x-i\varepsilon)] \varphi(x) \, dx \; = \; <\mu, \varphi>$$

*for any test function $\varphi \in \mathcal{D}$.*

**Proof:** (See, for example, H. Bremermann [4].) □

Every distribution $\mu \in \mathcal{O}'_{-1}$ is a tempered distribution and so there exists a continuous function $f$ of polynomial growth and an integer $k \in \mathbb{N}_0$ such that

$$\mu^\circ(z) = \frac{k!}{2\pi i} \int_{-\infty}^{+\infty} \frac{f(t)}{(t-z)^{k+1}} \, dt \, , \quad \forall z \in \mathbb{C} \backslash \mathbb{R}.$$

**Exercise 4.2 :**

1. Show that the Cauchy operator $S_r : \mathcal{O}'_{-1} \to \mathcal{H}(\mathbb{C} \backslash \mathbb{R})$ has the following properties:

    (a) $S_r[a\mu + b\nu] = aS_r[\mu] + bS_r[\nu]$,
    (b) $S_r[D^k \mu] = D_z^k S_r[\mu]$,
    (c) $S_r[\tau_\alpha \mu] = \tau_\alpha S_r[\mu]$,

    for all $\mu, \nu \in \mathcal{O}'_{-1}, a, b \in \mathbb{C}, \alpha \in \mathbb{R}, k \in \mathbb{N}_0$.

2. Find the following Cauchy transforms:

$$S_r[\exp(i\omega t)] \, ; \, S_r[\delta] \, ; \, S_r[\delta^{(k)}] \, ; \, S_r[\mathrm{Pf}\{t^{-1}\}].$$

3. If we define $\delta_+(x) = \lim_{\varepsilon \downarrow 0} \delta^\circ(x + i\varepsilon)$ and $\delta_-(x) = \lim_{\varepsilon \downarrow 0} \delta^\circ(x - i\varepsilon)$, then we have $\delta(x) = \delta_+(x) - \delta_-(x)$. Find $\delta_+^\circ(z)$ and $\delta_-^\circ(z)$ and show that

$$\delta_+(x) = +\frac{1}{2}\delta(x) - \frac{1}{2\pi i} \mathrm{Pf}\{x^{-1}\} \, , \quad \delta_-(x) = -\frac{1}{2}\delta(x) - \frac{1}{2\pi i} \mathrm{Pf}\{x^{-1}\}$$

so that

$$\delta_+(x) + \delta_-(x) = -\frac{1}{\pi i} \mathrm{Pf}\{x^{-1}\}.$$

### 4.1.3.4 The Cauchy transform in $\mathcal{E}'$

The space $\mathcal{E}'$ of all distributions of compact support is a linear subspace of $\mathcal{O}'_{-1}$. The Cauchy representation of a distribution of compact support has a particular behaviour at infinity according to the following:

**Theorem 4.10** *Let $\mu$ be a distribution of compact support in $\mathcal{E}'$. Then $|\mu^\circ(z)|$ is $O(|z|^{-1})$ as $|z|$ tends to infinity.*

**Proof:** (a) Let $\theta \in \mathcal{D}$ be such that $\theta(t) = 1$ on a neighbourhood of $\mathrm{supp}(\mu)$. Then $\theta_z(t) \equiv \theta(t)(t-z)^{-1}, t \in \mathbb{R}$, and each of its derivatives converges uniformly to zero on every compact subset of $\mathbb{R}$ as $|z| \to \infty$. Hence $\theta_z(t), t \in \mathbb{R}$, converges to zero in $\mathcal{E}$ as $|z| \to \infty$, and so also does $\mu^\circ(z)$.

(b) The function $\mu^o(z)$ has a Laurent expansion at infinity. The point at infinity is a removable singularity (in view of part (a) above) and so

$$\mu^o(z) = \frac{a_1}{z} + \frac{a_2}{z^2} + \cdots + \frac{a_n}{z^n} + \cdots$$

the series converging outside a circle, centre the origin, which contains $\text{supp}(\mu)$. Hence $|\mu^o(z)| = O(|z|^{-1})$ as $|z| \to \infty$. □

It is also of some interest to know how the Cauchy transform of a distribution of compact support behaves in a neighbourhood of its support.

**Theorem 4.11** *For every distribution $\mu \in \mathcal{E}'$ there exists $r \in \mathbb{N}_0$ and $C > 0$ such that the inequality*

$$|\mu^o(z)| \leq C/|\Im(z)|^r$$

*holds for any $z \in \mathbb{C}$ so that $\Re(z) \in \text{supp}(\mu)$ and $\Im(z) \neq 0$.*

**Proof:** If $\mu \in \mathcal{E}'$ then

$$\mu = \sum_{j=0}^{k} D^j f_j$$

where the $f_j$ are continuous functions of compact support contained in a neighbourhood of $\text{supp}(\mu)$, say $I_\varepsilon(\mu) \equiv \text{supp}(\mu) + [-\varepsilon, +\varepsilon]$. Hence

$$\mu^o(z) = \frac{1}{2\pi i} < \mu(t), \frac{1}{t-z} >$$

$$= \sum_{j=0}^{k} \frac{j!}{2\pi i} \int_{I_\varepsilon(\mu)} f_j(t)(t-z)^{-(j+1)} \, dt, \quad z \notin \text{supp}(\mu)$$

and so we get

$$|\mu^o(z)| \leq \sum_{j=0}^{k} C_j \sup_t |t-z|^{-(j+1)} \leq C/|\Im(z)|^{k+1}$$

for any $z \in \mathbb{C}$ such that $\Re(z) \in \text{supp}(\mu)$ and $|\Im(z)|$ sufficiently small (but $\neq 0$). The result follows when we set $r = k+1$. □

## 4.2 ANALYTIC REPRESENTATION OF DISTRIBUTIONS

The definition of the Cauchy transform as given in §4.1.3 has been restricted by the imposition of quite severe growth conditions on the distributions in $\mathcal{O}'_{-1}$. To remove these restrictions and extend the transform to a wider class of distributions we shall need to modify the original definition. The generalization thus obtained will be called the **Cauchy-Stieltjes transform** and denoted by $\mathbf{S}[\cdot]$. We consider first an extension of the Cauchy transform to the space $\mathcal{D}'_{\text{fin}}$ of all finite-order distributions.

### 4.2.1 Finite-order distributions

The definition of the Cauchy-Stieltjes transform of a finite order distribution can be based on lemma 3.4.1. For convenience we give again the essential result of that lemma here.

> For every locally bounded function $f : \mathbb{R} \to \mathbb{C}$ there exists an entire function $\Phi : \mathbb{C} \to \mathbb{C}$ in $\mathcal{H}(\mathbb{C})$, with no zeros, such that for all $t \in \mathbb{R}$ we have $\Phi(t) \geq |f(t)|$.

The following lemma will also be of importance in the sequel.

**Lemma 4.2.1** *Let $f : \mathbb{R} \to \mathbb{C}$ be an $L^1(\mathbb{R})$-function and $\Phi$ the restriction to $\mathbb{R}$ of an entire function with no real zeros such that the product $\Phi f$ is in $L^{i,1}(\mathbb{R})$; then $f$ also belongs to $L^{i,1}(\mathbb{R})$ and moreover*

$$\mathbf{S}_r[\Phi f](z) - \Phi(z)\mathbf{S}_r[f](z)$$

*is an entire function.*

**Proof:** That $f \in L^{i,1}(\mathbb{R})$ is immediate. For the rest consider

$$\mathbf{S}_r[\Phi f](z) - \Phi(z)\mathbf{S}_r[f](z) = \frac{1}{2\pi i}\int_{-\infty}^{+\infty}\frac{\Phi(t)-\Phi(z)}{t-z}f(t)\,dt = \int_{-\infty}^{+\infty}g(z,t)f(t)\,dt$$

where $g(z,t) = [\Phi(t) - \Phi(z)]/(t-z)$ is an entire function of two variables. If $\Gamma$ denotes a simple closed contour of the $z$-plane we get

$$\oint_\Gamma \frac{1}{2\pi i}\int_{-\infty}^{+\infty} g(z,t)f(t)\,dt\,dz = \int_{-\infty}^{+\infty} f(t)\left\{\frac{1}{2\pi i}\oint_\Gamma g(z,t)\,dz\right\}dt = 0$$

where the interchange between integrations is justified by the Fubini theorem. Since $\Gamma$ is arbitrary it follows that $\mathbf{S}_r[\Phi f](z) - \Phi(z)\mathbf{S}_r[f](z)$ is entire. □

Now let $f : \mathbb{R} \to \mathbb{C}$ be an absolutely continuous function of arbitrary growth such that $f' : \mathbb{R} \to \mathbb{C}$ is a locally bounded function. From lemma 3.4.1 there exists an entire function $\Phi : \mathbb{C} \to \mathbb{C}$, positive and without zeros on $\mathbb{R}$, such that $f/\Phi \in L^1(\mathbb{R})$. (In fact, if $\Phi_1$ is an entire function such that $\Phi_1(t) \geq |f(t)|$ on $\mathbb{R}$ then we have only to take $\Phi(t) = (1+t^2)\Phi_1(t)$). The Cauchy transform of $f/\Phi$ exists as

$$\mathbf{S}_r[f/\Phi](z) = \frac{1}{2\pi i}\int_{-\infty}^{+\infty} \frac{f(t)}{(t-z)\Phi(t)}\,dt, \quad \Im(z) \neq 0$$

and from theorem 4.7 it follows that the one-sided limits

$$g^\pm(x) = \lim_{\varepsilon\downarrow 0}\mathbf{S}_r[f/\Phi](x\pm i\varepsilon)$$

exist locally uniformly on $\mathbb{R}$ and moreover that

$$\frac{f(x)}{\Phi(x)} = g^+(x) - g^-(x) , \quad x \in \mathbb{R}$$

where $g^+$ and $g^-$ are continuous functions on $\mathbb{R}$. Hence

$$f(x) = f^+(x) - f^-(x) , \quad x \in \mathbb{R}$$

where

$$f^\pm(x) = \Phi(x)g^\pm(x)$$

are continuous functions on $\mathbb{R}$. Since $\Phi$ is an entire function on $\mathbb{R}$ it follows easily from the inequality

$$|\Phi(x+iy)\mathbf{S}_r[f/\Phi](x+iy) - \Phi(x)g^\pm(x)| \le$$

$$|\Phi(x+iy)||\mathbf{S}_r[f/\Phi](x+iy) - g^\pm(x)| + |g^\pm(x)||\Phi(x+iy) - \Phi(x)|$$

that

$$f^+(x) = \lim_{\varepsilon \downarrow 0}\{\Phi(x+i\varepsilon)\mathbf{S}_r[f/\Phi](x+i\varepsilon)\}$$

and

$$f^-(x) = \lim_{\varepsilon \downarrow 0}\{\Phi(x-i\varepsilon)\mathbf{S}_r[f/\Phi](x-i\varepsilon)\}$$

the convergence being locally uniform on $\mathbb{R}$. Thus,

$$f(x) = \lim_{\varepsilon \downarrow 0}\{\Phi(x+i\varepsilon)\mathbf{S}_r[f/\Phi](x+i\varepsilon) - \Phi(x-i\varepsilon)\mathbf{S}_r[f/\Phi](x-i\varepsilon)\} \quad (4.12)$$

locally uniformly on $\mathbb{R}$ and so the function

$$f_\Phi^\circ(z) = \Phi(z)\mathbf{S}_r[f/\Phi](z) , \quad \Im(z) \ne 0$$

is an analytic representation of the function $f$.

Note that if $\Psi$ is any other entire function with no real zeros such that $f/\Psi \in L^{i,1}(\mathbb{R}))$ then $f_\Phi^\circ(z) - f_\Psi^\circ$ is entire.

If $\mu$ is a finite-order distribution then without loss of generality we may assume that $\mu = D^k f$, where $k \in \mathbb{N}_0$ and $f$ is a function satisfying the conditions of theorem 4.7. (This follows from the fact that every $f \in \mathcal{C}^0(\mathbb{R})$ is the first derivative of an absolutely continuous function with locally bounded derivative on $\mathbb{R}$.) The function $D_z^k f_\Phi^\circ(z)$ will not, in general, have an inverse Cauchy transform in the sense of (4.12). However for any function $\varphi \in \mathcal{D}$ we have

$$\lim_{\varepsilon \downarrow 0}\int_{-\infty}^{+\infty}\left\{D_z^k f_\Phi^\circ(x+i\varepsilon) - D_z^k f_\Phi^\circ(x-i\varepsilon)\right\}\varphi(x)dx \;=\; <\mathbf{S}_r^{-1}[f_\Phi^\circ](t),(-1)^k\varphi^{(k)}(t)>$$

which means that the inverse Cauchy transform of the function

$$\mu_\Phi^\circ(z) \equiv D_z^k f_\Phi^\circ(z) \,,\; \Im(z) \neq 0$$

does exist in the sense of $\mathcal{D}'$ and that $\mu_\Phi^\circ(z)$ is an analytic representation for $\mu \equiv D^k f \in \mathcal{D}'_{\text{fin}}$. Moreover for any other such representative $\mu_\Psi^\circ$, where $\Psi \in \mathcal{H}(\mathbb{C})$, we would have $\mu_\Phi^\circ - \mu_\Psi^\circ \in \mathcal{H}(\mathbb{C})$. Accordingly we define an equivalence relation $\sim$ on the space $\mathcal{H}(\mathbb{C}\backslash\mathbb{R})$ of all functions analytic outside the real axis by setting

$$\mu^\circ \sim \nu^\circ \Leftrightarrow \mu^\circ - \nu^\circ \in \mathcal{H}(\mathbb{C}).$$

We denote by $\mathcal{H}_S$ the quotient $\mathcal{H}(\mathbb{C}\backslash\mathbb{R})/\mathcal{H}(\mathbb{C})$. Then we define the Cauchy-Stieltjes transform on $\mathcal{D}'_{\text{fin}}$ as follows:

**Definition 4.4** *The Cauchy-Stieltjes transform* $\mathbf{S}[\mu]$ *of a finite-order distribution* $\mu \equiv D^k f$, *where* $f$ *is an absolutely continuous function with locally bounded derivative on* $\mathbb{R}$, *is the equivalence class*

$$[\mu_\Phi^\circ] \in \mathcal{H}_S \equiv \mathcal{H}(\mathbb{C}\backslash\mathbb{R})/\mathcal{H}(\mathbb{C})$$

*where for some suitable* $\Phi \in \mathcal{H}(\mathbb{C})$, $\mu_\Phi^\circ = D_z^k(\Phi \mathbf{S}_r[f/\Phi])$.

Note that for any $j \in \mathbb{N}_0$ and all $z \in \mathbb{C}\backslash\mathbb{R}$ we have

$$|D_z^j \mathbf{S}_r[f/\Phi](z)| \leq C_j \cdot |\Im(z)|^{-(j+1)}.$$

Hence all functions of the form $\mu_\Phi^\circ(z) \equiv D_z^k f_\Phi^\circ(z), \Im(z) \neq 0$, have arbitrary growth to infinity and polynomial growth towards the real axis.

The Cauchy-Stieltjes transform $\mathbf{S} : \mathcal{D}'_{\text{fin}} \subset \mathcal{D}' \to \mathcal{H}_S$ is an injective mapping which extends the Cauchy transform $\mathbf{S}_r : \mathcal{O}'_{-1} \subset \mathcal{D}'_{\text{fin}} \to \mathcal{H}(\mathbb{C}\backslash\mathbb{R})$ in the sense that for any distribution $\mu \in \mathcal{O}'_{-1}$

$$\mathbf{S}[\mu] \equiv [\mathbf{S}_r[\mu]] \in \mathcal{H}_S.$$

If $\Phi\,|_{\mathbb{R}}$ is the restriction to $\mathbb{R}$ of a function $\Phi \in \mathcal{H}(\mathbb{C})$ then we have

$$\mathbf{S}[\Phi\,|_{\mathbb{R}}](z) = \epsilon(z)\Phi(z) + \mathcal{H}(\mathbb{C}) \,,\; \Im(z) \neq 0 \qquad (4.13)$$

where

$$\epsilon(z) = \begin{cases} 1 & \text{for } \Im(z) > 0 \\ 0 & \text{for } \Im(z) < 0 \end{cases} \qquad (4.14)$$

and we can express this by writing

$$\mathbf{S}\{\mathcal{H}\,|_{\mathbb{R}}\} \subset \mathcal{H}_S.$$

**Theorem 4.12** *If* $S[\mu] = [\mu_\Phi^\circ] \in \mathcal{H}_S$ *is the Cauchy-Stieltjes transform of a distribution of finite order* $\mu \equiv D^k f \in \mathcal{D}'_{\text{fin}}$ *then*

$$\lim_{\varepsilon \downarrow 0} \int_{-\infty}^{+\infty} \{\mu^\circ(x+i\varepsilon) - \mu^\circ(x-i\varepsilon)\}\varphi(x)\,dx = <\mu,\varphi>$$

*for any* $\varphi \in \mathcal{D}$ *and any representative* $\mu^\circ \in S[\mu]$.

**Proof:** (See Exercises 4.3, below) □

**Remark 4.2 :**

(1) *Let* $\mu^\circ$ *belong to* $S[\mu]$ *for some distribution* $\mu$ *of finite order. The integral*

$$I(\varepsilon) = \int_{-\infty}^{+\infty} \{\mu^\circ(x+i\varepsilon) - \mu^\circ(x-i\varepsilon)\}\varphi(x)\,dx$$

*is independent of the particular function* $\mu^\circ$ *chosen in* $S[\mu]$. *If we write*

$$<S^{-1}[[\mu^\circ]], \varphi> = \lim_{\varepsilon \downarrow 0} I(\varepsilon)$$

*then theorem 4.12 may be stated as*

$$\forall \varphi \in \mathcal{D} \;:\; <S^{-1}[[\mu^\circ]], \varphi> = <\mu, \varphi>\;.$$

(2) *Note that the product of two equivalence classes cannot generally be defined. In fact, for* $[\mu^\circ] \equiv \mu^\circ + \mathcal{H}(\mathbb{C})$ *and* $[\nu^\circ] \equiv \nu^\circ + \mathcal{H}(\mathbb{C})$, *if we write*

$$[\mu^\circ] \cdot [\nu^\circ] = \mu^\circ \nu^\circ + \{\mu^\circ \cdot \mathcal{H}(\mathbb{C}) + \nu^\circ \cdot \mathcal{H}(\mathbb{C})\} + \mathcal{H}(\mathbb{C})$$

*then* $\mu^\circ \cdot \mathcal{H}(\mathbb{C})$ *and* $\nu^\circ \cdot \mathcal{H}(\mathbb{C})$ *depend upon the particular function chosen in* $\mathcal{H}(\mathbb{C})$ *to make the products and are therefore indeterminate entities.*

**Exercise 4.3 :**

1. Let $f$ be an absolutely continuous function on $\mathbb{R}$ with locally bounded derivative. If $\Phi, \Psi$ are entire functions such that $f/\Phi$, $f/\Psi$ both belong to $L^{i,1}(\mathbb{R})$, prove that $f_\Phi^\circ(z) - f_\Psi^\circ(z)$ is entire.

2. Prove Theorem 4.12.

3. By considering $S[\delta] \cdot S[\delta]$ show that a possible meaning for the distributional product $\delta \cdot \delta = \delta^2$ is given by

$$\delta^2 = \frac{1}{2\pi i}\delta' + c\delta$$

   where $c$ is an arbitrary constant.

4. Show that $S[H] = [-\frac{1}{2\pi i}\log(-z)]$.

### 4.2.1.1  General distributions

Since a general distribution $\mu \in \mathcal{D}'$ is not a derivative of a continuous function on $\mathbb{R}$ the method described above to determine the Cauchy-Stieltjes transform of $\mu$ does not apply. Instead, following Bremermann [4], we shall describe the construction of an analytic representation for $\mu$ by expressing it as a locally finite sum of distributions of compact support.

**Theorem 4.13** *For any distribution $\mu \in \mathcal{D}'$ there exists a function $\mu^\circ \in \mathcal{H}(\mathbb{C}\backslash\mathbb{R})$ such that*

$$< \mathbf{S}^{-1}[[\mu^\circ]], \varphi > = <\mu, \varphi>$$

*for all functions $\varphi \in \mathcal{D}$.*

**Proof:** Let $(\theta_j)_{j\in\mathbb{Z}}$ be a partition of unity subordinate to the open covering $\{\Omega_j\}_{j\in\mathbb{Z}}$ of the real line defined, for each $j \in \mathbb{Z}$, by

$$\Omega_j \equiv \{t \in \mathbb{R} : |t - j| < 1\}.$$

The functions $\theta_j, j \in \mathbb{Z}$, belong to $\mathcal{D}$ and are such that for all $t \in \mathbb{R}$

$$\sum_{j=-\infty}^{+\infty} \theta_j(t) = 1$$

and so defining $\mu_j \equiv \theta_j\mu$ for every $j \in \mathbb{Z}$ we get

$$\lim_{n\to\infty} \sum_{j=-n}^{n} <\mu_j, \varphi> = <\mu, \varphi>$$

for all $\varphi \in \mathcal{D}$. For every $j \in \mathbb{Z}$ $\mu_j$ is a distribution of compact support and therefore it belongs to $\mathcal{O}'_{-1}$, that is, $\mu_j$ has a well defined Cauchy representation

$$\mu_j^\circ(z) \equiv \mathbf{S}_r[\mu_j](z), \quad z \notin \mathrm{supp}(\mu_j).$$

For any $n \in \mathbb{N}_0$ the sum

$$\sum_{j=-n}^{n} \mu_j^\circ(z)$$

is a well defined function analytic outside the support of $\mu_{-n} + \cdots + \mu_n \in \mathcal{E}'$. This sum may not converge as $n \to \infty$, but it is possible to construct a modified sum which does converge. Let $j \in \mathbb{Z}$ be such that $|j| > 1$. The support of $\mu_j$ does not intersect the disk $D_j = \{z \in \mathbb{C} : |z| \leq |j| - 1\}$ and therefore the function $\mu_j^\circ(z) \equiv \mathbf{S}_r[\mu_j](z)$ is analytic on $D_j$ and may be expanded into a power series

$$\mu_j^\circ(z) = \sum_{k=0}^{\infty} a_k^{(j)} z^k$$

which converges uniformly on $D_j$. Hence there exists a number $n_j \in \mathbb{N}_0$ such that

$$\left| \mu_j^\circ(z) - \sum_{k=0}^{n_j} a_k^{(j)} z^k \right| < 2^{-|j|} .$$

for all $z \in D_j$ and consequently the series

$$\mu^\circ(z) = \sum_{j=-1}^{1} \mu_j^\circ(z) + \sum_{|j| \geq 2} \left[ \mu_j^\circ(z) - \sum_{k=0}^{n_j} a_k^{(j)} z^k \right]$$

converges uniformly in every compact subset of the $z$-plane, after subtracting finitely many terms from each $\mu_j^\circ(z)$. That is, if $K$ is a compact of $\mathbb{C}$, then there is a number $n_0 \in \mathbb{N}$ such that $K \subset D_{n_0}$; then

$$\mu^\circ(z) = \sum_{j=-1}^{1} \mu_j^\circ(z) + \sum_{2 \leq |j| \leq n_0} \left[ \mu_j^\circ(z) - \sum_{k=0}^{n_j} a_k^{(j)} z^k \right] + \sum_{n_0 < |j|} \left[ \mu_j^\circ(z) - \sum_{k=0}^{n_j} a_k^{(j)} z^k \right]$$

where the first two summands are finite sums and the third summand is formed by functions analytic on $K$ such that, for all $z \in K$,

$$\left| \sum_{n_0 < |j|} \left[ \mu_j^\circ(z) - \sum_{k=0}^{n_j} a_k^{(j)} z^k \right] \right| < \sum_{j \geq n_0} \left( \frac{1}{2} \right)^{j-1} < \infty$$

and therefore converges uniformly on $K$. Hence $\mu^\circ(z)$ is analytic on $\mathbb{C}\backslash\mathbb{R}$ (more generally on $\mathbb{C}\backslash\mathrm{supp}(\mu)$).

Consider now a function $\varphi \in \mathcal{D}$. Let $n_\varphi \in \mathbb{N}_0$ be such that $\mathrm{supp}(\varphi) \subset D_{n_\varphi}$. Then, since

$$\forall t \in \mathrm{supp}(\varphi) : \sum_{|j| \leq n_\varphi} \theta_j(t) = 1$$

we have

$$\varphi(t) = \sum_{|j| \leq n_\varphi} \theta_j(t)\varphi(t)$$

and therefore

$$<\mu, \varphi> = \sum_{j=-n_\varphi}^{n_\varphi} <\mu_j, \varphi> .$$

Since, for every $j \in \mathbb{Z}$, $\mu_j \in \mathcal{O}'_{-1}$ then, by theorem 4.9, we have

$$<\mu_j, \varphi> = \lim_{\varepsilon \downarrow 0} \int_{-\infty}^{+\infty} \left[ \mu_j^\circ(x + i\varepsilon) - \mu_j^\circ(x - i\varepsilon) \right] \varphi(x)\, dx$$

and so

$$\sum_{j=-n_\varphi}^{n_\varphi} <\mu_j, \varphi> = \lim_{\varepsilon \downarrow 0} \int_{-\infty}^{+\infty} \left\{ \sum_{j=-n_\varphi}^{n_\varphi} \left[ \mu_j^\circ(x + i\varepsilon) - \mu_j^\circ(x - i\varepsilon) \right] \right\} \varphi(x)\, dx.$$

Finally we have

$$\sum_{j=-n_\varphi}^{n_\varphi} \mu_j^\circ(z) = \mu^\circ(z) - \sum_{|j|>n_\varphi} \left[\mu_j^\circ(z) - \sum_{k=0}^{n_j} a_k^{(j)} z^k\right] + \sum_{1<|j|\le n_\varphi}\sum_{k=0}^{n_\varphi} a_k^{(j)} z^k$$
$$= \mu^\circ(z) + \Phi_\varphi(z)$$

where $\Phi_\varphi$ is analytic on $\mathrm{supp}(\varphi)$, and so

$$<\mu,\varphi> = \sum_{j=-n_\varphi}^{n_\varphi} <\mu_j,\varphi>$$
$$= \lim_{\varepsilon\downarrow 0}\int_{-\infty}^{+\infty} [\mu^\circ(x+i\varepsilon) - \mu^\circ(x-i\varepsilon)]\,\varphi(x)\,dx$$

as was to be proved. □

**Corollary 4.2.1** *If $\mu$ is a distribution in $\mathcal{D}'$ then $\mu^\circ(z), \Im(z) \ne 0$, has locally polynomial growth to the real axis, that is, for every compact $K \subset \mathbb{R}$ there exist $C_K > 0$ and $r_K \in \mathbb{N}_0$ such that*

$$|\mu^\circ(z)| \le C_K/|\Im(z)|^{r_K}$$

*for all $z \in \mathbb{C}$ such that $\Re(z) \in K$ and sufficiently small $\Im(z) \ne 0$.*

**Proof:** Let $K$ be a compact of the real line. For some $j_K \in \mathbb{N}_0$ we have that $K \subset D_{j_K}$ and so for all complex numbers $z \in D_{j_K}$ such that $\Re(z) \in K$ and $\Im(z) \ne 0$ we have

$$\mu^\circ(z) = \sum_{j=-j_K}^{j_K} \mu_j^\circ(z) + \Phi(z)$$

where $\Phi \in \mathcal{H}(\mathbb{C})$. Since $\mu_j \in \mathcal{E}'$ for every $j \in \mathbb{Z}$ then, taking theorem 4.11 into account, for every $j \in \{-j_K,\ldots,j_K\}$ there exists $r_j \in \mathbb{N}_0$ such that

$$|\mu_j^\circ(z)| \le C_j/|\Im(z)|^{r_j}$$

for all $z \in \mathbb{C}$ such that $\Re(z) \in \mathrm{supp}(\mu_j)$ and sufficiently small $\Im(z) \ne 0$. Hence, defining

$$r_K = \max\{r_j \in \mathbb{N}_0 : j = -j_K,\ldots,+j_K\}$$

there exists $C_K > 0$ such that the inequality

$$|\mu^\circ(z)| \le C_K/|\Im(z)|^{r_K}$$

holds for all $z \in \mathbb{C}$ such that $\Re(z) \in K$ and sufficiently small $\Im(z) \ne 0$. □

We may now define the Cauchy-Stieltjes transform in $\mathcal{D}'$

$$\mathbf{S}: \mathcal{D}' \to \mathcal{H}_S \equiv \mathcal{H}(\mathbb{C}\setminus\mathbb{R})/\mathcal{H}(\mathbb{C})$$

by setting $\mathbf{S}[\mu] = [\mu^\circ]$, where $\mu^\circ$ is any analytic representation of $\mu$. The Cauchy-Stieltjes transform is easily seen to enjoy the following properties

(a) $S[a\mu + b\nu] = aS[\mu] + bS[\nu]$,
(b) $S[D^k \mu] = D^k S[\mu]$,
(c) $S[\tau_\alpha \mu] = \tau_\alpha S[\mu]$,

for all $\mu, \nu \in \mathcal{D}'$, $a, b \in \mathbb{C}$, $\alpha \in \mathbb{R}$ and $k \in \mathbb{N}_0$.

Denote by $\mathcal{H}^{p,loc}(\mathbb{C}\backslash\mathbb{R}) \subset \mathcal{H}(\mathbb{C}\backslash\mathbb{R})$ the space of all functions analytic on $\mathbb{C}\backslash\mathbb{R}$ with arbitrary growth to infinity and locally of polynomial growth to the real axis. Then the following result has been established by Tillmann [111]:

**Theorem 4.14 (Tillmann theorem)** *The mapping*

$$S : \mathcal{D}' \to \mathcal{H}^{p,loc}(\mathbb{C}\backslash\mathbb{R})/\mathcal{H}(\mathbb{C}) \subset \mathcal{H}_S$$

*is a topological isomorphism between linear spaces.*

Using Tillmann's theorem we can generalise the Painlevé theorem to the following

**Theorem 4.15 (Generalised Painlevé theorem)** *Let $\mu^\circ(z)$ be a function in $\mathcal{H}^{p,loc}(\mathbb{C}\backslash\mathbb{R})$ such that*

$$\lim_{\varepsilon \downarrow 0} < \mu^\circ(x + i\varepsilon), \varphi(x) > = \lim_{\varepsilon \downarrow 0} < \mu^\circ(x - i\varepsilon), \varphi(x) >$$

*for every $\varphi \in \mathcal{D}$. Then $\mu^\circ \in \mathcal{H}(\mathbb{C})$.*

**Proof:** From Tillmann's theorem it follows that $[\mu^\circ]$ is the Cauchy-Stieltjes transform of a distribution $\mu$. Let $(K_n)_{n \in \mathbb{N}}$ be the exhausting family of compacts of $\mathbb{R}$ defined by $K_n = \{x \in \mathbb{R} : |x| \le n\}$. Then for every $n \in \mathbb{N}$ there exist $f_n^\circ \in \mathcal{H}^{p,loc}(\mathbb{C}\backslash\mathbb{R})$ and an integer $r_n \in \mathbb{N}_0$ such that for all $\varphi \in \mathcal{D}_{K_n}$ we have

$$< \mu^\circ(x \pm i\varepsilon), \varphi(x) > = < f_n^\circ(x \pm i\varepsilon), (-1)^{r_n} \varphi^{(r_n)}(x) >$$

and such that the functions

$$f_{\pm,n}(x) \equiv \lim_{\varepsilon \downarrow 0} f_n^\circ(x \pm i\varepsilon)$$

are continuous on $K_n$ (the limits being uniform). Hence

$$\lim_{\varepsilon \downarrow 0} f_n^\circ(x + i\varepsilon) = \lim_{\varepsilon \downarrow 0} f_n^\circ(x - i\varepsilon)$$

and thus, from the Painlevé theorem 4.1, $f_n^\circ$ extends analytically into the strip $\{z \in \mathbb{C} : |\Re(z)| < n\}$. Consequently $\mu^\circ$ also extends analytically into the same strip. Since the result is valid for every $n \in \mathbb{N}$ then $\mu^\circ$ is analytic over the whole of $\mathbb{C}$. $\square$

If $\mu \in \mathcal{D}'$ is a distribution of finite order then the construction described above also produces a function $\mu^\circ$ defined and analytic on $\mathbb{C}\backslash\mathbb{R}$ such that

$$\lim_{\varepsilon \downarrow 0} \int_{-\infty}^{+\infty} [\mu^\circ(x+i\varepsilon) - \mu^\circ(x-i\varepsilon)]\varphi(x)\,dx = <\mu,\varphi>$$

for every $\varphi \in \mathcal{D}$. From the generalised Painlevé theorem it thus follows at once that $\mu^\circ \in \mathbf{S}[\mu]$, which shows that the Cauchy-Stieltjes transform in $\mathcal{D}'$ is an extension of the transform previously defined in $\mathcal{D}'_{\text{fin}}$.

## 4.3 SATO HYPERFUNCTIONS ON THE LINE

The space $\mathcal{H}_S \equiv \mathcal{H}(\mathbb{C}\backslash\mathbb{R})/\mathcal{H}(\mathbb{C})$ of all equivalence classes of analytic functions of the form

$$[\vartheta] \equiv \{v \in \mathcal{H}(\mathbb{C}\backslash\mathbb{R}) : v - \vartheta \in \mathcal{H}(\mathbb{C})\}$$

will be called the space of **Sato hyperfunctions**. It is convenient to refer to the analytic function $\vartheta$ as the **defining function** for the hyperfunction $[\vartheta] \in \mathcal{H}_S$. If $\mu$ is any distribution in $\mathcal{D}'$ then, as shown in the preceding section, there exists a function $\mu^\circ : \mathbb{C}\backslash\mathbb{R} \to \mathbb{C}$ such that

$$\lim_{\varepsilon \downarrow 0} < \mu^\circ(x+i\varepsilon) - \mu^\circ(x-i\varepsilon), \varphi(x) > = <\mu,\varphi>$$

for all $\varphi \in \mathcal{D}$. Conversely for any function $\vartheta \in \mathcal{H}^{p,loc}(\mathbb{C}\backslash\mathbb{R})$ there exists some distribution $\nu \in \mathcal{D}'$ for which $\vartheta$ will be the Cauchy-Stieltjes transform, $\nu^\circ$, of $\nu$. It follows that we can identify $\nu \in \mathcal{D}'$ with the hyperfunction in $\mathcal{H}_S$ represented by the equivalence class $[\nu^\circ]$. For example, the delta function $\delta \in \mathcal{D}'$ corresponds to, and may be identified with, the hyperfunction $[-1/2\pi i z]$; similarly the distribution $\mathrm{Pv}\{x^{-1}\}$ may be identified with the hyperfunction $[\mathrm{sgn}(\Im(z))/2z]$. It will be convenient hereafter to use the same symbol $\mu$ for the distribution in $\mathcal{D}'$ and for the corresponding hyperfunction $[\mu^\circ]$ in $\mathcal{H}_S$.

Suppose, however, that $\vartheta : \mathbb{C}\backslash\mathbb{R} \to \mathbb{C}$ is an analytic function for which the limit

$$\lim_{\varepsilon \downarrow 0}[\vartheta(x+i\varepsilon) - \vartheta(x-i\varepsilon)]$$

does not exist in the sense of $\mathcal{D}'$. It is certainly true that $\vartheta \in \mathcal{H}(\mathbb{C}\backslash\mathbb{R})/\mathcal{H}(\mathbb{C})$ but it does not follow from Tillmann's theorem that there is any corresponding distribution in $\mathcal{D}'$. Consider, for example, the function

$$\nu^\circ(z) = \exp\left(-\frac{1}{z^2}\right), \quad z \neq 0.$$

This is analytic on $\mathbb{C}\backslash\mathbb{R}$ but has exponential growth in the neighbourhood of the origin. If it were the Cauchy-Stieltjes transform of some distribution $\nu \in \mathcal{D}$ then $\nu$ would have support concentrated on the set $\{0\}$ and would therefore be a finite linear combination of derivatives of $\delta(x)$. Hence, for some $N \in \mathbb{N}$, we should have

$$\nu^\circ(z) = \sum_{n=1}^{N} a_n z^{-n}$$

whereas

$$\exp(-1/z^2) = \sum_{n=0}^{\infty} \frac{(-1)^n}{n! z^{2n}}.$$

It follows that if we write $f(z) = \exp(-1/z^2) - \nu^\circ(z)$ then we should get

$$\lim_{\varepsilon \downarrow 0} \int_{-\infty}^{+\infty} [f(x+i\varepsilon) - f(x-i\varepsilon)] \varphi(x) dx = 0$$

for all $\varphi \in \mathcal{D}$. This is not the case, as can be seen by taking a counter-example of the form

$$\varphi(x) = \alpha_r(x) x^{2N-1}$$

where $\alpha_r \in \mathcal{D}$ and $\alpha_r(x) = 1$ for $|x| \leq r$. (A detailed discussion is to be found in Bremermann [4, p.70]).

The hyperfunction $\nu \in \mathcal{H}_S$ is sometimes represented as a boundary value of the form

$$\nu(x) \equiv \nu^\circ(x+i0) - \nu^\circ(x-i0)$$

where the right-hand side is to be interpreted a limit, in some appropriate sense, as $\varepsilon \downarrow 0$. This is equivalent to saying that the hyperfunction is to be represented in terms of the jump of a function analytic on $\mathbb{C}\backslash\mathbb{R}$ from just below to just above the real axis. From this point of view it is only the behaviour of $\nu^\circ(z)$, $\Im(z) \neq 0$, in the vicinity of $\mathbb{R}$ which matters. This raises the following question : *could we take another (smaller) complex neighbourhood of* $\mathbb{R}$*, say* $\Lambda$ *(such that* $\mathbb{R} \subset \Lambda \subset \mathbb{C}$*) to define* $\mathcal{H}_S \equiv \mathcal{H}_S(\mathbb{R})$ *as the factor-space* $\mathcal{H}(\Lambda\backslash\mathbb{R})/\mathcal{H}(\Lambda)$*?* The answer is in the affirmative since the restriction map

$$R_\Lambda : \mathcal{H}(\mathbb{C}\backslash\mathbb{R}) \to \mathcal{H}(\Lambda\backslash\mathbb{R})$$

defined by $R_\Lambda(\nu^\circ) = \nu^\circ|_{\Lambda\backslash\mathbb{R}}$ induces a natural mapping

$$\xi : \mathcal{H}(\mathbb{C}\backslash\mathbb{R})/\mathcal{H}(\mathbb{C}) \to \mathcal{H}(\Lambda\backslash\mathbb{R})/\mathcal{H}(\Lambda)$$

such that $\xi([\nu^\circ]) = [R_\Lambda(\nu^\circ)]$, and this can easily be seen to be a linear space isomorphism. The rigorous definition of the space $\mathcal{H}_S$ of Sato hyperfunctions on the line can therefore be given as in Kaneko [48] as follows. Let $\{\Lambda\}_{\Lambda \supset \mathbb{R}}$ be the family of all complex neighbourhoods of $\mathbb{R}$. To every $\Lambda$ associate the linear space $\mathcal{H}_{S,\Lambda} \equiv \mathcal{H}(\Lambda \backslash \mathbb{R})/\mathcal{H}(\Lambda)$ and to any two neighbourhoods in the family $\Lambda_i, \Lambda_j$ such that $\mathbb{R} \subset \Lambda_j \subset \Lambda_i$ associate a restriction map $R_{j,i} : \mathcal{H}_{S,\Lambda_i} \to \mathcal{H}_{S,\Lambda_j}$ such that

(a) $R_{i,i} = id$,

(b) $\mathbb{R} \subset \Lambda_k \subset \Lambda_j \subset \Lambda_i \Rightarrow R_{k,i} = R_{k,j} \circ R_{j,i}$.

Then $\{\mathcal{H}_{S,\Lambda}\}_{\Lambda \supset \mathbb{R}} \equiv \{\mathcal{H}(\Lambda \backslash \mathbb{R})/\mathcal{H}(\Lambda)\}_{\Lambda \supset \mathbb{R}}$ is an inductive system of linear spaces. The space of hyperfunctions is the inductive limit of this system

$$\mathcal{H}_S = \lim_{\Lambda \supset \mathbb{R}} \mathcal{H}(\Lambda \backslash \mathbb{R})/\mathcal{H}(\Lambda) .$$

**Remark 4.3** *More precisely the space $\mathcal{H}_S$ of Sato hyperfunctions on the line can be defined as follows : let $E = \cup_{\Lambda \supset \mathbb{R}} \mathcal{H}_{S,\Lambda}$ be the disjoint union of $\{\mathcal{H}_{S,\Lambda}\}_{\Lambda \supset \mathbb{R}}$ (that is, $E$ is the union of all $\mathcal{H}_{S,\Lambda}$ regarded as mutually unrelated). Introduce in $E$ the equivalence relation $\sim$ defined for any two classes $[\nu^\circ], [\sigma^\circ] \in E$ (such that $[\nu^\circ] \in \mathcal{H}_{S,\Lambda_i}$, and $[\sigma^\circ] \in \mathcal{H}_{S,\Lambda_j}$) by*

$$[\nu^\circ] \sim [\sigma^\circ] \Leftrightarrow \exists \Lambda_k \subset \Lambda_i \cap \Lambda_j : \mathbb{R} \subset \Lambda_k \wedge R_{k,i}([\nu^\circ]) = R_{k,j}([\sigma^\circ]) .$$

*Then*

$$\mathcal{H}_S = E/\sim .$$

As can be seen, there exists a natural mapping $R_\Lambda : \mathcal{H}_{S,\Lambda} \to \mathcal{H}_S \equiv E/\sim$ and, conversely, every element of this space can be written as $R_\Lambda([\nu^\circ])$ with $[\nu^\circ] \in \mathcal{H}_{S,\Lambda}$ for some $\Lambda \supset \mathbb{R}$. This induces a natural $\mathbb{C}$-linear space structure into the inductive limit $\mathcal{H}_S$, and each $R_\Lambda$ becomes a $\mathbb{C}$-linear mapping with respect to it. Indeed, the linear combination of two elements $R_{\Lambda_i}([\nu^\circ])$ and $R_{\Lambda_j}([\sigma^\circ])$ is given by

$$\alpha R_{\Lambda_i}([\nu^\circ]) + \beta R_{\Lambda_j}([\sigma^\circ]) = R_{\Lambda_k}(\alpha R_{k,i}([\nu^\circ]) + \beta R_{k,j}([\sigma^\circ]))$$

where $\alpha, \beta$ are two scalars and $\Lambda_k \subset \Lambda_i \cap \Lambda_j$ is a complex neighbourhood of $\mathbb{R}$ and we use the vector space structure of $\mathcal{H}_{S,\Lambda_k}$. That is, we need only to restrict the defining functions to a common domain and compute the linear combination there.

The **zero** hyperfunction, 0, is represented by any function in $\mathcal{H}(\mathbb{C})$.

Now let $\Omega$ be an open subset of $\mathbb{R}$ and define $\mathbb{C}_\Omega = \{z \in \mathbb{C} : \Re(z) \in \Omega\}$. Then

$$\mathcal{H}_S(\Omega) = \mathcal{H}(\mathbb{C}_\Omega \backslash \Omega)/\mathcal{H}(\mathbb{C}_\Omega)$$

is the space of hyperfunctions on $\Omega \subset \mathbb{R}$. The restriction to $\Omega$ of a hyperfunction $\nu \in \mathcal{H}_S(\mathbb{R})$ is the hyperfunction on $\Omega$, $\nu|_\Omega \in \mathcal{H}_S(\Omega)$, defined by

$$\nu|_\Omega \equiv \left[\nu^\circ \,|_{\mathbf{C}_\Omega \backslash \Omega}(z)\right]$$

where $\nu^\circ(z), \Im(z) \neq 0$ is a defining function of $\nu \in \mathcal{H}_S(\mathbb{R})$. A hyperfunction $\nu \in \mathcal{H}_S(\mathbb{R})$ is said to be null on the open set $\Omega \subset \mathbb{R}$ if $\nu|_\Omega$ is the null hyperfunction on $\mathcal{H}_S(\Omega)$, that is, its defining function belongs to $\mathcal{H}(\mathbf{C}_\Omega)$.

**Definition 4.5** *The support of a hyperfunction $\nu \in \mathcal{H}_S(\mathbb{R})$, denoted by* $\mathrm{supp}(\nu)$, *is the complement in $\mathbb{R}$ of the largest open subset of $\mathbb{R}$ on which $\nu$ is the zero hyperfunction.*

The support of a hyperfunction on $\mathbb{R}$ is always a closed subset of $\mathbb{R}$.
Let $K \sqsubset \mathbb{R}$ be any compact. We denote by $\mathcal{H}_S[K]$ the space of hyperfunctions of compact support on $K$, it being easy to see that $\mathcal{H}_S[K] \equiv \mathcal{H}(\mathbf{C}\backslash K)/\mathcal{H}(\mathbf{C})$.

**Exercise 4.4 :**

1. Confirm the following representation of distributions as hyperfunctions

    (a) for any $\lambda \in \mathbb{R}$, $(x+i0)^\lambda$ is the hyperfunction with defining function

    $$\nu_+^\circ(z) = \begin{cases} +z^\lambda & \text{for } \Im(z) > 0, \\ 0 & \text{for } \Im(z) < 0, \end{cases}$$

    and $(x-i0)^\lambda$ is the hyperfunction with defining function

    $$\nu_-^\circ(z) = \begin{cases} 0 & \text{for } \Im(z) > 0, \\ -z^\lambda & \text{for } \Im(z) < 0, \end{cases}$$

    (b) for every $\lambda \notin \mathbb{Z}$,

    $$x_+^\lambda = \left[\frac{-(-z)^\lambda}{2i\sin(\pi\lambda)}\right] \;;\; x_-^\lambda = \left[\frac{z^\lambda}{2i\sin(\pi\lambda)}\right],$$

    and for any $n \in \mathbb{Z}$,

    $$\mathrm{Pf}\left\{\frac{H_0(x)}{x^n}\right\} = \left[\frac{-z^n}{2\pi i}\log(-z)\right],$$

    $$\mathrm{Pf}\left\{\frac{H_0(-x)}{x^n}\right\} = \left[\frac{(-z)^n}{2\pi i}\log(-z)\right],$$

2. Find the supports of the hyperfunctions $H, \delta, (x+i0)^{-1}$.

### 4.3.1 Basic operations

Two hyperfunctions $\nu \equiv [\nu^\circ], \sigma \equiv [\sigma^\circ] \in \mathcal{H}_S$ are added in the natural way, that is

$$\nu + \sigma = [\nu^\circ + \sigma^\circ] \in \mathcal{H}_S,$$

and the product of a hyperfunction $\nu \equiv [\nu^\circ] \in \mathcal{H}_S$ by a scalar $a \in \mathbb{C}$ is defined by

$$a\nu = [a\nu^\circ] \in \mathcal{H}_S.$$

With these operations $\mathcal{H}_S \equiv \mathcal{H}_S(\mathbb{R})$ is a $\mathbb{C}$-linear space.

If $\alpha$ is an entire function in $\mathcal{H}(\mathbb{C})$ then we can multiply any hyperfunction $\nu \equiv [\nu^\circ]$ by $\alpha$ in the following way

$$\alpha\nu = [\alpha(z)\nu^\circ(z)] \in \mathcal{H}_S. \tag{4.15}$$

More generally, bearing in mind the inductive limit structure of the space $\mathcal{H}_S$, we may define the product $\alpha\nu$ for any real analytic function $\alpha$ (that is, a function analytically continuable into some complex neighbourhood of $\mathbb{R}$).

**Example 4.3.1** If $\alpha$ is an entire function then we have

$$\alpha(z) = \alpha(0) + \alpha'(0)z + \cdots + \left(\alpha^{(n)}(0)/n!\right)z^n + \cdots$$

the series converging for all $z \in \mathbb{C}$. Thus we get, for example, the following

$$\begin{aligned}\alpha(x)\delta &= \left[\frac{-\alpha(0)}{2\pi i z} + \sum_{n=1}^{\infty} \frac{\alpha^{(n)}(0)}{2\pi i n!} z^{n-1}\right] \\ &= \left[\frac{-\alpha(0)}{2\pi i z}\right] \equiv \alpha(0)\delta\end{aligned}$$

a result which agrees with the Schwartz product.

The differentiation of a hyperfunction

$$\nu \equiv [\nu^\circ] \in \mathcal{H}_S$$

is defined naturally by

$$D\nu = [D_z \nu^\circ] \in \mathcal{H}_S$$

and, more generally, if $P(x, D)$ is a linear differential operator of finite order with analytic coefficients, we have

$$P(x, D)\nu = [P(z, D_z)\nu^\circ] \in \mathcal{H}_S.$$

**Example 4.3.2** 1. from exercise 4.3-(4) we get

$$DH = \left[\frac{d}{dz}\frac{(-1)}{2\pi i}\log(-z)\right] = \left[\frac{(-1)}{2\pi i z}\right] \equiv \delta,$$

2. for any $\lambda \in \mathbb{C}\backslash\mathbb{Z}$

$$Dx_+^\lambda = \left[\frac{d}{dz}\frac{-(-z)^\lambda}{2i\sin(\pi\lambda)}\right]$$
$$= \left[\frac{\lambda(-z)^{\lambda-1}}{2i\sin(\pi\lambda)}\right] = \lambda\left[\frac{d}{dz}\frac{-(-z)^{\lambda-1}}{2i\sin(\pi(\lambda-1))}\right] \equiv \lambda x_+^{\lambda-1}$$

and for $\lambda = -n$, $(n \in \mathbb{N})$,

$$D\text{Pf}\left\{\frac{H_0(x)}{x^n}\right\} = -n\left[\frac{-z^{-n-1}}{2\pi i}\log(-z) + \frac{-1}{2\pi i z^{n+1}}\right]$$
$$\equiv -n\text{Pf}\left\{\frac{H_0(x)}{x^{n+1}}\right\} + \frac{(-1)^n}{n!}\delta^{(n)}.$$

### 4.3.2 Local operators

Consider for example a hyperfunction $\nu$ with support at the origin. Then its defining function $\nu^\circ$ extends analytically into $\mathbb{C}\backslash\{0\}$ and therefore has a Laurent expansion at the origin of the form

$$\sum_{k=-\infty}^{+\infty} a'_k z^k.$$

Since, in the sense of hyperfunctions, only the negative powers of $z$ are meaningful we may consider without any loss of generality that

$$\nu^\circ(z) = \sum_{k=0}^{+\infty} a_k / z^{k+1}, \quad (a_k \equiv a'_{-k}, \ k \in \mathbb{N}_0)$$

and therefore since $1/z^{k+1} = \left(2\pi i(-1)^{k-1}/k!\right) D_z^k(-1/2\pi i z)$ we obtain

$$\nu^\circ(z) = \sum_{k=0}^\infty b_k D_z^k\left(-\frac{1}{2\pi i z}\right)$$

where $b_k = 2\pi i(-1)^{k-1}a_k/k!$. Hence $\nu \in \mathcal{H}_S[\{0\}]$ may be formally expressed as an infinite-order derivative of the delta-function as follows

$$\nu \equiv \sum_{k=0}^\infty b_k \delta^{(k)}.$$

Note that since the Laurent expansion of $\nu^\circ(z)$ has infinite radius of convergence then $\limsup_{k\to\infty} \sqrt[k]{|a_k|} = 0$ and therefore the sequence $(b_k)_{k\in\mathbb{N}_0}$ must be such that $\limsup_{k\to\infty} \sqrt[k]{|b_k|k!} = 0$. An infinite-order operator of the form

$$J(D) = \sum_{k=0}^{\infty} b_k D^k$$

where the sequence of numbers $(b_k)_{k\in\mathbb{N}_0}$ is such that $\limsup_{k\to\infty} \sqrt[k]{|b_k|k!}$ is called a **local operator** of constant coefficients. The following result is proved in Kaneko [48].

**Theorem 4.16** *Every hyperfunction $\nu \in \mathcal{H}_S$ can be obtained by applying a local operator with constant coefficients to a continuous function on $\mathbb{R}$.*

This theorem emphasises the difference between distributions (considered as members of a special sub-class of $\mathcal{H}_S$) and hyperfunctions in general. Every distribution can be expressed locally as the derivative to some finite order of a continuous function. In particular if $\mu$ is a distribution with support concentrated at the origin then $\mu$ must be a finite linear combination of finite-order derivatives of $\delta$ (and therefore of finite-order derivatives of the continuous function $x_+$). By contrast there exist hyperfunctions with support concentrated at the origin which are infinite-order derivatives of the delta function. If we interpret the term "derivative" in the sense of local operators then we can say that **every** hyperfunction can be expressed as the derivative of a continuous function.

**Exercise 4.5 :**

1. Find representations of $x_+^{-n}$ and of $x_-^{-n}$ as hyperfunctions (n =1,2,3, ...).

2. Obtain an expression for the hyperfunction with defining function $\exp(-1/z)$ as an infinite-order derivative of a delta function.

## 4.4 ULTRADISTRIBUTIONS

As already remarked above (§4.1.2) Carleman was able to extend the classical definition of Fourier transform from $L^1(\mathbb{R})$ to functions of polynomial growth (and, in fact, to general tempered distributions) by representing them as pairs of analytic functions on $\mathbb{C}^+$ and $\mathbb{C}^-$ respectively. In our notation this corresponds to replacing every such function (or any finite-order derivative of it) by its Cauchy representation.

The Carleman idea leads, on the one hand, to the representation by analytic functions not only of ordinary functions but also of general distributions, and ultimately to the concept of hyperfunction. On the other hand the same idea may be used to extend the domain of application of the Fourier transform and leads to the definition of the so-called Fourier-Carleman Transform. The Fourier-Carleman transform of a pair of functions, analytic on $\mathbb{C}^+$ and $\mathbb{C}^-$ respectively, which represents a tempered distribution is again a pair of functions which represents a tempered distribution. This agrees with the well known fact that $\mathcal{F}\{\mathcal{S}'\} = \mathcal{S}'$. However if we consider a pair of functions analytic on $\mathbb{C}^+$ and $\mathbb{C}^-$ respectively which represents, for example, a distribution of exponential type then we cannot expect the transform to be a pair of functions analytic on $\mathbb{C}^+$ and $\mathbb{C}^-$. In general we will obtain a pair of functions analytic on open half-planes of the form

$$\mathbb{C}_b^+ = \{z \in \mathbb{C} : \Im(z) > +b\}$$

$$\mathbb{C}_b^- = \{z \in \mathbb{C} : \Im(z) < -b\}$$

for some $b \geq 0$. In the case when $b$ must be taken strictly positive we have a pair of functions which does not represent a distribution or even a hyperfunction; its interpretation must be sought for in terms of other objects, such as ultradistributions.

### 4.4.1 Silva tempered ultradistributions

#### 4.4.1.1 The Cauchy-Stieltjes transform in $\mathcal{U}'$

Let $\hat{\mu}$ denote a Silva tempered ultradistribution in $\mathcal{U}'$. Then there exists a distribution of exponential type $\mu \in \mathcal{K}'$ such that

$$<\hat{\mu}, \varphi> = <\mu, \hat{\varphi}> \tag{4.16}$$

for all $\varphi \in \mathcal{U}$. From the representation theorem 3.31 it follows that there exist numbers $b \geq 0$ and $r \in \mathbb{N}_0$ such that

$$\mu(\omega) = D^r[\exp(b|\omega|)f(\omega)] \tag{4.17}$$

in the sense of $\mathcal{K}'$, where $f$ is a continuous bounded function on $\mathbb{R}$. Hence we get

$$<\hat{\mu}, \varphi> = \int_{-\infty}^{+\infty} \exp(b|\omega|)f(\omega)(-1)^r \hat{\varphi}^{(r)}(\omega) d\omega \tag{4.18}$$

for all $\varphi \in \mathcal{U}$ (and therefore for all $\hat{\varphi} \in \mathcal{K}$). Now

$$\hat{\varphi}(\omega) = \int_{-\infty}^{+\infty} \exp(-i\omega t)\varphi(t)dt \quad , \quad \omega \in \mathbb{R}$$

and, since $\varphi$ is an entire function which belongs to $\mathcal{S}$ on every horizontal line, then from the Cauchy Residue Theorem we may write

$$\varphi(t) = \frac{1}{2\pi i} \oint_{\Gamma_{b'}} \frac{\varphi(z)}{z-t} dz$$

where, for every $b' > 0$, $\Gamma_{b'}$ is the counterclockwise oriented boundary of the strip $\Lambda_{b'} = \{z \in \mathbb{C} : |\Im(z)| < b'\}$. Thus we obtain

$$\hat{\varphi}(\omega) = \int_{-\infty}^{+\infty} \exp(-i\omega t) \left\{ \frac{1}{2\pi i} \oint_{\Gamma_{b'}} \frac{\varphi(t)}{z-t} dz \right\} dt$$

$$= -\oint_{\Gamma_{b'}} \varphi(z) \left\{ \frac{1}{2\pi i} \int_{-\infty}^{+\infty} \frac{\exp(-i\omega t)}{t-z} dt \right\} dz$$

the interchange of integrations being justified by invoking Fubini's theorem and taking into account the growth properties of $\varphi(z)$ on horizontals. Using the Cauchy representation of $\exp(-i\omega t)$, $t \in \mathbb{R}$ we have

$$\omega < 0 : \qquad \hat{\varphi}(\omega) = -\int_{\Gamma_{b'}^+} \varphi(z) \exp(-i\omega z) dz$$

$$\omega > 0 : \qquad \hat{\varphi}(\omega) = +\int_{\Gamma_{b'}^-} \varphi(z) \exp(-i\omega z) dz$$

where $\Gamma_{b'}^+ = \Gamma_{b'} \cap \mathbb{C}^+$ and $\Gamma_{b'}^- = \Gamma_{b'} \cap \mathbb{C}^-$. We can differentiate under the integral sign as many times as we wish. Hence, substituting $\hat{\varphi}^{(r)}(\omega)$ into (4.18), we get

$$<\hat{\mu}, \varphi> = \left\{ \int_{-\infty}^{0} + \int_{0}^{+\infty} \right\} \exp(b|\omega|) f(\omega) (-1)^r \hat{\varphi}^{(r)}(\omega) d\omega$$

$$= -\int_{-\infty}^{0} \exp(-b\omega) f(\omega) \int_{\Gamma_{b'}^+} (iz)^r \varphi(z) \exp(-i\omega z) dz d\omega$$

$$+ \int_{0}^{+\infty} \exp(+b\omega) f(\omega) \int_{\Gamma_{b'}^-} (iz)^r \varphi(z) \exp(-i\omega z) dz d\omega$$

$$= -\int_{\Gamma_{b'}^+} (iz)^r \varphi(z) \int_{-\infty}^{0} f(\omega) \exp\left(-i\omega(z-ib)\right) d\omega dz$$

$$+ \int_{\Gamma_{b'}^-} (iz)^r \varphi(z) \int_{0}^{-\infty} f(\omega) \exp\left(-i\omega(z+ib)\right) d\omega dz \quad (4.19)$$

where the interchange of order of integrations is justified under the constraint $b' > b$. The integrals

$$\int_{-\infty}^{0} f(\omega) \exp\left(-i\omega(z-ib)\right) d\omega$$

and
$$\int_0^{+\infty} f(\omega) \exp\left(-i\omega(z+ib)\right) d\omega$$
converge uniformly on compacts of $\mathbb{C}_b^+$ and $\mathbb{C}_b^-$, respectively, and we define
$$\hat{f}_+(z) = +\int_{-\infty}^0 f(\omega) \exp\left(-i\omega(z-ib)\right) d\omega \;, \quad z \in \mathbb{C}_b^+$$
$$\hat{f}_-(z) = -\int_0^{+\infty} f(\omega) \exp\left(-i\omega(z+ib)\right) d\omega \;, \quad z \in \mathbb{C}_b^-. \qquad (4.20)$$

From the above remarks it also follows that we can differentiate under the integral sign with respect to $z$ and therefore that $\hat{f}_+$ and $\hat{f}_-$ are analytic functions on their respective domains (whether or not $\hat{f}_+$ and $\hat{f}_-$ extend beyond $\mathbb{C}_b^+$ and $\mathbb{C}_b^-$ depends on the particular behaviour of the function $f(\omega)$ on the half-lines $\omega < 0$ and $\omega > 0$, respectively.) The function
$$\hat{\mu}^\circ(z) = \begin{cases} (iz)^r \hat{f}_+(z) & \text{if } \Im(z) > +b \\ (iz)^r \hat{f}_-(z) & \text{if } \Im(z) < -b \end{cases} \qquad (4.21)$$
is an analytic function outside the closed strip $\bar{\Lambda}_b = \{z \in \mathbb{C} : |\Im(z)| \leq b\}$; moreover it is easily seen that $\hat{\mu}^\circ(z)$ grows no faster than a polynomial on every closed domain $\bar{\mathbb{C}}_{b'}^\circ \equiv \mathbb{C}\backslash\Lambda_{b'}$ with $b' > b$. Taking (4.19) into account we have therefore proved the following

**Theorem 4.17** *For every Silva tempered ultradistribution $\hat{\mu} \in \mathcal{U}'$ there exists a function $\hat{\mu}^\circ(z)$, defined and analytic on $\mathbb{C}_b^\circ \equiv \mathbb{C}\backslash\bar{\Lambda}_b$ for some $b \geq 0$ and of polynomial growth on $\bar{\mathbb{C}}_{b'}^\circ$ for every $b' > b$, such that*
$$<\hat{\mu}, \varphi> = -\oint_{\Gamma_{b'}} \hat{\mu}^\circ(z)\varphi(z)\, dz$$
*for all test functions $\varphi \in \mathcal{U}$.*

The function $\hat{\mu}^\circ(z)$ has all its singularities in the strip $\bar{\Lambda}_b$ and in this sense it will be said that the Silva tempered ultradistribution $\hat{\mu} \in \mathcal{U}'$ is **concentrated** on $\bar{\Lambda}_b$. The intersection of $\bar{\Lambda}_b$ with the imaginary axis is a closed imaginary interval $[-ib, ib]$ which we call the **imaginary support** of $\hat{\mu} \in \mathcal{U}'$. Silva tempered ultradistributions are thus ultradistributions of **compact imaginary support** (a fact which might have been expected since test-functions in $\mathcal{U}$ have arbitrary growth along vertical lines in the $z$-plane). The converse is not true : there are ultradistributions with compact imaginary support which are not Silva tempered ultradistributions.

**Definition 4.6** *The function $\hat{\mu}^\circ : \mathbb{C}_b^\circ \to \mathbb{C}$ as defined in (4.21) is called an* **analytic representation** *of the ultradistribution $\hat{\mu} \in \mathcal{U}'$.*

Theorem 4.17 above concerning the analytic representation of a Silva tempered ultradistribution may be refined somewhat to give the following result:

**Theorem 4.18** *For every Silva tempered ultradistribution $\hat{\mu} \in \mathcal{U}'$ there exists a continuous function $\hat{f} \in L^1 \cap L^2(\mathbb{R})$ such that if $\hat{f}^\circ(z) = \mathbf{S}_r[\hat{f}](z)$, $\Im(z) \neq 0$, then*

$$\hat{\mu}^\circ(z) = \begin{cases} (iz)^r \hat{f}^\circ(z - ib) & \text{if } \Im(z) > +b \\ (iz)^r \hat{f}^\circ(z + ib) & \text{if } \Im(z) < -b \end{cases}$$

*is an analytic representation of that ultradistribution.*

**Proof:** The number $b \geq 0$ in (4.17) can be chosen so that $f$ is a $L^1 \cap L^2(\mathbb{R})$ function; then, by the Plancherel theorem, we can assert that there is a continuous bounded $L^2(\mathbb{R})$-function $\hat{f}(t), t \in \mathbb{R}$, such that

$$f(\omega) = \frac{1}{2\pi} \int_{-\infty}^{+\infty} \exp(i\omega t)\hat{f}(t)dt \quad , \quad \omega \in \mathbb{R}. \tag{4.22}$$

We can even choose $b \geq 0$ and $r \in \mathbb{N}_0$ in such a way that $f$ belongs to $C^2(\mathbb{R})$ with $f'$ and $f''$ integrable; under these circumstances the function

$$\hat{f}(t) = \int_{-\infty}^{+\infty} \exp(-i\omega t)f(\omega)d\omega \quad , \quad t \in \mathbb{R} \tag{4.23}$$

is $O(t^{-2})$ as $|t| \to \infty$ and therefore $\hat{f}$ is also a continuous bounded $L^1 \cap L^2(\mathbb{R})$-function. Substituting (4.22) into (4.20) gives

$$\begin{aligned}
\hat{f}_+(z) &= \int_{-\infty}^{0} \exp(-i\omega(z - ib)) \left\{ \frac{1}{2\pi} \int_{-\infty}^{+\infty} \exp(i\omega t)\hat{f}(t)dt \right\} d\omega \\
&= \frac{1}{2\pi} \int_{-\infty}^{+\infty} \hat{f}(t) \int_{-\infty}^{0} \exp(i\omega(t - (z - ib))) d\omega dt \\
&= \frac{1}{2\pi i} \int_{-\infty}^{+\infty} \frac{\hat{f}(t)}{t - (z - ib)} dt = \hat{f}^\circ(z - ib)
\end{aligned}$$

where, in view of the above remarks, the interchange between integrations may be justified by Fubini's theorem provided that $\Im(z) > b$. Similarly we would obtain

$$\hat{f}_-(z) = \frac{1}{2\pi i} \int_{-\infty}^{+\infty} \frac{\hat{f}(t)}{t - (z + ib)} dt = \hat{f}^\circ(z + ib)$$

provided that $\Im(z) < -b$. Taking the last two results into (4.21) we get

$$\hat{\mu}^\circ(z) = (iz)^r \hat{f}^\circ(z - i\,\mathrm{sgn}(\Im(z))b) \quad, \quad |\Im(z)| > b$$

where
$$\hat{f}^\circ(z) = \frac{1}{2\pi i} \int_{-\infty}^{+\infty} \frac{\hat{f}(t)}{t-z} dt \;,\; \Im(z) \neq 0$$
is the Cauchy representation of the function $\hat{f} \in L^1 \cap L^2(\mathbb{R})$. $\square$

The analytic representation of $\hat{\mu} \in \mathcal{U}'$ is not unique. From the integral representation of $\hat{\mu}$ in theorem 4.17 it follows that any other function which differs from $\hat{\mu}^\circ(z)$ by a polynomial also represents the same ultradistribution. Conversely every function $\hat{\mu}^\circ(z)$ defined and analytic on $\mathbf{C}_b^\circ$ for some $b \geq 0$ with (at most) polynomial growth on every $\bar{\mathbf{C}}_{b'}^\circ$, $b' > b$, defines through that integral representation a Silva tempered ultradistribution.

For every fixed $b \geq 0$ let $\mathcal{H}(\mathbf{C}_b^\circ)$ be the space of all functions defined and analytic outside $\bar{\Lambda}_b$ and then consider
$$\mathcal{H}^\circ \equiv \bigcup_{b \geq 0} \mathcal{H}(\mathbf{C}_b^\circ). \tag{4.24}$$

$\Pi[z]$, the space of all polynomial functions, is a subspace of $\mathcal{H}^\circ$. Then define $\mathbf{S}[\hat{\mu}]$, the Cauchy-Stieltjes transform of $\hat{\mu} \in \mathcal{U}'$, to be the class in $\mathcal{H}^\circ/\Pi[z]$ determined by any Cauchy representation $\hat{\mu}^\circ \in \mathcal{H}^\circ$ of $\hat{\mu}$. The map $\mathbf{S}: \mathcal{U}' \to \mathcal{H}^\circ/\Pi[z]$ is an into (but not onto) mapping.

Representing the inverse transformation, as usual, by $\mathbf{S}^{-1}$ we have
$$<\mathbf{S}^{-1}[[\hat{\mu}^\circ]], \varphi> = -\oint_{\Gamma_{b'}} \hat{\mu}^\circ(z)\varphi(z) dz$$
where $\Gamma_{b'}$ is a counterclockwise oriented boundary of a horizontal strip containing in its interior all singularities of $\hat{\mu}^\circ$.

All the usual properties of the Cauchy-Stieltjes transform hold. In particular it make sense within the ultradistribution theory to define complex translations. Let $\hat{\mu} \in \mathcal{U}'$ be such that $\mathbf{S}[\hat{\mu}] = [\hat{\mu}^\circ]$, and let $\alpha$ denote any complex number. Then we have
$$<\tau'_\alpha \hat{\mu}, \varphi> = -\oint_{\Gamma_{b'+|\Im(\alpha)|}} \hat{\mu}^\circ(z-\alpha)\varphi(z) dz$$
for any $\varphi \in \mathcal{U}$ and therefore
$$\mathbf{S}[\tau'_\alpha \hat{\mu}] = \tau_\alpha \mathbf{S}[\hat{\mu}] . \tag{4.25}$$

**Exercise 4.6 :**

1. Let $\hat{\mu}$ be an ultradistribution in $\mathcal{U}'$ concentrated on $\bar{\Lambda}_b$ (for some $b \geq 0$).

(a) Show that there exists a polynomial $p \in \Pi[z]$ such that
$$\hat{\mu}_p^{\circ}(z) = \frac{p(z)}{2\pi i} < \hat{\mu}(\lambda), \frac{1}{p(\lambda)(\lambda - z)} >$$
is a function analytic for $|\Im(z)| > b$ which belongs to $S[\hat{\mu}]$.

(b) Show that if $p$ and $q$ are two polynomials such that both $\hat{\mu}_p^{\circ}$ and $\hat{\mu}_q^{\circ}$ are well defined as above then
$$\hat{\mu}_p^{\circ} - \hat{\mu}_q^{\circ}$$
is a polynomial in $\Pi[z]$.

2. Prove the following

(a) $S[D\hat{\mu}] = DS[\hat{\mu}]$,

(b) $S[\tau_\alpha \hat{\mu}] = \tau_\alpha S[\hat{\mu}]$,

(c) $S[p\hat{\mu}] = pS[\hat{\mu}]$,

where $\hat{\mu} \in \mathcal{U}', \alpha \in \mathbb{C}$ and $p \in \Pi[z]$.

### 4.4.1.2 Tempered distributions

Recall that if $\hat{\mu} \in \mathcal{S}'$ then there exists $\mu \in \mathcal{S}'$ such that $\hat{\mu} = \mathcal{F}[\mu]$ in the sense of $\mathcal{S}'$. Then as seen in chapter 3 we have that
$$\mu = D^r[(1 + |\omega|)^j f(\omega)]$$
where $f$ is a continuous bounded function on $\mathbb{R}$. Since for any $\varepsilon > 0$
$$\mu(\omega) = D^r \left\{ \exp(\varepsilon|\omega|) \left[ (1 + |\omega|)^j \exp(-\varepsilon|\omega|) f(\omega) \right] \right\}$$
it follows that $\mu \in \mathcal{K}'$. Substituting into (4.20) gives
$$\hat{f}_+(z) = +\int_{-\infty}^{0} (1 - \omega)^j f(\omega) \exp(-i\omega z) dz \quad , \quad \Im(z) > 0$$
$$\hat{f}_-(z) = -\int_{0}^{+\infty} (1 + \omega)^j f(\omega) \exp(-i\omega z) dz \quad , \quad \Im(z) < 0$$
and therefore there exists $C > 0$ such that
$$|\hat{f}_\pm(z)| \leq C/|\Im(z)|^{j+1}$$
for all $z \in \mathbb{C}_0^{\circ} \equiv \mathbb{C} \backslash \mathbb{R}$. Hence
$$\hat{\mu}^{\circ}(z) = \begin{cases} (iz)^r \hat{f}_+(z) & \text{for } \Im(z) > 0 \\ (iz)^r \hat{f}_-(z) & \text{for } \Im(z) < 0 \end{cases}$$

is an analytic representation of $\hat{\mu} \in \mathcal{S}'$. Any other analytic representation of the tempered distribution $\hat{\mu}$ differs from $\hat{\mu}^\circ$ by a polynomial. The characterization of the Cauchy-Stieltjes transform of tempered distributions has been made in J. S. Silva [104] with the following theorem

**Theorem 4.19** *A function $\hat{\mu}^\circ(z)$ defined and analytic on $\mathbb{C}\backslash\mathbb{R}$ belongs to the Cauchy-Stieltjes transform of a tempered distribution in $\mathcal{S}'$ if and only if there exist $C > 0$ and $m, n \in \mathbb{N}$ such that*

$$|\hat{\mu}^\circ(z)| \leq C(1+|z|)^m/|\Im(z)|^n$$

*for all $z \in \mathbb{C}\backslash\mathbb{R}$. Moreover the limits*

$$\hat{\mu}_+(x) = \lim_{\varepsilon \downarrow 0} \hat{\mu}^\circ(x+i\varepsilon) \quad , \quad \hat{\mu}_-(x) = \lim_{\varepsilon \downarrow 0} \hat{\mu}^\circ(x-i\varepsilon)$$

*exist in the sense of $\mathcal{S}'$ and $\hat{\mu}(x) = \hat{\mu}_+(x) - \hat{\mu}_-(x)$.*

#### 4.4.1.3 Ultradistributions of compact support

Recall from theorem 4.17 that for any Silva tempered ultradistribution $\hat{\mu} \in \mathcal{U}'$ we have

$$<\hat{\mu}, \varphi> = -\oint_{\Gamma_{b'}} \hat{\mu}^\circ(z)\varphi(z)\,dz \quad , \quad \forall \varphi \in \mathcal{U}$$

where $\Gamma_{b'}$ is the counterclockwise oriented contour $|\Im(z)| = b' > b \geq 0$ and $\hat{\mu}^\circ(z)$ is any Cauchy representation of $\hat{\mu}$. Suppose that $\hat{\mu}^\circ(z)$ is analytically continuable to the strip $\{z \in \mathbb{C} : c < \Re(z) < d\}$ and then for $c < c' < d' < d$ consider the decomposition of the contour $\Gamma_{b'}$ as shown in the figure.

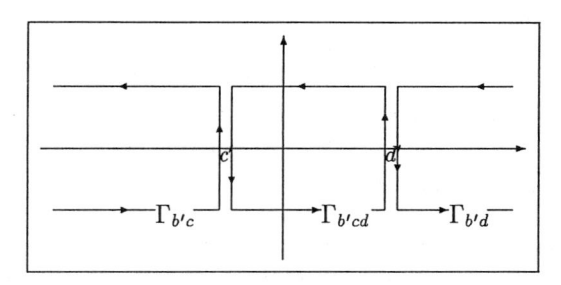

For any $\varphi \in \mathcal{U}$ we have

$$\oint_{\Gamma_{b'}} \hat{\mu}^\circ(z)\varphi(z)dz = \left\{\oint_{\Gamma_{b'c}} + \oint_{\Gamma_{b'cd}} + \oint_{\Gamma_{b'd}}\right\} \hat{\mu}^\circ(z)\varphi(z)dz$$

and since $\hat{\mu}^\circ(z)\varphi(z)$ is analytic within and on $\Gamma_{b'cd}$ we get

$$\oint_{\Gamma_{b'}} \hat{\mu}^\circ(z)\varphi(z)dz = \left\{\oint_{\Gamma_{b'c}} + \oint_{\Gamma_{b'd}}\right\} \hat{\mu}^\circ(z)\varphi(z)dz,$$

which means that the value of $< \hat{\mu}, \varphi >$ is independent of the values of $\hat{\mu}^\circ(z)$ in open vertical strips in which it is analytically continuable. Then we have

**Definition 4.7** *A Silva tempered ultradistribution $\hat{\mu} \in \mathcal{U}'$ is said to be* **null** *on the open interval $(c, d)$, $(c, d \in \bar{\mathbb{R}})$, if every function in the Cauchy-Stieltjes transform of $\hat{\mu}$ is analytically continuable into the strip $\{z \in \mathbb{C} : c < \Re(z) < d\}$ (where we may have $c = -\infty$ and/or $d = +\infty$).*

**Definition 4.8** *Two Silva tempered ultradistributions $\hat{\mu}, \hat{\sigma} \in \mathcal{U}'$ are said to be equal over the open interval $(c, d)$, $(c, d \in \bar{\mathbb{R}})$ if and only if $\hat{\mu} - \hat{\sigma}$ is the null ultradistribution over that interval.*

**Definition 4.9** *The complement (with respect to $\mathbb{R}$) of the union of all subsets of $\mathbb{R}$ where $\hat{\mu} \in \mathcal{U}'$ is null is called the* **real support** *of $\hat{\mu}$.*

A Silva tempered ultradistribution with compact real support is said to be **compactly concentrated**. In this case there exists a compact $K$ of the complex-plane such that

$$< \hat{\mu}, \varphi > = - \oint_\Gamma \hat{\mu}^\circ(z)\varphi(z)dz \qquad (4.26)$$

for any $\varphi \in \mathcal{U}$, where $\hat{\mu}^\circ(z)$ is a function analytic outside $K \sqsubset \mathbb{C}$ and $\Gamma$ is any simple closed contour, counterclockwise oriented, which encloses $K$ in its interior.

Denote by $\mathcal{U}'_0$ the subspace of all Silva tempered ultradistributions which are compactly concentrated. Then we have

**Theorem 4.20** *The space $\mathcal{U}'_0$ of all Silva tempered ultradistributions of compact real support may be identified with the dual $\mathcal{H}'$ of the space $\mathcal{H}(\mathbb{C})$ of all entire functions, with the topology generated by the uniform convergence on compacts of $\mathbb{C}$.*

**Proof:** (1) Let $\hat{\mu} \in \mathcal{U}'_0$ be compactly concentrated on a compact $K \sqsubset \mathbb{C}$. It follows from (4.26) that $\hat{\mu}$ may be extended to all functions in $\mathcal{H}(\mathbb{C})$ so that we have $\mathcal{U}'_0 \subset \mathcal{H}'$.

(2) Now let $L$ be a continuous linear functional on $\mathcal{H}(\mathbb{C})$. For any $\varphi \in \mathcal{H}(\mathbb{C})$ there is a sequence $(\sum_{j=0}^n a_j z^j)_{n \in \mathbb{N}_0}$ which converges in $\mathcal{H}(\mathbb{C})$ to $\varphi$ and which is therefore such that

$$\lim_{j \to +\infty} \sup \sqrt[j]{|a_j|} = 0 \, .$$

Then the sequence $(L[\sum_{j=0}^{n} a_j z^j])_{n \in \mathbb{N}_0}$ converges in $\mathbb{C}$ to $L[\varphi] \in \mathbb{C}$. For $j = 0, 1, 2, \ldots$, define $\lambda_j \equiv L[z^j]$; then the sequence $(\sum_{j=0}^{n} a_j \lambda_j)_{n \in \mathbb{N}_0}$ converges to the (finite) complex number $L[\varphi]$. Since we must have $\limsup_{j \to +\infty} \sqrt[j]{|a_j \lambda_j|} < 1$ there must exist $C > 0$ and $\tau \geq 0$, (depending on $L$), such that

$$|\lambda_j| \leq C \tau^j , \quad j = 0, 1, 2, \ldots .$$

The function $\hat{\mu}^\circ$ defined by

$$\hat{\mu}^\circ(z) = -\frac{1}{2\pi i} \sum_{j=0}^{\infty} \frac{\lambda_j}{z^{j+1}}$$

is analytic outside the disc $\bar{D}_\tau = \{z \in \mathbb{C} : |z| \leq \tau\}$ and therefore defines an ultradistribution $\hat{\mu} \in \mathcal{U}_0'$ through

$$<\hat{\mu}, \varphi> = -\oint_\Gamma \hat{\mu}^\circ(z) \varphi(z) dz$$

where $\Gamma$ is a simple closed contour enclosing $\bar{D}_\tau \sqsubset \mathbb{C}$.
It remains to show that $L[\varphi] = <\hat{\mu}, \varphi>$. But, since polynomials are dense in $\mathcal{H}(\mathbb{C})$, this follows from the fact that

$$\begin{aligned}<\hat{\mu}, z^n> &= -\oint_\Gamma \hat{\mu}^\circ(z) z^n dz \\ &= \sum_{j=0}^{\infty} \lambda_j \frac{1}{2\pi i} \oint_\Gamma \frac{z^n}{z^{j+1}} dz = \lambda_n = L[z^n].\end{aligned}$$

Therefore we also have $\mathcal{H}' \subset \mathcal{U}_0'$. □

Since the function $\mathbf{1}$ belongs to $\mathcal{H}(\mathbb{C})$ we can now define the integral of an ultradistribution of compact support $\hat{\mu} \in \mathcal{H}'$ by setting

$$\int_{-\infty}^{+\infty} \hat{\mu}(t) dt \equiv <\hat{\mu}, 1> = -\oint_\Gamma \hat{\mu}^\circ(z) dz \qquad (4.27)$$

where $\hat{\mu}^\circ$ is any analytic representation of $\hat{\mu}$ compactly concentrated on $\bar{D}_\tau \sqsubset \mathbb{C}$ and $\Gamma$ is any simple closed counterclockwise oriented contour enclosing $\bar{D}_\tau$ in its interior. In particular we have for example

$$\int_{-\infty}^{+\infty} \delta(t) dt = -\frac{1}{2\pi i} \oint_\Gamma \frac{dz}{z} = 1$$

as expected.

Let $\hat{\mu}$ be any Silva tempered ultradistribution in $\mathcal{U}'$. Then $\hat{\mu}$ is the Fourier transform of a distribution $\mu$ of exponential type in $\mathcal{K}'$ which is such that

$$<\mu, \hat{\varphi}> = (-1)^r \int_{-\infty}^{+\infty} F(\omega) \hat{\varphi}^{(r)}(\omega) d\omega , \quad \forall \hat{\varphi} \in \mathcal{K},$$

where $F$ is a continuous function of exponential type $b \geq 0$. Since we also have

$$\begin{aligned}
<\mu, \hat{\varphi}> \; &= \; <\hat{\mu}, \varphi> \; = \; -\oint_{\Gamma_{b'}} \hat{\mu}^\circ(z)\varphi(z)dz \\
&= \; -\oint_{\Gamma_{b'}} \hat{\mu}^\circ(z) \left\{ \frac{1}{2\pi} \int_{-\infty}^{+\infty} \hat{\varphi}(\omega)\exp(i\omega z)d\omega \right\} dz \\
&= \; -\oint_{\Gamma_{b'}} \frac{\hat{\mu}^\circ(z)}{(iz)^r} \left\{ \frac{(-1)^r}{2\pi} \int_{-\infty}^{+\infty} \hat{\varphi}^{(r)}(\omega)\exp(i\omega z)d\omega \right\} dz \\
&= \; (-1)^r \int_{-\infty}^{+\infty} \left\{ -\frac{1}{2\pi} \oint_{\Gamma_{b'}} \frac{\hat{\mu}^\circ(z)}{(iz)^r} \exp(i\omega z)dz \right\} \hat{\varphi}^{(r)}(\omega)d\omega
\end{aligned}$$

then we get

$$F(\omega) \; = \; -\frac{1}{2\pi} \oint_{\Gamma_{b'}} \frac{\hat{\mu}^\circ(z)}{(iz)^r} \exp(i\omega z)dz \;, \; \omega \in \mathbb{R}$$

which is effectively a formula giving the inverse Fourier transform in $\mathcal{U}'$.
Now let $\hat{\mu}$ be an ultradistribution in $\mathcal{H}' \subset \mathcal{U}'$ with support contained in the compact

$$\{z \in \mathbb{C} : |z| \leq \tau\} \subset \mathbb{C}.$$

Then the inverse Fourier transform of $\hat{\mu}$ is a finite-order derivative of the function $F(\omega)$ defined by

$$F(\omega) \; = \; -\frac{1}{2\pi} \int_{\Gamma_\tau} \frac{\hat{\mu}^\circ(z)}{(iz)^r} \exp(i\omega z)dz \;, \; \omega \in \mathbb{R}$$

where $\Gamma_\tau$ is any circle, centre the origin and radius $\tau+\varepsilon$ for some $\varepsilon > 0$; hence $F$ may extend into the complex $\lambda$-plane as an entire function. Consequently $\mu = \mathcal{F}^{-1}[\hat{\mu}]$ is a regular distribution of exponential type generated by the function

$$f(\lambda) \; = \; D^r F(\lambda) \; = \; -\frac{1}{2\pi} \oint_{\Gamma_\tau} \hat{\mu}^\circ(z)\exp(i\lambda z)dz \;, \; \lambda \in \mathbb{C}.$$

Since for any $\varepsilon > 0$ we have

$$|f(\lambda)| \; \leq \; \left\{ \sup_{z \in \Gamma_\tau} |\exp(i\lambda z)| \right\} \frac{1}{2\pi} \oint_{\Gamma_\tau} |\hat{\mu}^\circ(z)||dz| \; = \; A_\varepsilon \exp\left((\tau+\varepsilon)|\lambda|\right)$$

it follows that $f$ is an entire function of exponential type $\leq \tau$.
The converse of this theorem also holds and we have

**Theorem 4.21 (Paley-Wiener-Gel'fand)** *The Fourier transform establishes a topological and algebraic isomorphism between the space $\mathcal{H}'$, of all compactly concentrated ultradistributions, and the space $\mathcal{H}_e$ of all entire functions of exponential type.*

**Proof:** It has already been shown that $\mathcal{F}^{-1}\{\mathcal{H}'\} \subseteq \mathcal{H}_e$. It remains to show that $\mathcal{H}_e \subseteq \mathcal{F}^{-1}\{\mathcal{H}'\}$ also holds.

Suppose that $f \in \mathcal{H}_e$. Then there exists a constant $\tau \geq 0$ such that for every $\varepsilon > 0$ there exists $A_\varepsilon > 0$ so that

$$|f(\lambda)| \leq A_\varepsilon \exp\left((\tau + \varepsilon)|\lambda|\right), \quad \lambda \in \mathbb{C}.$$

The restriction of $f$ to $\mathbb{R}$ is a regular distribution in $\mathcal{K}'$ and for each $\hat{\varphi} \in \mathcal{K}$ we have

$$<f, \hat{\varphi}> \;=\; <\sum_{n=0}^{\infty} a_n \omega^n, \hat{\varphi}> \;=\; \sum_{n=0}^{\infty} a_n <\omega^n, \hat{\varphi}(\omega)>$$

$$= \sum_{n=0}^{\infty} \frac{2\pi}{i^n} a_n \varphi^{(n)}(0) \;=\; \sum_{n=0}^{\infty} \frac{2\pi}{i^n} a_n \frac{n!}{2\pi i} \oint_\Gamma \frac{\varphi(z)}{z^{n+1}} dz,$$

where $\Gamma$ is any circle centred at the origin, and $(a_n)_{n \in \mathbb{N}_0}$ is the sequence of the Taylor coefficients of the function $f \in \mathcal{H}_e$. Then, provided $\Gamma$ has an appropriate radius, we may write

$$<f, \hat{\varphi}> \;=\; -\oint_\Gamma \left[-\frac{1}{2\pi i} \sum_{n=0}^{\infty} 2\pi i^{-n} a_n z^{-n-1}\right] \varphi(z) dz$$

$$= -\oint_\Gamma \left[-\frac{1}{2\pi i} \sum_{n=0}^{\infty} \lambda_n z^{-n-1}\right] \varphi(z) dz \;=\; -\oint_\Gamma \hat{\mu}^\circ(z) \varphi(z) dz,$$

where

$$\hat{\mu}^\circ(z) \;=\; -\frac{1}{2\pi i} \sum_{n=0}^{\infty} \lambda_n z^{-n-1}$$

and $\lambda_n = 2\pi i^{-n} n! a_n$, $n = 0, 1, 2, \ldots$.

We now claim that $\hat{\mu}^\circ(z)$ represents an ultradistribution compactly concentrated on a circle centred at the origin. In fact, for any $n = 0, 1, 2, \ldots$

$$a_n \;=\; \frac{1}{2\pi i} \oint_C \frac{f(z)}{z^{n+1}} dz,$$

where $C$ is any circular contour centred at the origin of any radius $R > 0$. Thus

$$|a_n| \;\leq\; \frac{1}{2\pi} \oint_C \frac{|f(z)|}{|z|^{n+1}} |dz|$$

$$\leq\; \frac{A_\varepsilon}{2\pi} \frac{\exp((\tau + \varepsilon)R)}{R^{n+1}} 2\pi R \;=\; \frac{A_\varepsilon}{R^n} \exp((\tau + \varepsilon)R).$$

This bound holds for any $R > 0$ and, therefore, for $n = 1, 2, \ldots$, taking $R_\varepsilon$ so that $(\tau + \varepsilon)R_\varepsilon = n$, we obtain

$$|a_n| \;\leq\; A_\varepsilon \frac{(\tau + \varepsilon)^n}{\exp(-n) n^n}.$$

Using the Stirling formula $n! = O\left(\sqrt{2\pi n}\, n^n \exp(-n)\right)$ as $n \to \infty$, after some easy manipulations, we get
$$\sqrt[n]{n!|a_n|} \leq \sqrt[n]{A'_\varepsilon}\,(\tau + \varepsilon),$$
where $A'_\varepsilon$, for each fixed $\varepsilon > 0$, is a positive constant.

Hence, for every $\varepsilon > 0$, there exists a constant $C_\varepsilon > 0$ such that
$$|\lambda_n| \leq C_\varepsilon (\tau + \varepsilon)^n, \quad \forall n \in \mathbb{N}_0.$$

Consequently the function $\hat{\mu}^\circ(z)$, as given above, is analytic on
$$\{z \in \mathbb{C} : |z| > \tau + \varepsilon\},$$
for any $\varepsilon > 0$ and thus represents an ultradistribution $\hat{\mu} \in \mathcal{H}'$ concentrated on the compact $\{z \in \mathbb{C} : |z| < \tau\}$. □

### 4.4.2 Ultradistributions of exponential type

Let $\hat{\sigma}$ be an ultradistribution of exponential type. Then, from (3.74), it follows that there exists $j_0 \in \mathbb{N}$ such that $\hat{\sigma}$ belongs to $\mathcal{Z}'_{\exp,j}$ for all $j \geq j_0$. This means that $\hat{\sigma}$ is concentrated on a strip $\bar{\Lambda}_b$ for some $b \in [0, j_0)$ and may extend to functions analytic on $\Lambda_j$ decreasing there more rapidly than $\exp(-j|\Re(z)|)$. By appealing to Cauchy's Residue Theorem we have, for any $\varphi \in \mathcal{Z}_{\exp}$ and any $\zeta \in \mathbb{C}$ such that $|\Im(\zeta)| \leq b$,

$$\varphi(\zeta) = \frac{\cosh^{-\nu}(a\zeta)}{2\pi i} \oint_{\Gamma_{b'}} \frac{\varphi(z)\cosh^\nu(az)}{z - \zeta}\, dz, \quad |\Im(\zeta)| < b', \qquad (4.28)$$

where

(a) $\Gamma_{b'}$ is a counterclockwise oriented contour of the form $|\Im(z)| = b'$, and

(b) $a \geq 0$ and $\nu \in \mathbb{N}_0$ are two numbers chosen so that an inequality of the form
$$\cosh^\nu(az) \geq C \cdot \exp(j|\Re(z)|), \quad \forall z \in \bar{\Lambda}_b$$
holds for some constant $C > 0$.

Then, taking $b' > b$, $2ab > \pi$ and $\nu a \geq j$, since $<\hat{\sigma}, \varphi>$ depends only on the values that the test-function $\varphi$ assumes on $\bar{\Lambda}_b$, we have

$$<\hat{\sigma}, \varphi> = <\hat{\sigma}(\zeta), \frac{\cosh^{-\nu}(a\zeta)}{2\pi i} \oint_{\Gamma_{b'}} \frac{\varphi(z)\cosh^\nu(az)}{z - \zeta}\, dz>$$
$$= -\oint_{\Gamma_{b'}} \hat{\sigma}^\circ(z)\varphi(z)\, dz \qquad (4.29)$$

where
$$\hat{\sigma}^{\circ}(z) = \frac{\cosh^{\nu}(az)}{2\pi i} < \hat{\sigma}(\zeta), 1/[(\zeta - z)\cosh^{\nu}(a\zeta)] > \qquad (4.30)$$
is a function analytic on $\Lambda_b^{\circ} \equiv \mathbf{C}\setminus\bar{\Lambda}_b$. The function $\hat{\sigma}^{\circ}(z), z \in \Lambda_b^{\circ}$ is an **analytic representation** of the ultradistribution of exponential type $\hat{\sigma} \in \mathcal{Z}'_{\exp}$. Such representation, however, is not unique. Adding to $\hat{\sigma}^{\circ}$ an entire function which does not grow more than an exponential of first-order of growth and finite type on horizontals does not modify the value of the integral on the right-hand side of (4.29).

For each $b \geq 0$ let $\mathcal{H}_{e,p}(\mathbf{C}_b^{\circ})$ be the space of all functions defined and analytic on $\mathbf{C}_b^{\circ}$ which are bounded by an exponential of first order of growth and finite type on horizontals and by a polynomial on verticals of $\bar{\mathbf{C}}_{b'}^{\circ}$, with $b' > b$. Then define
$$\mathcal{H}_{e,p}^{\circ} \equiv \bigcup_{b \geq 0} \mathcal{H}_{e,p}(\mathbf{C}_b^{\circ}) \subset \mathcal{H}^{\circ}.$$

The space $\mathcal{H}_{e,p}(\mathbf{C}) \subset \mathcal{H}(\mathbf{C})$ of all entire functions of exponential type on horizontals and polynomial growth on verticals is a subspace of $\mathcal{H}_{e,p}^{\circ}$. Hence we can finally define the quotient space
$$\mathcal{H}_{S,\exp} \equiv \mathcal{H}_{e,p}^{\circ}/\mathcal{H}_{e,p}(\mathbf{C})$$
and then define the Cauchy-Stieltjes transform $\mathbf{S}[\hat{\mu}]$ of $\hat{\mu} \in \mathcal{Z}'_{\exp}$ to be the class in $\mathcal{H}_{S,\exp}$ determined by any Cauchy representation $\mu^{\circ} \in \mathcal{H}_{e,p}^{\circ}$ of $\hat{\mu}$. The space $\mathbf{S}\{\mathcal{U}'\}$ of Cauchy-Stieltjes transforms of all Silva tempered ultradistributions may be embedded into $\mathcal{H}_{S,\exp}$ associating each class $[\hat{\mu}^{\circ}]$ in $\mathbf{S}\{\mathcal{U}'\}$ with the class in $\mathcal{H}_{S,\exp}$ represented by the same function $\hat{\mu}^{\circ}$.

## 4.5 THE FOURIER-CARLEMAN TRANSFORM

Let $\mathbf{S}[\sigma] = [\sigma^{\circ}], \mathbf{S}[\hat{\sigma}] = [\hat{\sigma}^{\circ}]$ be the respective Cauchy-Stieltjes transforms of the ultradistributions $\sigma, \hat{\sigma} \in \mathcal{Z}'_{\exp}$. The mapping
$$\mathcal{FC} : \mathcal{H}_{S,\exp} \to \mathcal{H}_{S,\exp} \qquad (4.31)$$
which maps the Cauchy-Stieltjes transform $\mathbf{S}[\sigma] \in \mathcal{H}_{S,\exp}$ of the ultradistribution $\sigma$ in $\mathcal{Z}'_{\exp}$ into the Cauchy-Stieltjes transform $\mathbf{S}[\hat{\sigma}] \in \mathcal{H}_{S,\exp}$ of its Fourier transform $\hat{\sigma} \in \mathcal{Z}'_{\exp}$ is called the Fourier-Carleman transform. The inverse Fourier-Carleman transform
$$\mathcal{FC}^{-1} : \mathcal{H}_{S,\exp} \to \mathcal{H}_{S,\exp} \qquad (4.32)$$
is defined similarly.

Let $\hat{\sigma}^\circ \in \mathcal{H}_{e,p}^\circ$ be a function belonging to the Cauchy-Stieltjes transform of an ultradistribution of exponential type $\hat{\sigma} \in \mathcal{Z}'_{\exp}$. Then there exist $b \geq 0$ and $a > 0$ such that $\hat{\sigma}$ is concentrated in the strip $\bar{\Lambda}_b = \{z \in \mathbb{C} : |\Im(z)| \leq b\}$ and $\hat{\sigma}^\circ$ is bounded by $\exp(a|\Re(z)|)$ on every horizontal line of $\mathbb{C}_b^\circ = \mathbb{C} \setminus \bar{\Lambda}_b$. We have that

$$<\hat{\sigma}, \varphi> = -\oint_{\Gamma_{b'}} \hat{\sigma}^\circ(z) \varphi(z) dz$$

for any $\varphi \in \mathcal{Z}_{\exp}$ and any counterclockwise oriented boundary $\Gamma_{b'}$ of $\Lambda_{b'}$ for $b' > b$. But, according to lemma 3.6.2, we get

$$\varphi(z) = \frac{1}{2\pi} \int_{-\infty+i\eta}^{+\infty+i\eta} \hat{\varphi}(\zeta) \exp(i\zeta z) d\zeta \ , \ z \in \mathbb{C}$$

for any $\eta \in \mathbb{R}$, and therefore

$$<\hat{\sigma}, \varphi> = -\int_{\Gamma_{b',l}} \hat{\sigma}^\circ(z) \left\{ \frac{1}{2\pi} \int_{-\infty+i\eta}^{+\infty+i\eta} \hat{\varphi}(\zeta) \exp(i\zeta z) d\zeta \right\} dz$$

$$- \int_{\Gamma_{b',r}} \hat{\sigma}^\circ(z) \left\{ \frac{1}{2\pi} \int_{-\infty+i\eta}^{+\infty+i\eta} \hat{\varphi}(\zeta) \exp(i\zeta z) d\zeta \right\} dz = I_1 + I_2$$

where $\Gamma_{b',l} = \Gamma_{b'} \cap \{z \in \mathbb{C} : \Re(z) < 0\}$ and $\Gamma_{b',r} = \Gamma_{b'} \cap \{z \in \mathbb{C} : \Re(z) > 0\}$. Taking into account the growth properties of the functions $\hat{\sigma}^\circ(\zeta)$ and $\hat{\varphi}(\zeta)$ then, by Fubini's theorem, we can interchange the order of integrations in $I_1$ provided that $\eta = \Im(\zeta) < -a$, and in $I_2$ provided that $\eta = \Im(\zeta) > a$. Hence, for any $a' > a$, we obtain

$$<\hat{\sigma}, \varphi> = \int_{-\infty-ia'}^{+\infty-ia'} \hat{\varphi}(\zeta) \left\{ \frac{1}{2\pi} \int_{\Gamma_{b',l}} \hat{\sigma}^\circ(z) \exp(i\zeta z) dz \right\} d\zeta$$

$$- \int_{-\infty+ia'}^{+\infty+ia'} \hat{\varphi}(\zeta) \left\{ \frac{-1}{2\pi} \int_{\Gamma_{b',r}} \hat{\sigma}^\circ(z) \exp(i\zeta z) dz \right\} d\zeta.$$

Defining

$$\sigma_+^\circ(\zeta) = \frac{-1}{2\pi} \int_{\Gamma_{b',r}} \hat{\sigma}^\circ(z) \exp(i\zeta z) dz \tag{4.33}$$

$$\sigma_-^\circ(\zeta) = \frac{1}{2\pi} \int_{\Gamma_{b',l}} \hat{\sigma}^\circ(z) \exp(i\zeta z) dz \tag{4.34}$$

and

$$\sigma^\circ(\zeta) = \begin{cases} \sigma_-^\circ(\zeta) & \text{for } \Im(\zeta) < -a \\ \sigma_+^\circ(\zeta) & \text{for } \Im(\zeta) > +a \end{cases} \tag{4.35}$$

we arrive at

$$<\hat{\sigma}, \varphi> = -\oint_{\Gamma_{b'}} \hat{\sigma}^\circ(z) \varphi(z) dz$$

$$= -\oint_{\Gamma_{a'}} \sigma^\circ(\zeta) \hat{\varphi}^\circ(\zeta) d\zeta = <\sigma, \hat{\varphi}>. \tag{4.36}$$

From (4.33) and (4.34) it is easy to see that there exists a constant $C > 0$ such that
$$|\sigma^\circ(\zeta)| \leq C \cdot \exp\left(b'|\Re(\zeta)|\right) \qquad (4.37)$$
for all $\zeta \in \mathbf{C}_a^\circ$ and all $b' > b$. Thus $\hat{\sigma}^\circ(\zeta), \zeta \in \mathbf{C}_a^\circ$, is a function in $\mathcal{H}_{e,p}^\circ$ which belongs to the Cauchy-Stieltjes transform of the ultradistribution $\sigma \in \mathcal{Z}'_{\exp}$.

Suppose now that we had considered a different representative $\hat{\sigma}_1^\circ \in \mathcal{H}_{e,p}^\circ$ of the ultradistribution $\hat{\sigma} \in \mathcal{Z}'_{\exp}$. Then there exists an entire function $\hat{\theta}$ in $\mathcal{H}_{e,p}(\mathbf{C})$ such that
$$\hat{\sigma}_1^\circ = \hat{\sigma}^\circ + \hat{\theta}.$$
The function $\hat{\sigma}_1^\circ$ after being transformed according to (4.33) and (4.34) gives rise to a new function $\sigma_1^\circ(\zeta), \zeta \in \mathbf{C}_{a_1}^\circ$, where $\sigma_1^\circ(\zeta) = \sigma_{1+}^\circ(\zeta)$ for $\Im(\zeta) > a_1$, and $\sigma_1^\circ(\zeta) = \sigma_{1-}^\circ(\zeta)$ for $\Im(\zeta) < -a_1$ for some $a_1 \geq 0$. Then we obtain
$$\sigma_{1+}^\circ(\zeta) - \sigma_+^\circ(\zeta) = -\frac{1}{2\pi} \int_{\Gamma_{b',r}} \hat{\theta}(z) \exp(i\zeta z) dz$$
where the integral converges for $\Im(\zeta) > \max\{a_1, a\}$.

Now, since $\hat{\theta}(z) \exp(i\zeta z)$ is an entire function of $z \in \mathbf{C}$ for each fixed $\zeta \in \mathbf{C}$, we have
$$\oint_{\Gamma_\alpha^+} \hat{\theta}(z) \exp(i\zeta z) \, dz = 0$$
where $\Gamma_\alpha^+$ is a contour of the form shown below

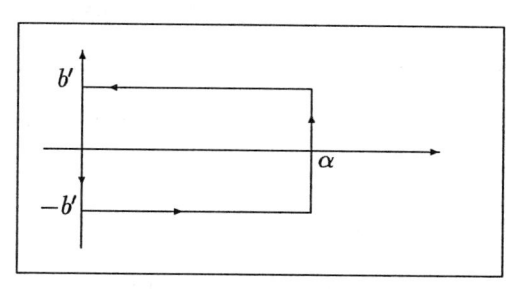

Hence
$$\begin{aligned}
\sigma_{1+}^\circ(\zeta) - \sigma_+^\circ(\zeta) &= -\frac{1}{2\pi} \int_{\Gamma_{b',r}} \hat{\theta}(z) \exp(i\zeta z) dz \\
&= -\frac{i}{2\pi} \int_{-b'}^{+b'} \hat{\theta}(iy) \exp(-y\zeta) dy \\
&\quad + \lim_{\alpha \to +\infty} \int_{\alpha - ib'}^{\alpha + ib'} \hat{\theta}(\alpha + iy) \exp\left(i(\alpha + iy)\zeta\right) dy \\
&= -\frac{i}{2\pi} \int_{-b'}^{+b'} \hat{\theta}(iy) \exp(-y\zeta) dy
\end{aligned}$$

since the limit term vanishes for $\Im(\zeta) > \max\{a, a_1\}$. Similarly we can show that

$$\sigma_{1-}^\circ(\zeta) - \sigma_-^\circ(\zeta) = \frac{1}{2\pi}\int_{\Gamma_{b',l}} \hat{\theta}(z)\exp(i\zeta z)dz = -\frac{i}{2\pi}\int_{-b'}^{+b'} \hat{\theta}(iy)\exp(-y\zeta)dy$$

provided that $\Im(\zeta) < -\max\{a, a_1\}$. Consequently, for $|\Im(\zeta)| > \max\{a, a_1\}$

$$\sigma_1^\circ(\zeta) - \sigma^\circ(\zeta) = \frac{i}{2\pi}\int_{b'}^{+b'} \hat{\theta}(iy)\exp(-y\zeta)dy \tag{4.38}$$

and since $\hat{\theta} \in \mathcal{H}_{e,p}(\mathbf{C})$ is certainly continuous on the segment $[-ib', +ib']$ the above equality shows that $\sigma_1^\circ(\zeta) - \sigma^\circ(\zeta)$ may extend into the complex $\zeta$-plane as an entire function in $\mathcal{H}_{e,p}(\mathbf{C})$. On the other hand this means that if the functions $\hat{\sigma}^\circ, \hat{\sigma}_1^\circ \in \mathcal{H}_{e,p}^\circ$ belong to the same class in $\mathcal{H}_{S,\exp}$ then the corresponding transforms $\sigma^\circ, \sigma_1^\circ \in \mathcal{H}_{e,p}^\circ$ obtained through (4.33) and (4.34) also belong to the same class in $\mathcal{H}_{S,\exp}$. In a similar way we can get $[\hat{\sigma}^\circ] = \mathcal{FC}\left[[\sigma^\circ]\right]$ where

$$\hat{\sigma}_+^\circ(z) = -\int_{\Gamma_{a',l}} \sigma^\circ(\zeta)\exp(-i\zeta z)d\zeta \quad , \quad \Im(z) > +b \tag{4.39}$$

$$\hat{\sigma}_-^\circ(z) = +\int_{\Gamma_{a',r}} \sigma^\circ(\zeta)\exp(-i\zeta z)d\zeta \quad , \quad \Im(z) < -b \tag{4.40}$$

and

$$\hat{\sigma}^\circ(z) = \begin{cases} \hat{\sigma}_-^\circ(z) & \text{for } \Im(z) < -b \\ \hat{\sigma}_+^\circ(z) & \text{for } \Im(z) > +b \end{cases} \tag{4.41}$$

Summarising, we can state the result:

**Theorem 4.22** *The inverse Fourier-Carleman transform of the class $[\hat{\sigma}^\circ] \in \mathcal{H}_{S,\exp}$ is the class $[\sigma^\circ] \equiv \mathcal{FC}^{-1}[\hat{\sigma}^\circ] \in \mathcal{H}_{S,\exp}$ as defined by (4.33) and (4.34). Similarly (4.39) and (4.40) define the direct Fourier-Carleman transform. Moreover $\mathcal{FC} \circ \mathcal{FC}^{-1} = \mathcal{FC}^{-1} \circ \mathcal{FC} = id$.*

**Proof:** It is only necessary to prove the the second part. From (4.39) we have

$$\left(\mathcal{FC} \circ \mathcal{FC}^{-1}[\hat{\sigma}^\circ]\right)_+(z) = -\int_{\Gamma_{a',l}} \mathcal{FC}^{-1}[\hat{\sigma}^\circ](\zeta)\exp(-i\zeta z)d\zeta$$

and, using (4.33) and (4.34), we can write this (after some manipulations) in the form

$$-\int_{-\infty}^0 \left\{\frac{1}{2\pi}\int_{\Gamma_{b',l}} \hat{\sigma}^\circ(\lambda)e^{i(\xi-ia')\lambda}d\lambda\right\} e^{-i(\xi-ia')z}d\xi$$

$$-\int_{-\infty}^0 \left\{-\frac{1}{2\pi}\int_{\Gamma_{b',r}} \hat{\sigma}^\circ(\lambda)e^{i(\xi+ia')\lambda}d\lambda\right\} e^{-i(\xi+ia')z}d\xi$$

and since
$$|\hat{\sigma}^\circ(\alpha \pm ib')\exp(i(\xi \pm ia')(\lambda - z))| = O\left(\exp[a|\alpha| \mp a'(\alpha - x) - \xi(b' - y)]\right)$$

then, for $a' > a$ and $y = \Im(z) > b' > b$, by Fubini's theorem, we may interchange the order of integrations to get

$$\begin{aligned}\left(\mathcal{FC} \circ \mathcal{FC}^{-1}[\hat{\sigma}^\circ]\right)_+(z) &= -\frac{1}{2\pi}\int_{\Gamma_{b',l}}\hat{\sigma}^\circ(\lambda)\int_{-\infty}^0 e^{i(\xi - ia')(\lambda - z)}d\xi d\lambda \\ &\quad -\frac{1}{2\pi}\int_{\Gamma_{b',l}}\hat{\sigma}^\circ(\lambda)\int_{-\infty}^0 e^{i(\xi - ia')(\lambda - z)}d\xi d\lambda \\ &= -\frac{1}{2\pi i}\int_{\Gamma_{b',l}}\frac{\hat{\sigma}^\circ(\lambda)\exp(a'(\lambda - z))}{\lambda - z}d\lambda \\ &\quad -\frac{1}{2\pi i}\int_{\Gamma_{b',l}}\frac{\hat{\sigma}^\circ(\lambda)\exp(-a'(\lambda - z))}{\lambda - z}d\lambda.\end{aligned}$$

After evaluating the above integrals by contour integration using contours $C_{R,l}, C_{R,r}$, respectively, as shown below

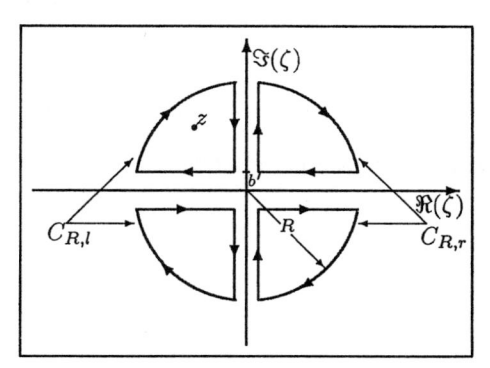

we obtain finally (as $R \to +\infty$)

$$\left(\mathcal{FC} \circ \mathcal{FC}^{-1}[\hat{\sigma}^\circ]\right)_+(z) = \hat{\sigma}^\circ(z), \quad \Im(z) > b.$$

Similarly we would get

$$\left(\mathcal{FC} \circ \mathcal{FC}^{-1}[\hat{\sigma}^\circ]\right)_-(z) = \hat{\sigma}^\circ(z), \quad \Im(z) < -b$$

and

$$\left(\mathcal{FC}^{-1} \circ \mathcal{FC}[\sigma^\circ]\right)_+(\zeta) = \sigma^\circ(\zeta), \quad \Im(\zeta) > +a$$

$$\left(\mathcal{FC}^{-1} \circ \mathcal{FC}[\sigma^\circ]\right)_-(\zeta) = \sigma^\circ(\zeta), \quad \Im(\zeta) < -a$$

which completes the proof. $\square$

The Fourier-Carleman Transform in $\mathcal{H}_{S,\text{exp}}$ enjoys the usual properties of the Fourier transform in $\mathcal{Z}_{\text{exp}}$. Namely, for any function $\theta \in \mathcal{H}_{e,p}(\mathbb{C})$, we have

$$\mathcal{FC}[\theta(-iD)\sigma^\circ](z) = \theta(z)\mathcal{FC}[\sigma](z),$$
$$\mathcal{FC}^{-1}[\theta(iD)\hat{\sigma}^\circ](z) = \theta(\zeta)\mathcal{FC}^{-1}[\sigma](\zeta)$$

and

$$\mathcal{FC}^{-1}[\theta(\cdot)\hat{\sigma}^\circ](\zeta) = \theta(-iD)\mathcal{FC}^{-1}[\sigma](\zeta),$$
$$\mathcal{FC}[\theta(\cdot)\sigma^\circ](z) = \theta(iD)\mathcal{FC}[\sigma](z).$$

Moreover, for any $\alpha \in \mathbb{C}$, we also obtain

$$\mathcal{FC}[\tau_\alpha \sigma^\circ](z) = \exp(-i\alpha z)\mathcal{FC}[\sigma^\circ](z),$$
$$\mathcal{FC}^{-1}[\tau_\alpha \hat{\sigma}^\circ](\zeta) = \exp(i\alpha\zeta)\mathcal{FC}^{-1}[\sigma](\zeta)$$

and

$$\mathcal{FC}[\exp(i\alpha \cdot)\sigma^\circ](z) = \tau_\alpha \circ \mathcal{FC}[\sigma^\circ](z),$$
$$\mathcal{FC}^{-1}[\exp(-i\alpha \cdot)\hat{\sigma}^\circ](\zeta) = \tau_\alpha \circ \mathcal{FC}^{-1}[\sigma](\zeta).$$

**Exercise 4.7 :**

1. Confirm the properties of the Fourier-Carleman transform stated above.

2. Find the Fourier-Carleman transform of $[-1/2\pi i z]$.

# 5

# Irregular Operations and Colombeau Generalised Functions

## 5.1 INTRINSIC PRODUCTS OF DISTRIBUTIONS

### 5.1.1 Intrinsic products and normal products

Some of the problems associated with the definition of a useful and comprehensive multiplicative product of distributions have been discussed briefly in §2.4.3. In that section we defined an intrinsic product to be one in which the product of two distributions, whenever it is defined at all, is always itself a distribution. That is to say, an intrinsic product is a mapping $M$ defined on some subset of $\mathcal{D}' \times \mathcal{D}'$ into $\mathcal{D}'$. The restriction of the domain of such a map to a (proper) subset of $\mathcal{D}' \times \mathcal{D}'$ seems to be unavoidable if important and useful properties of multiplication are to be preserved. Thus the Schwartz product was defined originally only for products of the form $\theta\mu$ where $\theta \in \mathcal{C}^\infty$ and $\mu \in \mathcal{D}'$:

$$<\theta\mu,\varphi> = <\mu,\theta\varphi> \ , \ \forall \varphi \in \mathcal{D}.$$

(Throughout all that follows we shall continue to denote the Schwartz product by simple juxtaposition of the factors. Other definitions of product will be distinguished, as necessary, by recourse to special, often temporary, notations.) We can weaken the infinite differentiability of the factor $\theta$ provided that the singularities of the other factor $\mu$ are appropriately restricted. Moreover even at the level of the basic Schwartz product itself it was found necessary to abandon the property of associativity.

In a fundamental study of the general multiplication problem Itano [38] listed those properties which would seem to be most desirable for an intrinsic product of distributions to satisfy: A product $M(\mu, \nu) \equiv \mu \circ \nu$ is said to be **normal** if it is well defined on a subset of $\mathcal{D}' \times \mathcal{D}'$ which contains $\mathcal{C}(\mathbb{R}) \times \mathcal{C}(\mathbb{R})$ and is such that,

**NP1** If $f, g \in \mathcal{C}(\mathbb{R})$ then
$$\mu_{(f)} \circ \mu_{(g)} = \mu_{(fg)}.$$

**NP2** The product is commutative, and distributive with respect to addition of distributions
$$\mu \circ \nu = \nu \circ \mu \; ; \; (\mu + \nu) \circ \sigma = \mu \circ \sigma + \nu \circ \sigma.$$

**NP3** If $\mu \circ \nu$ exists and if $\theta$ is a function in $\mathcal{C}^\infty(\mathbb{R})$ then the product $(\theta\mu) \circ \nu$ exists and,
$$(\theta\mu) \circ \nu = \theta(\mu \circ \nu).$$

**NP4** If $(D\mu) \circ \nu$ exists then so also do $\mu \circ \nu$ and $\mu \circ (D\nu)$, and we have
$$D(\mu \circ \nu) = (D\mu) \circ \nu + \mu \circ (D\nu).$$

**NP5** If $\Omega$ is any open subset of $\mathbb{R}$ and if $\mu|_\Omega$ denotes the restriction of $\mu$ to $\Omega$, then
$$(\mu|_\Omega) \circ (\nu|_\Omega) = (\mu \circ \nu)|_\Omega.$$

**NP6** If $\mathbb{R} = \bigcup_{j \in J} \Omega_j$, where the $\Omega_j$ are non-empty open subsets of $\mathbb{R}$, and if the product $(\mu|_{\Omega_j}) \circ (\nu|_{\Omega_j})$ exists for each $j \in J$, then the product $\mu \circ \nu$ exists.

**NP7** The product is invariant under diffeomorphisms of $\mathbb{R}$ into $\mathbb{R}$.

Clearly, any normal product will agree with the Schwartz product; that is we always have $\theta \circ \mu = \theta\mu$ for any $\theta \in \mathcal{C}^\infty(\mathbb{R})$ and any $\mu \in \mathcal{D}'$. Also it is easy to establish that the Leibniz formula is valid for any normal product:

$$\mu \circ L^p \nu = \sum_{q=0}^{p}(-1)^q \binom{p}{q} D^{p-q}\left[(D^q\mu) \circ \nu\right].$$

Associativity holds for a normal product only in the weak form given under **NP3**. Nevertheless this weak associativity has important implications. In the first place if $\mu \circ \nu$ is a well defined normal product then for any $\theta \in C^\infty(\mathbb{R})$ we have

$$\theta(\mu \circ \nu) = (\theta\mu) \circ \nu = \mu \circ (\theta\nu).$$

Hence if $\theta \in \mathcal{D}$ has support contained in the complement of $\text{supp}(\mu)$ or in the complement of $\text{supp}(\nu)$, then either $\theta\mu = 0$ or $\theta\nu = 0$, and therefore certainly $\theta(\mu \circ \nu) = 0$. It follows then that

$$\text{supp}(\mu \circ \nu) \subset \text{supp}(\mu) \cap \text{supp}(\nu).$$

Thus suppose for example that the normal product $\mu \circ \delta$ exists for some $\mu \in \mathcal{D}'$. Then $\mu \circ \delta$ must be concentrated on $\{0\}$, and so

$$\mu \circ \delta = a_0 \delta + a_1 \delta' + \ldots + a_n \delta^{(n)}$$

for some $n \in \mathbb{N}_0$, where the $a_k$ are numerical constants. Now from the established properties of the Schwartz product we have

$$x \circ \delta \equiv x\delta = 0 \quad \text{and} \quad x \circ \delta^{(j)} \equiv x\delta^{(j)} = -j\delta^{(j-1)}$$

for $j \geq 1$, and this implies that

$$0 = \mu \circ (x\delta) = x(\mu \circ \delta) = -\sum_{j=1}^{n} j a_j \delta^{(j-1)}.$$

It follows that we must have $a_1 = \ldots = a_n = 0$. Hence we obtain the following result

**Theorem 5.1** *If $\mu \in \mathcal{D}'$ and $\mu \circ \delta$ exists as a normal product then we must have*

$$\mu \circ \delta = c\delta \tag{5.1}$$

*where $c$ is a numerical constant.*

In particular if the expression $\delta \circ \delta \equiv \delta^2$ can be interpreted as a normal product then for some number $c$ we should have $\delta^2 = c\delta$. But, if $a$ is any positive number, then under the diffeomorphism $x \to ax$ we have $\delta(ax) = a^{-1}\delta(x)$, and it follows from the diffeomorphism condition **NP7** for a normal product that then

$$(a^{-1}\delta(x)) \circ (a^{-1}\delta(x)) = ca^{-1}\delta(x)$$

so that
$$\delta(x) \circ \delta(x) = ca\delta(x)$$
for any $a > 0$. This means that the only possible value for the constant $c$ is zero; if $\delta^2$ is well defined as a normal product then it can only be as the null distribution. Expressions equivalent to a product of the form $\delta^2$ arise frequently in the equations of physics and have always presented problems of interpretation. The result $\delta^2 = 0$ is often inappropriate on physical grounds, and other (non-zero) values have been assigned to the expression $\delta^2$ in some theories of intrinsic multiplication. Güttinger [27] suggests that we should have $\delta^2 = c\delta$, where $c$ is an arbitrary constant (and not necessarily zero); in certain situations in quantum field theory there is even the possibility that $\delta^2$ should be interpreted as some infinite multiple of $\delta$.

The fact that we have only a weak form of associativity for normal products does not entirely free us from the problem posed by the Schwartz counter example cited in §2.4.3. If a normal product of the distributions $\text{Pf}\{x^{-1}\} \equiv x^{-1}$ and $\delta(x)$ were to exist then we would get

$$x\left(\text{Pf}\{x^{-1}\} \circ \delta(x)\right) = \left(x\text{Pf}\{x^{-1}\}\right) \circ \delta(x) = 1 \circ \delta(x) = \delta(x)$$

and

$$x\left(\text{Pf}\{x^{-1}\} \circ \delta(x)\right) = \text{Pf}\{x^{-1}\} \circ \{x\delta(x)\} = 0,$$

so that there is still a contradiction. Thus although it may be possible in principle to define $\delta^2$ as a normal product (even though the result may not always be desirable) we are obliged to conclude that no normal product of the form $x^{-1} \circ \delta(x)$ can exist. This suggests already that it may be necessary to look for an extended definition of product for which some at least of the listed properties of a normal product may be invalid. We shall consider first some definitions of distributional product which can be shown to be normal.

**Exercise 5.1 :**

1. Confirm that the Leibniz rule holds for a normal product.

2. Prove that if the normal product $\delta \circ \delta^k$ exists for any value of $k \geq 0$ then $\delta \circ \delta^k = 0$.

3. Prove that if the normal product $\delta^p \circ \delta^q$ exists for some $p \geq 0$ and $q \geq 0$, then $\delta^r \circ \delta^s = 0$ for $0 \leq r + s \leq p + q$.

## 5.1.2 Sequential theories of intrinsic products

Many different forms of definition have been proposed for an intrinsic product. For example, if $\theta \in \mathcal{S}$ and $\mu \in \mathcal{S}'$ then the Schwartz product $\theta\mu$ exists and the exchange formula holds

$$\theta\mu = \mathcal{F}^{-1}[\hat{\theta} * \hat{\mu}].$$

Now let $\mu, \nu$ be arbitrary distributions in $\mathcal{D}'$ such that for each $x \in \mathbb{R}$ there exists a neighbourhood $\Omega(x)$ and a function $\theta_x$ in $\mathcal{D}$ with $\theta_x(y) \equiv 1$ on $\Omega(x)$ for which the convolution $\mathcal{F}[\theta_x\mu] * \mathcal{F}[\theta_x\nu]$ is well defined on $\mathcal{S}'$. Then the so-called **Ambrose product** $\mu \cdot \nu$ is defined as the distribution for which

$$<\mu \cdot \nu, \varphi> = <\mathcal{F}^{-1}[\mathcal{F}[\theta_x\mu] * \mathcal{F}[\theta_x\nu]], \varphi> \qquad (5.2)$$

for any $\varphi \in \mathcal{D}(\Omega(x))$. It can be shown that there does exist a (unique) member of $\mathcal{D}'$ which coincides with the local distribution defined by (5.2).

In many respects however it is definitions based on some form of sequential approach which have attracted the most attention. In §2.2.2.1 a strict delta sequence was defined as a sequence $(\rho_n)_{n \in \mathbb{N}}$ of functions in $\mathcal{D}$ such that

**A** $\rho_n(x) \geq 0$ for all $x \in \mathbb{R}$ and for each $n \in \mathbb{N}$,

**B** $\int_{-\infty}^{+\infty} \rho_n(x)dx = 1$, for each $n \in \mathbb{N}$,

**C** $\mathrm{supp}(\rho_n) \to \{0\}$, as $n \to \infty$.

Sequential methods for defining multiplicative products of distributions rely in general on the fact that we can represent an arbitrary distribution $\mu \in \mathcal{D}'$ as the limit of a sequence of infinitely smooth functions. Given any distributions $\mu, \nu \in \mathcal{D}'$ the convolutions $\mu_n = \mu * \rho_n$ and $\nu_n = \nu * \rho_n$ always exist as infinitely smooth functions for each $n$ if $(\rho_n)_{n \in \mathbb{N}}$ is any strict delta sequence. Hence the elementary Schwartz products

$$(\mu * \rho_n)\nu, \quad \mu(\nu * \rho_n), \quad (\mu * \rho_n)(\nu * \rho_n),$$

are always well defined. This suggests four possible definitions of product for $\mu$ and $\nu$. Provided the limits exist in $\mathcal{D}'$, independently of the choice of the $\rho_n$, we write:

$$\begin{aligned}
\mathbf{MHO-1} \quad & \mathrm{SP1}\{\mu, \nu\} \equiv [\mu]\nu &&= \lim_{n \to \infty}(\mu * \rho_n)\nu, \\
\mathbf{MHO-2} \quad & \mathrm{SP2}\{\mu, \nu\} \equiv \mu[\nu] &&= \lim_{n \to \infty}\mu(\nu * \rho_n), \\
\mathbf{MHO-3} \quad & \mathrm{SP3}\{\mu, \nu\} \equiv [\mu][\nu] &&= \lim_{n \to \infty}(\mu * \rho_{1n})(\nu * \rho_{2n}), \\
\mathbf{MHO-4} \quad & \mathrm{SP4}\{\mu, \nu\} \equiv [\mu\nu] &&= \lim_{n \to \infty}(\mu * \rho_n)(\nu * \rho_n).
\end{aligned}$$

We shall refer to these as (strict) **Mikusinski-Hirata-Ogata (MHO) products**. The first two definitions, due to Hirata and Ogata [31], and the third, due to

Mikusinski [74], can be shown to be equivalent. The fourth, proposed in the text by Antosik, Mikusinski and Sikorski, [1], is more general. For example, we have

$$\mathrm{SP4}\{x^{-1}, \delta(x)\} \equiv [\mathrm{Pf}\{x^{-1}\}\delta(x)] = -\frac{1}{2}\delta'(x)$$

whereas the SP3 product $[\mathrm{Pf}\{x^{-1}\}][\delta(x)]$ does not exist.

Note that the non-negative constraint **A** on the functions $\rho_n$ can be replaced by the condition:

**A$_1$** $\rho_n$ is a function in $\mathcal{D}$ such that for each $p \in \mathbb{N}$ there exists $M_p > 0$, independent of $n$, for which

$$\varepsilon_n^p \int_{-\infty}^{+\infty} |D^p \rho_n(x)| dx \leq M_p,$$

where $\mathrm{supp}(\rho_n) \subset \mathcal{B}(0, \varepsilon_n)$ and $\varepsilon_n \to 0$ as $n \to \infty$.

With this modified condition $(\rho_n)_{n \in \mathbb{N}}$ is usually called a **Kaminski $\wedge$-sequence**. The Kaminski $\wedge$-product $\mu \wedge \nu$ is then defined by

$$\mu \wedge \nu = \lim_{n \to \infty} (\mu * \rho_{1n})(\nu * \rho_{2n})$$

provided that the limit exists in the sense of $\mathcal{D}'$ for every $\wedge$-sequence $(\rho_{1n})_{n \in \mathbb{N}}$ and every $\wedge$-sequence $(\rho_{2n})_{n \in \mathbb{N}}$. Whenever the Mikusinski product $\mathrm{SP3}\{\mu, \nu\} \equiv [\mu][\nu]$ exists so also does the Kaminski product and then we have

$$[\mu][\nu] = \mu \wedge \nu.$$

On the other hand if we take, for example, $\mu = \delta$ and $\nu(x) = \sum_{m=1}^{\infty} m^{-r} \delta(x - 1/m)$, where $r > 2$, then the Kaminski product $\delta \wedge \nu$ is well defined as the null distribution whereas the Mikusinski strict product $[\delta][\nu]$ does not exist.

**Remark 5.1** *Suppose that for each $x \in \mathbb{R}$ there exists an open neighbourhood $\Omega(x)$ such that for all functions $\theta, \psi \in \mathcal{D}(\Omega(x))$ we have*

$$\mathcal{F}[\theta \mu] \mathcal{F}^{-1}[\psi \nu] \in L^1(\mathbb{R}).$$

*Then the Ambrose product $\mu \cdot \nu$ exists and $\mu \cdot \nu = [\mu][\nu]$. On the other hand, suppose that $f$ is a function in $L^\infty(\mathbb{R})$ which is continuous over any neighbourhood of 0; then the Mikusinski product $[f][\delta]$ certainly exists whereas the Ambrose product $f \cdot \delta$ does not.*

### 5.1.2.1 Model MHO-products

Some increase in generality can be obtained if the strict delta sequences used in the **MHO** definitions are further restricted by imposing additional constraints. In particular we can define products MP1 - MP4 using the limits given in **MHO1 - 4** above but with the strict delta sequences $(\rho_n)_{n \in \mathbb{N}}$ constrained to be model delta sequences; that is we require

$$\rho_n(x) = n\rho(nx) \ , \ n = 1, 2, 3, \ldots,$$

where $\rho$ is some function in $\mathcal{D}$ such that

$$\int_{-\infty}^{+\infty} \rho(x) dx = 1.$$

The limits in **MHO1/4** are then required to exist independently of the particular function chosen in $\mathcal{D}$. In the terminology introduced in §2.2.2.1 such delta sequences are described as **model** delta sequences and the corresponding products will therefore be referred to as **model MHO-products**. Once again it can be shown that the model products MP1, MP2 and MP3 are equivalent to each other (cf. Itano [37], Shiraishi and Itano [100], Kaminski [47]).

The model products MP1, MP2, MP3 are more general than the corresponding strict products SP1, SP2 and SP3, while the model product MP4 appears to be the most general of all. Thus if $\nu$ is again the distribution $\sum_{m=1}^{\infty} m^{-r}\delta(x - 1/m)$ then the model product $MP4\{\delta, \nu\} \equiv [\delta\nu]$ exists for all $r \geq 2$ being equal to the null distribution for any $r > 2$ and equal to $\frac{1}{2}\delta$ for $r = 2$. (Nevertheless it fails to exist for any value of $r < 2$)

It is proved in Itano [37] that the model **MHO**-product MP3 is a normal product. It can be extended to apply to more than two factors in an obvious way by setting

$$MP3\{\mu, \nu, \sigma\} = \lim_{n \to \infty} (\mu * \rho_{1n})(\nu * \rho_{2n})(\sigma * \rho_{3n})$$

provided the limits exist and are independent of the choice of the sequences $(\rho_{1n})_{n \in \mathbb{N}}$, $(\rho_{2n})_{n \in \mathbb{N}}$ and $(\rho_{3n})_{n \in \mathbb{N}}$. It then follows that whenever $[\nu][\sigma]$ and $[\mu][\nu][\sigma]$ both exist we have

$$[\mu][\nu][\sigma] = [\mu]([\nu][\sigma]).$$

### 5.1.3 Extended forms of sequential products

The limitations of a classical sequential definition of distributional product can be clearly seen from a celebrated example discussed by Mikusinski [75]. The following equivalence arises naturally in quantum mechanics:

$$\delta^2(x) - \frac{1}{\pi^2}\left(\frac{1}{x}\right)^2 = -\frac{1}{\pi^2}\frac{1}{x^2}. \tag{5.3}$$

The term $x^{-2}$ appearing on the right-hand side of (5.3) admits the usual interpretation as the (singular) distribution which is the second derivative of the regular distribution $-\log|x|$. The problem is to find a meaning for the terms on the left-hand side. We have already seen (exercise 2.4) that there exist strict delta sequences $(\rho_n)_{n\in\mathbb{N}}$ such that

$$\lim_{n\to\infty}\int_{-\infty}^{+\infty}\rho_n^2(x)dx = +\infty$$

and therefore such that

$$\lim_{n\to\infty}\int_{-\infty}^{+\infty}\rho_n^2(x)\varphi(x)dx = \pm\infty$$

for any test function $\varphi$ for which $\varphi(0) \neq 0$. Mikusinski concluded from this that $\delta^2$ cannot exist as a distribution. In the same way if by $(x^{-1})^2$ we are to understand a limit of the form

$$\left(x^{-1}\right)^2 = \left(x^{-1}\right)\left(x^{-1}\right) = \lim_{n\to\infty}\left(\rho_n * x^{-1}\right)^2$$

then $(x^{-1})^2$ also cannot exist as a distribution since there exist strict delta sequences for which

$$\lim_{n\to\infty}\int_{-\infty}^{+\infty}\left(x^{-1}\right)_n^2\varphi(x)dx$$

is again generally infinite. Nevertheless Mikusinski was able to show that the expression $\{\delta^2(x) - \pi^{-2}(x^{-1})^2\}$ on the left-hand side of (5.3) does have a well-defined meaning in $\mathcal{D}'$ if we treat it as a whole, without trying to assign meaning to the individual terms. Thus we may consider the distributional limit of

$$\rho_n^2(x) - \frac{1}{\pi^2}\left(x^{-1} * \rho_n(x)\right)^2 \equiv \left\{\rho_n(x) - \frac{1}{i\pi}\left(x^{-1} * \rho_n(x)\right)\right\}^2$$
$$+ \frac{2}{i\pi}\rho_n(x)\left(x^{-1} * \rho_n(x)\right).$$

For each $n \in \mathbb{N}$ let $\mu_n(x) = \rho_n(x) - (i\pi)^{-1}(x^{-1} * \rho_n(x))$. Taking Fourier transforms we have

$$\mathcal{F}[\mu_n](y) \equiv \hat{\mu}_n(y) = \hat{\rho}_n(y) - [1 - 2H(y)]\hat{\rho}_n(y) \equiv 2H(y)\hat{\rho}_n(y),$$

while for any $\varphi \in \mathcal{S}$,

$$\lim_{n\to\infty} <\hat{\mu}_n(y),\varphi(y)> = <2H(y),\varphi(y)> = 2\int_0^{+\infty}\varphi(y)dy.$$

Hence
$$\lim_{n\to\infty} (\hat{\mu}_n * \hat{\mu}_n)(y) = (2H * 2H)(y) = 4yH(y).$$

Now
$$\mathcal{F}\left[-\frac{1}{i\pi}\frac{1}{x^2}\right] = \mathcal{F}\left[D_x \frac{1}{i\pi x}\right](y) = iy\{1 - 2H(y)\}$$

and since $iy$ is the Fourier transform of $\delta'(x)$ we get

$$4yH(y) = -\frac{2}{i}\mathcal{F}\left[-\delta'(x) - \frac{1}{i\pi x^2}\right].$$

But, if $\mu = \lim_{n\to\infty} \mu_n$ then the exchange formula gives

$$4yH(y) = (\hat{\mu} * \hat{\mu})(y) = 2\pi \mathcal{F}\left[\delta(x) - \frac{1}{i\pi x}\right]^2 (y)$$

and so we have shown the equivalence

$$\left\{\delta(x) - \frac{1}{i\pi x}\right\}^2 = \frac{1}{i\pi}\delta'(x) - \frac{1}{\pi^2 x^2}. \tag{5.4}$$

To establish the equivalence (5.3) it remains to show that the sequence $u_n(x) = \rho_n(x)(x^{-1} * \rho_n(x))$ converges distributionally to $-\frac{1}{2}\delta'(x)$. This is easily shown if $(\rho_n)_{n\in\mathbb{N}}$ is a (strict delta) sequence of *even* functions, so that $\rho_n(-x) = \rho_n(x)$ for each $n$. In that case if we write

$$I_n = \int_{-\infty}^{+\infty} u_n(x)dx = \int_{-\infty}^{+\infty}\int_{-\infty}^{+\infty} \frac{\rho_n(x)\rho_n(t)}{x-t} dx dt$$

and

$$K_n = \int_{-\infty}^{+\infty} xu_n(x)dx = \int_{-\infty}^{+\infty}\int_{-\infty}^{+\infty} \frac{x\rho_n(x)\rho_n(t)}{x-t} dx dt$$

(where the integrals are understood in the Cauchy principal value sense) we can obtain explicit values for $I_n$ and $K_n$ by simply interchanging $x$ and $t$ in the integrands. This would cause a change of sign in $I_n$ and so we must have $I_n = 0$ for every $n$. To evaluate $K_n$ we first write

$$K_n = 1 - \int_{-\infty}^{+\infty}\int_{-\infty}^{+\infty} \frac{t\rho_n(x)\rho_n(t)}{x-t} dx dt.$$

Interchanging $x$ and $t$ in the integrand here gives $K_n = 1 - K_n$, which implies that $K_n = 1/2$. It now follows easily that the sequence $(F_n)_{n\in\mathbb{N}}$, where

$$F_n(x) = \int_{-\infty}^{x} (x-t)u_n(t)dt = x\int_{-\infty}^{x} u_n(t)dt - \int_{-\infty}^{x} tu_n(t)dt$$

tends to 0 for $x < 0$ and to $-1/2$ for $x > 0$ as $n \to \infty$, and is bounded by a constant function independent of $n$. Therefore $F_n$ tends distributionally to $-\frac{1}{2}H$. Since $F_n'' = u_n$ it follows that $u_n$ tends distributionally to $-\frac{1}{2}\delta'$.

Thus the equivalence (5.3) is confirmed, together with the result that $x^{-1} \cdot \delta(x)$ is to be interpreted as $-\frac{1}{2}\delta'(x)$. This has been achieved at the expense of defining the product in terms of an even more restricted class of delta sequences. There are very many ways of imposing such extra constraints (the assumption that the functions of the sequence are all even being among the simplest) and so arriving at a wider range of possible products. The price to be paid is that we no longer have a normal product, and so that some of the desirable properties of multiplication will not generally be valid. It would not be possible to give anything other than a very incomplete review of the huge field of attempts to provide a (partial) solution of the multiplication problem in this way. Instead we shall trace the application of classical sequential methods and their subsequent generalisation in a particular, well defined, context. The work of Brian Fisher throughout the last two decades offers a nice illustration of the way in which an increasingly wide class of distributional products can be defined by generalising the limiting process. In the sections which follow we consider in particular the development of Fisher's approach to the problem of defining products of delta functions and generalised powers of $x$.

### 5.1.3.1 The Fisher classical product

For $r = 1, 2, \ldots$, the product $x^r \delta^{(r)}(x)$ presents no problems since it is well defined by the ordinary Schwartz product:

$$< x^r \delta^{(r)}(x), \varphi(x) > = < \delta^{(r)}(x), x^r \varphi(x) > = (-1)^r r! \varphi(0). \qquad (5.5)$$

Hence $x^r \delta^{(r)}(x) = (-1)^r r! \delta(x)$.

Products of the form $x_+^r \delta^{(r)}(x)$ and $x_-^r \delta^{(r)}(x)$ can be defined using the **MP4** sequential definition together with the additional assumption that the generating functions $\rho(x)$ are even. For clarity we repeat here the specific conditions imposed by Fisher [18] in his original (classical) definition of sequential product: Let $\rho$ be a function in $\mathcal{D}$ such that

**(A)** $\rho(x) \geq 0$ for all $x \in \mathbb{R}$;
**(B)** $\int_{-\infty}^{+\infty} \rho(x)dx = 1$;
**(C)** $\rho(x) = 0$ for $|x| \geq 1$;
**(D)** $\rho(-x) = \rho(x)$ for every $x$.

Then any delta sequence $(\rho_n)_{n\in\mathbb{N}}$ where $\rho_n(x) = n\rho(nx)$ will be called a **symmetric model sequence**. For any given $\mu \in \mathcal{D}'$ we set $\mu_n = \mu * \rho_n$. Thus, for example, if $r > 0$ we write

$$\left(x_+^r\right)_n = x_+^r * \rho_n(x) = \int_{-1/n}^{x} (x-t)^r \rho_n(t)\,dt \quad \text{for } x \geq -1/n,$$

$$\left(x_-^r\right)_n = x_-^r * \rho_n(x) = \int_{x}^{+1/n} (t-x)^r \rho_n(t)\,dt \quad \text{for } x \leq +1/n.$$

Fisher begins with a product $\mu \cdot \nu \equiv [\mu, \nu]$ of **MP4** type, using a symmetric model sequence.

**Definition 5.1 (classical Fisher product)** *The product $\mu \cdot \nu$ exists and is equal to $\sigma \in \mathcal{D}'$ if and only if*

$$\lim_{n\to\infty} <\mu_n \nu_n, \varphi> = <\sigma, \varphi>, \quad \forall \varphi \in \mathcal{D}.$$

Under this definition the products $x_+^r \cdot \delta^{(r)}(x)$ and $x_-^r \cdot \delta^{(r)}(x)$ are easily established using a simple inductive argument. First note that if $H_n(x) = \int_{-\infty}^{x} \rho_n(t)\,dt$ then $\lim_{n\to\infty} H_n^2 = H$ and we have immediately

$$\lim_{n\to\infty} 2 H_n \rho_n = \delta.$$

Hence for the symmetric Fisher product we have, unequivocally,

$$H(x) \cdot \delta(x) = \frac{1}{2}\delta(x). \tag{5.6}$$

Now assume that for some $r > 0$ we have

$$x_+^r \cdot \delta^{(r)}(x) = \frac{1}{2}(-1)^r r! \delta(x).$$

Then $\lim_{n\to\infty} \left(x_+^{r+1}\right)_n \left(\delta^{(r)}\right)_n = 0$, and differentiation of this gives

$$\lim_{n\to\infty} \left(x_+^{r+1}\right)_n \left(\delta^{(r+1)}\right)_n = -\lim_{n\to\infty}(r+1)\left(x_+^r\right)_n \left(\delta^{(r)}\right)_n = \frac{1}{2}(-1)^{r+1}(r+1)!\delta(x).$$

Therefore we have

$$x_+^r \cdot \delta^{(r)}(x) = \frac{1}{2}(-1)^r r!\, \delta(x) \tag{5.7}$$

for all $r \geq 0$. Using (5.5) and the fact that $x_+^r + (-1)^r x_-^r = x^r$, we get

$$x_-^r \cdot \delta^{(r)}(x) = \frac{1}{2} r!\, \delta(x) \tag{5.8}$$

for all $r \geq 0$.

### 5.1.3.2  The Fisher product $x^{-r} \cdot \delta^{(r-1)}$

The Fisher product $x^{-1} \cdot \delta(x)$ exists and is equal to $-\frac{1}{2}\delta'(x)$, as shown by the Mikusinski analysis quoted above. Fisher [19] was able to obtain the following generalisation

$$x^{-r} \cdot \delta^{(r-1)} = (-1)^r \frac{(r-1)!}{2(2r-1)!} \delta^{(2r-1)}(x) \qquad (5.9)$$

where $r = 0, 1, 2, \ldots$, by working with symmetric generating functions $\rho(x)$ which satisfy the additional constraint

(E)  $\rho^{(r)}(x)$ has only $r$ changes of sign for $r = 1, 2, \ldots$.

This constraint is certainly satisfied by the (symmetric) model sequence obtained by choosing

$$\rho(x) = \begin{cases} A \cdot \exp[1/(x^2 - 1)] & \text{if } |x| < 1, \\ 0 & \text{if } |x| \geq 1, \end{cases}$$

where

$$A^{-1} = \int_{-1}^{+1} \exp[-(1-x^2)^{-1}] dx.$$

Now for $r = 1, 2, \ldots$, the singular distribution $x^{-r}$ is defined as

$$x^{-r} = \frac{(-1)^{r-1}}{(r-1)!} D^r \log|x|$$

so that

$$(x^{-r})_n = \frac{(-1)^{r-1}}{(r-1)!} \int_{-1/n}^{+1/n} \rho_n^{(r)}(t) \log|x-t| dt.$$

If $\Im f(x)$ denotes the integration operator $\Im f(x) \equiv \int_{-\infty}^{x} f(t) dt$ then iterating this operator $2r-1$ times will give

$$\Im^{2r-1}\left[(x^{-r})_n \rho_n^{(r-1)}(x)\right] = \frac{1}{(2r-2)!} \int_{-1/n}^{x} (t^{-r})_n \rho_n^{(r-1)}(t)(x-t)^{2r-2} dt$$

while a direct computation (see Remark 5.2) shows that

$$\int_{-1/n}^{+1/n} (x^{-r})_n \rho_n^{(r-1)}(x) x^m dx = \begin{cases} -\frac{1}{2}(-1)^r(r-1)! & \text{if } m = 2r-1, \\ 0 & \text{if } m = 0, 1, 2, \ldots, 2r-2. \end{cases}$$

Hence

$$\int_{-1/n}^{+1/n} (t^{-r})_n \rho_n^{(r-1)}(t)(1/n - t)^{2r-2} dt = 0$$

and since, by definition, $\rho_n^{(r-1)}(x)$ is zero for $|x| \geq 1/n$ it follows that

$$\Im^{2r-1}\left[(x^{-r})_n \rho_n^{(r-1)}(x)\right] = 0$$

for $|x| \geq 1/n$, and therefore has support contained in $(-1/n, +1/n)$. Since $\rho^{(r)}(x)$ has only $r$ changes of sign it follows that $(x^{-r})_n$ has only $r$ changes of sign and therefore that $(x^{-r})_n \rho_n^{(r-1)}(x)$ has at most $(2r-1)$ changes of sign. Thus the $(2r-1)$th primitive of $(x^{-r})_n \rho_n^{(r-1)}(x)$ is either $\geq 0$ everywhere or else $\leq 0$ everywhere. Finally

$$\int_{-1/n}^{+1/n} \mathfrak{I}^{2r-1}[(x^{-r})_n \rho_n^{(r-1)}(x)]dx = \frac{1}{(2r-1)!} \int_{-1/n}^{+1/n} (t^{-r})_n \rho_n^{(r-1)}(t)(\frac{1}{n} - t)^{2r-1} dt$$

$$= (-1)^r \frac{(r-1)!}{2(2r-1)!}.$$

Hence if $F_n(x) \equiv \mathfrak{I}^{2r-1}[(x^{-r})_n \rho_n^{(r-1)}(x)]$ then $(F_n)_{n \in \mathbb{N}}$ is a regular sequence converging distributionally to the limit $(-1)^r[(r-1)!/2(2r-1)!]\delta(x)$. Differentiating $(2r-1)$ times shows that the distributional limit of the sequence $((x^{-r})_n \rho_n^{(r-1)}(x))_{n \in \mathbb{N}}$ is $(-1)^r[(r-1)!/2(2r-1)!]\delta^{(2r-1)}(x)$, $r = 1, 2, \ldots$. In particular for $r = 1$ we have the Mikusinski result $x^{-1} \cdot \delta(x) = -\frac{1}{2}\delta'(x)$.

**Remark 5.2** To evaluate $J_n \equiv \int_{-1/n}^{+1/n} (x^{-r})_n \rho_n^{(r-1)}(x) x^m dx$ note first that a change in order of integration gives

$$J_n = (-1)^r \frac{1}{(r-1)!} \int_{-1/n}^{+1/n} \rho^{(r)}(t) \left\{ \int_{-1/n}^{+1/n} \rho_n^{(r-1)}(x) x^m \log|x-t| dx \right\} dt.$$

Integration by parts reduces the integral within the braces to

$$\frac{1}{m+1} \int_{-1/n}^{+1/n} \rho_n^{(r-1)}(x) \log|x-t| d(x^{m+1} - t^{m+1})$$

$$= \frac{(-1)}{m+1} \int_{-1/n}^{+1/n} \left\{ \rho_n^{(r)}(x)(x^{m+1} - t^{m+1}) \log|x-t| + \rho_n^{(r-1)}(x) \sum_{j=0}^{m} x^{m-j} t^j \right\} dx$$

and so we get

$$J_n = \frac{(-1)^r}{(m+1)(r-1)!} \left\{ \int_{-1/n}^{+1/n} \int_{-1/n}^{+1/n} f(x,t) dx dt \right.$$

$$\left. + \sum_{j=0}^{m} \left[ \int_{-1/n}^{+1/n} \rho_n^{(r)}(t) t^j dt \right] \left[ \int_{-1/n}^{+1/n} \rho_n^{(r-1)}(x) x^{m-j} dx \right] \right\}$$

where $f(x,t) \equiv \rho_n^{(r)}(x)\rho_n^{(r)}(t)(x^{m+1} - t^{m+1}) \log|x-t|$. Since $f(x,t) = -f(t,x)$ it follows that the first double integral on the right-hand side is zero and therefore that

$$J_n = \frac{(-1)^r}{(m+1)(r-1)!} \sum_{j=0}^{m} \left[ \int_{-1/n}^{+1/n} \rho_n^{(r)}(t) t^j dt \right] \left[ \int_{-1/n}^{+1/n} \rho_n^{(r-1)}(x) x^{m-j} dx \right].$$

*Integration by parts shows that*

$$\int_{-1/n}^{+1/n} \rho_n^{(r)}(t) t^j \, dt = \begin{cases} (-1)^r r! & \text{if } j = r, \\ 0 & \text{if } j = 0, 1, \ldots, r-1, \end{cases}$$

*while*

$$\int_{-1/n}^{+1/n} \rho_n^{(r-1)}(x) x^{m-j} \, dx = \begin{cases} (-1)^{r-1}(r-1)! & \text{if } m - j = r - 1, \\ 0 & \text{if } m - j = 0, 1, \ldots, r-2. \end{cases}$$

*Therefore we get*

$$\int_{-1/n}^{+1/n} (x^{-r})_n \rho_n^{(r-1)}(x) x^m \, dx = \begin{cases} -\frac{1}{2}(-1)^r(r-1)! & \text{if } m = 2r - 1, \\ 0 & \text{if } m = 0, 1, \ldots, 2r-2, \end{cases}$$

*as asserted.*

**Exercise 5.2 :**

1. If $\alpha > -1$ and $\beta > -1$ show that for any symmetric model delta sequence $(\rho_n)_{n \in \mathbb{N}}$ we have

$$(x_+^\alpha)_n (x_-^\beta)_n = \int_{-1/n}^{+1/n} \rho_n(t) \int_t^{+1/n} \rho_n(\tau)(\tau - t)^{\alpha+\beta+1} \mathrm{B}(\alpha+1, \beta+1) d\tau dt$$

where $\mathrm{B}(\cdot, \cdot)$ is the Beta function. Hence show that

   (a) if $\alpha + \beta > -1$ then $x_+^\alpha \cdot x_-^{-1-\alpha} = 0$,
   
   (b) if $\alpha + \beta = -1$ then $x_+^\alpha \cdot x_-^{-1-\alpha} = -\frac{\pi}{2}\operatorname{cosec}(\pi\alpha)\delta(x)$, for $-1 < \alpha < 0$.

2. Show that the Fisher product $x_+^\alpha \cdot x_-^\beta$ does not generally exist when $\alpha > -1$ and $\beta > -1$, but $\alpha + \beta < -1$.

3. Show that $[|x|^\alpha \operatorname{sgn}(x)] \cdot \delta(x) = 0$ for all $\alpha > -1$.

### 5.1.3.3  Fisher neutrix products

All Fisher products considered so far have been established using only classical limiting processes. These are inadequate for general products of the form $x_+^{-r} \cdot \delta^{(p)}(x)$ and $x_-^{-r} \cdot \delta^{(p)}(x)$, where $p = 0, 1, 2, \ldots, r = 1, 2, \ldots$. To see this it is enough to consider products $\mu \cdot \nu$ given by the limits

$$<\mu[\nu], \varphi> = \lim_{n \to \infty} <\mu\nu_n, \varphi> = \lim_{n \to \infty} <\mu, \nu_n\varphi>.$$

Thus, for example, we have

$$\begin{aligned}
< x_+^{-1}, \rho_n^{(p)}(x)\varphi(x) > &= \int_0^1 \frac{\rho_n^{(p)}(x)\varphi(x) - \rho_n^{(p)}(0)\varphi(0)}{x} dx \\
&= \int_0^{+1/n} \rho_n^{(p)}(x) \frac{\varphi(x) - \varphi(0)}{x} dx \\
&\quad + \varphi(0) \int_0^{+1/n} \frac{\rho_n^{(p)}(x) - \rho_n^{(p)}(0)}{x} dx - \rho_n^{(p)}(0)\varphi(0) \int_{+1/n}^1 \frac{dx}{x} \\
&\equiv I_1 + \varphi(0) I_2 + \rho_n^{(p)}(0)\varphi(0) I_3.
\end{aligned}$$

Now

$$I_1 = \int_0^{+1/n} \rho_n^{(p)}(x) \left\{ \sum_{m=1}^{p+1} \frac{x^{m-1}}{m!} \varphi^{(m)}(0) + \frac{x^{p+1}}{(p+2)!} \varphi^{(p+2)}(\xi x) \right\} dx,$$

where $0 < \xi < 1$. Hence we get,

$$\begin{aligned}
I_1 &= \sum_{m=1}^p \frac{n^{p-m+1}}{m!} \varphi^{(m)}(0) \int_0^1 t^{m-1} \rho^{(p)}(t) dt \\
&\quad + \frac{\varphi^{(p+1)}(0)}{(p+1)!} \int_0^1 t^p \rho^{(p)}(t) dt + \frac{n^{-1}}{(p+2)!} \int_0^1 t^{p+1} \rho^{(p)}(t) \varphi^{(p+2)}(\xi t/n) dt \\
I_2 &= \varphi(0) n^{p+1} \int_0^1 t^{-1} \{\rho^{(p)}(t) - \rho^{(p)}(0)\} dt, \\
I_3 &= -\rho_n^{(p)}(0)\varphi(0) \log(n) \equiv -\rho^{(p)}(0) n^{p+1} \log(n).
\end{aligned}$$

The limit as $n$ tends to infinity does not exist because of the presence of terms in each of $I_1, I_2$ and $I_3$ which become infinite. If there were a rationale to allow the discard of these infinite terms in the limit and to extract a finite part then a meaning for the product $x_+^{-1} \cdot \delta^{(p)}(x)$ could be given by setting

$$< x_+^{-1} \cdot \delta^{(p)}(x), \varphi(x) > = \frac{\varphi^{(p+1)}(0)}{(p+1)!} \int_0^1 t^p \rho^{(p)}(t) dt.$$

Since the generating function $\rho(x)$ is even this would give

$$x_+^{-1} \cdot \delta^{(p)} = -\frac{1}{2(p+1)} \delta^{(p+1)}(x) \tag{5.10}$$

for $p = 0, 1, 2, \ldots$ (see Fisher [23]).

The required rationale for such an extended meaning of product is to be found in the use of the so-called neutrix calculus of van der Corput described earlier in §1.5.4. A neutrix was defined there as an additive commutative subgroup $\mathcal{N}$ of mappings of a nonempty set $X$ into a group $G$ such that the only constant function in $\mathcal{N}$ is the zero map. The members of $\mathcal{N}$ are usually called the negligible functions (with respect to the neutrix in question). In the present context it is enough to take $X = \mathbb{N}$ and $G = \mathbb{R}$, and to define the negligible functions to be all linear sums of

$$n^\lambda \log^{r-1}(n) \ , \ \log^r(n)$$

for $\lambda > 0, r = 1, 2, \ldots$, together with all real valued functions on $\mathbb{N}$ which converge to zero as $n \to \infty$. If $f \in \mathbb{R}^{\mathbb{N}}$ is such that there exists some constant $c$ for which $f(n) - c$ belongs to $\mathcal{N}$ then we say that the neutrix limit of $f(n)$ as $n$ tends to infinity is the number $c$. The Fisher neutrix product is then defined as follows:

**Definition 5.2 (Fisher neutrix product)** *With the neutrix $\mathcal{N} \subset \mathbb{R}^{\mathbb{N}}$ defined as above the neutrix product $\mu \odot \nu$ of distributions $\mu$ and $\nu$ is given by the neutrix limit*

$$<\mu \odot \nu, \varphi> = \mathcal{N}\!-\!\lim_{n \to \infty} <\mu\nu_n, \varphi> = \mathcal{N}\!-\!\lim_{n \to \infty} <\mu, \nu_n\varphi>.$$

The following theorems have been proved by Fisher [22], [23]:

**Theorem 5.2** *Let $\mu$ and $\nu$ be distributions for which the product $\mu \cdot \nu$ is well defined as the (classical) limit*

$$<\mu \cdot \nu, \varphi> = \lim_{n \to \infty} <\mu\nu_n, \varphi> = \lim_{n \to \infty} <\mu, \nu_n\varphi>.$$

*Then the neutrix product $\mu \odot \nu$ exists and $\mu \cdot \nu = \mu \odot \nu$.*

**Theorem 5.3** *If $\mu$ and $\nu$ are distributions for which both the neutrix products $\mu \odot \nu$ and $\mu \odot (D\nu)$ exist, then the neutrix product $(D\mu) \odot \nu$ also exists, and $D(\mu \odot \nu) = (D\mu) \odot \nu + \mu \odot (D\nu)$.*

From equation (5.10) and theorem 5.3 it follows that

$$x_+^{-r} \odot \delta^{(p)}(x) = \frac{(-1)^r p!}{2(p+r)!} \delta^{(p+r)}(x) \tag{5.11}$$

for $p = 0, 1, 2, \ldots$, and $r = 1, 2, \ldots$.

The neutrix product will not generally be commutative. Since

$$(x_+^{-r})_n = \frac{(-1)^r}{(r-1)!} \int_{-1/n}^{x} \log(x-t) \rho_n^{(r)}(t) dt$$

we get

$$< \delta(x), (x_+^{-r})_n \varphi(x) > = \frac{(-1)^r \varphi(0)}{(r-1)!} \int_{-1/n}^0 \log(-t) \rho_n^{(r)}(t) dt$$

and

$$\int_{-1/n}^0 \log(-t) \rho_n^{(r)}(t) dt = (-1)^r n^r \int_0^1 \log(t/n) \rho^{(r)}(t) dt$$
$$= (-1)^r n^r \int_0^1 \rho^{(r)}(t) \log(t) dt - (-1)^r n^r \log(n) \int_0^1 \rho^{(r)}(t) dt,$$

all terms becoming infinite, or else vanishing, as $n \to \infty$. In the sense of the neutrix this gives

$$\delta^{(p)}(x) \odot x_+^{-r} = 0,$$

for $p = 0, 1, 2, \ldots$, and $r = 1, 2, \ldots$.

Note that when both neutrix products $\mu \odot \nu$ and $\nu \odot \mu$ exist we can define a commutative product $\mu \cdot \nu$ by symmetrization:

$$\mu \cdot \nu = \frac{1}{2}(\mu \odot \nu + \nu \odot \mu).$$

**Exercise 5.3 :**

1. Find the neutrix products $x_-^{-r} \odot \delta^{(p)}(x)$ and $\delta^{(p)}(x) \odot x_-^{-r}$.

2. For $r, p = 0, 1, 2, \ldots$, prove that

   (a) $x_+^r \odot \delta^{(r+p)}(x) = \delta^{(r+p)}(x) \odot x_+^r = \frac{(-1)^r (r+p)!}{2p!} \delta^{(p)}(x),$
   
   (b) $x_-^r \odot \delta^{(r+p)}(x) = \delta^{(r+p)}(x) \odot x_-^r = \frac{(r+p)!}{2p!} \delta^{(p)}(x).$

## 5.2 GENERALISED FUNCTIONS AND OPERATORS

### 5.2.1 Distributions as operators

If we abandon the idea that the product of distributions must always be a distribution then it becomes possible to consider the construction of *algebras* of objects which are more general than distributions but within which $\mathcal{D}'$ may be embedded. This means that we expect combinations like $x(\text{Pf}\{x^{-1}\}\delta(x))$ and $(x\text{Pf}\{x^{-1}\})\delta(x)$ to co-exist and to be equal, though they will not represent distributions. As discussed in §2.4.4 some early proposals for such a non-intrinsic theory of multiplication were due to Güttinger and König. But the most effective and comprehensive theory of multiplication has been developed in recent years in the context of the so-called **New Generalised Functions** of J. F. Colombeau. Colombeau's original was framed in terms of infinitely differentiable functions defined on infinite dimensional spaces (of functions) [9], [10]. This involves concepts which are relatively unfamiliar and which (as Colombeau himself remarks) must be described as non-elementary. Colombeau later produced a more direct and elementary development of the theory [11]. In this he worked in terms of equivalence classes of mappings of the form

$$T: \Phi \times \mathbb{R} \to \mathbb{C}$$
$$(\varphi, x) \to T(\varphi, x)$$

such that $T(\varphi, x)$ is infinitely differentiable in $x$ for each fixed function $\varphi$ in a certain subset $\Phi$ of $\mathcal{D}$. For simplicity and ease of presentation we prefer to restate Colombeau's definition in an equivalent form which actually corresponds to that given in the account of the Colombeau theory in the text by Rosinger [92]. This presents generalised functions in terms of operators rather than in terms of functionals as in the familiar Schwartz approach to the theory of distributions. To give some motivation for this change of emphasis we begin by re-examining the concept of distribution itself in order to extend the sense of the term *generalised function*.

In §1.4 we gave a brief description of the axiomatic formulation of a theory of (finite order) distributions proposed by J. S. Silva. For convenience we restate the axiomatic scheme here: (finite order) distributions on $\mathbb{R}$ are defined as the members of a linear space $E$ for which linear maps

$$\iota : \mathcal{C}(\mathbb{R}) \to E , \quad D : E \to E$$

exist such that

> **Axiom 1** $\iota$ is injective; that is, every function in $\mathcal{C}(\mathbb{R})$ is a distribution on $\mathbb{R}$.
>
> **Axiom 2** To each distribution $\mu$ on $\mathbb{R}$ there corresponds another distribution $D\mu$, called the derivative of $\mu$, such that if $\mu = \iota(f)$ with $f \in \mathcal{C}^1(\mathbb{R})$ then $D\mu = \iota(f')$.
>
> **Axiom 3** If $\mu$ is a distribution on $\mathbb{R}$ then there exists a continuous function $f \in \mathcal{C}(\mathbb{R})$ and a number $r \in \mathbb{N}_0$ such that $\mu = D^r\iota(f)$.
>
> **Axiom 4** Given two functions $f, g \in \mathcal{C}(\mathbb{R})$, and a natural number $r$, the equality $D^r\iota(f) = D^r\iota(g)$ holds if and only if $(f - g)$ is a polynomial function of degree $< r$.

It was shown in §1.4.4 that $\mathcal{D}'_{\text{fin}}$, the linear subspace of all (finite-order) Schwartz distributions on $\mathbb{R}$, is a model for these axioms. Indeed this is the most familiar quasi-convolution operators $T_{[f]}$ and the latter could equally well be used as the basis for a model for a theory of distributions using the Silva axioms. But we now have a direct relation between operator and functional in the sense that

$$< \mu_{(f)}(y), \varphi(y) > \; = \; T_{[f]}\varphi(0). \qquad (5.14)$$

**Remark 5.3** *From the point of view of physical significance the conventional convolution operator $T_{(f)}$ is to be preferred since it defines a linear, translation-invariant mapping of $\mathcal{D}$ into $\mathcal{C}^\infty(\mathbb{R})$. This is particularly appropriate for example when $T_{(f)}$ is interpreted as a system operator carrying "input signals" from $\mathcal{D}$ into "output signals" in $\mathcal{C}^\infty(\mathbb{R})$. The operator $T_{[f]}$ on the other hand is linear but not translation-invariant. Indeed it has the contrasting property of being "translation-reflecting"; that is to say a system defined by a quasi-convolution operator is such that the response to a translated input $\tau_a \varphi(x) = \varphi(x-a)$ is always the reflected translate*

$$\tau_{-a}\{T_{[f]}\varphi(x)\} \; = \; T_{[f]}\varphi(x+a)$$

*of the system response to the input signal $\varphi(x)$. Note also that we have*

$$DT_{[f]}\varphi(x) \; = \; -\int_{-\infty}^{+\infty} f(y)\varphi'(y-x)dy \; = \; T_{[f]}\{-\varphi'\}(x),$$

*or, more generally,*

$$D^p T_{[f]}\varphi(x) \; = \; (-1)^p \int_{-\infty}^{+\infty} f(y)\varphi^{(p)}(y-x)dy \; = \; T_{[f]}\{(-1)^p \varphi^{(p)}\}(x)$$

*so that $T_{[f]}$ does not commute with the differential operator. In spite of this the use of the quasi-convolution operator in this context is preferred by most authorities currently researching in generalised function theory, because of (5.14). For the sake of consistency with the available literature we will generally follow this convention in the sequel. This means, for example, that as an operator representing the delta function we will usually consider the quasi-convolution operator $T_{[\delta]}$ given by*

$$T_{[\delta]}\varphi(x) \; = \; <\delta(y), \varphi(y-x)> \; = \; \varphi(-x),$$

*rather than the identity mapping $T_{(\delta)}$.*

### 5.2.2 The algebra $\mathcal{E}[\mathcal{D}, \mathcal{C}^\infty]$

Let $\mathcal{E}[\mathcal{D}, \mathcal{C}^\infty]$ denote the linear space of all operators $T$ which map $\mathcal{D}$ into $\mathcal{C}^\infty(\mathbb{R})$. Then $\mathcal{E}[\mathcal{D}, \mathcal{C}^\infty]$ is a differential algebra under the natural definitions for the algebraic operations and for derivations:

$$\begin{aligned}(T_1+T_2)\varphi &= T_1\varphi+T_2\varphi \\ (aT)\varphi &= a(T\varphi), \quad a \in \mathbb{C} \\ (T_1 T_2)\varphi &= (T_1\varphi)(T_2\varphi) \\ (DT)\varphi &= D(T\varphi).\end{aligned}$$

For $f \in \mathcal{C}^\infty(\mathbb{R}) \subset \mathcal{C}(\mathbb{R})$ let $T_f$ denote the constant map which carries each $\varphi \in \mathcal{D}$ into the function $f$:

$$T_f\varphi(x) = f(x), \quad \forall \varphi \in \mathcal{D}.$$

Then $T_f$ is a member of $\mathcal{E}[\mathcal{D},\mathcal{C}^\infty]$ for every such $f$, and identifying $f$ with $T_f$ gives an embedding of $\mathcal{C}^\infty(\mathbb{R})$ in $\mathcal{E}[\mathcal{D},\mathcal{C}^\infty]$ which is an embedding of algebras. On the other hand for each $f \in \mathcal{C}(\mathbb{R})$ the convolution operator $T_{(f)}$ and the quasi-convolution operator $T_{[f]}$ both belong to $\mathcal{E}[\mathcal{D},\mathcal{C}^\infty]$ and, as remarked above, we can choose to identify $f$ with either of these to obtain alternative models for a theory of distributions. So far as linear operations are concerned there would then be agreement between those carried out on the functions $f$ and on the corresponding operators $T_{(f)}$ or $T_{[f]}$. For example, for any functions $g, h \in \mathcal{C}(\mathbb{R})$ and any $\varphi \in \mathcal{D}$ we have

$$\begin{aligned}T_{[g+h]}\varphi &= T_{[g]}\varphi + T_{[h]}\varphi, \\ T_{[ag]}\varphi &= a\{T_{[g]}\varphi\}, \quad \forall a \in \mathbb{C}.\end{aligned}$$

This coherence fails for multiplication. The pointwise product $f = gh$, where $g$ and $h$ are functions in $\mathcal{C}(\mathbb{R})$, defines a quasi-convolution operator $T_{[f]}$ which will generally differ from the product of the operators $T_{[g]}$ and $T_{[h]}$. Identifying $f$ with $T_{[f]}$ would give an embedding of $\mathcal{C}(\mathbb{R})$ in $\mathcal{E}[\mathcal{D},\mathcal{C}^\infty]$ which is an embedding of linear spaces but not of algebras.

Again, the delta function would be identified with the operator $T_{[\delta]}$ which carries each function $\varphi \in \mathcal{D}$ into the function $\check{\varphi}$:

$$T_{[\delta]}\varphi(x) = <\delta(y),\varphi(y-x)> = \varphi(-x).$$

Consider the product of the operator $T_{[\delta]}$ with itself:

$$\begin{aligned}\left(T_{[\delta]}T_{[\delta]}\right)\varphi(x) &= (<\delta(y),\varphi(y-x)>)(<\delta(y),\varphi(y-x)>) \\ &= T_{[\delta^2]}\varphi(x) = \varphi^2(-x)\end{aligned}$$

where we define the functional $\delta^2$ by

$$<\delta^2(y),\varphi(y)> \equiv T_{[\delta^2]}\varphi(0) = \varphi^2(0).$$

This is a nonlinear functional on $\mathcal{D}$ and hence cannot be identified with any Schwartz distribution. Therefore if $T_{[\delta^2]}$ represents a generalised function in any sense at all then that generalised function must be something other than a distribution.

The essential step made by Colombeau is to identify functions (and ultimately distributions) not with individual operators in $\mathcal{E}[\mathcal{D}, \mathcal{C}^\infty]$ but with suitably defined *equivalence classes* of such operators. What is needed is an equivalence relation which ensures that, for any $f \in \mathcal{C}^\infty(\mathbb{R})$, the operators $T_f$ and $T_{[f]}$ belong to the same equivalence class, and which preserves the embedding of $\mathcal{C}^\infty(\mathbb{R})$ as an algebra. This would allow the identification of the ordinary function $f$ with the *generalised function* defined by the equivalence class $[T_f] \equiv \left[T_{[f]}\right]$. For an arbitrary function $f$ in $\mathcal{C}^0(\mathbb{R})$, not necessarily belonging to $\mathcal{C}^\infty(\mathbb{R})$, there will not generally be a map $T_f \in \mathcal{E}[\mathcal{D}, \mathcal{C}^\infty]$ but there will still exist a corresponding generalised function defined by the equivalence class $\left[T_{[f]}\right]$; similarly, for $\mu \in \mathcal{D}'$, there will be an equivalence class $\left[T_{[\mu]}\right]$. The Colombeau theory gives a differential algebra of equivalence classes of operators each of which defines a generalised function; distributions appear as members of a certain special sub-class of this algebra. The product of two distributions in the Colombeau sense is then always well defined as a generalised function but it will not generally be a distribution.

### 5.2.3 The basic differential algebra of Colombeau

To define the basic algebra from which the Colombeau theory of generalised functions is derived consider first the subset $\Phi$ of the space $\mathcal{D}$ comprising all functions $\varphi$ such that
$$\int_{-\infty}^{+\infty} \varphi(x)dx = 1.$$
Then, for each $m \in \mathbb{N}$, define $\Phi_m$ to be the set of all functions $\varphi \in \Phi \subset \mathcal{D}$ such that
$$\int_{-\infty}^{+\infty} x^k \varphi^{(k)}(x) dx = 0, \text{ for } 1 \leq k \leq m.$$
It is not difficult to show that $\Phi_m \neq \emptyset$ for each $m \in \mathbb{N}$, and it is obvious that we have
$$\mathcal{D} \supset \Phi \supset \Phi_1 \supset \Phi_2 \supset \ldots \supset \Phi_m \supset \ldots.$$
It will be convenient to adopt the following conventional notation: for any function $\varphi \in \mathcal{D}$ we shall now write
$$\varphi_{(\varepsilon)}(x) = \varepsilon^{-1} \varphi(x/\varepsilon), \quad 0 < \varepsilon < 1,$$
and
$$\varphi_n(x) \equiv \varphi_{(1/n)}(x) = n\varphi(nx), \quad n = 1, 2, \ldots$$

Colombeau's definition of his **New Generalised Functions** is given in terms of equivalence classes of mappings of the form

$$T : \Phi \times \mathbb{R} \to \mathbb{C}$$
$$(\varphi, x) \to T(\varphi, x)$$

which are infinitely differentiable with respect to $x$ for each $\varphi \in \Phi$. Instead of mappings defined on a product space $\Phi \times \mathbb{R}$ and taking (complex) numerical values we prefer to work with operators carrying functions in $\Phi \subset \mathcal{D}$ into complex-valued functions in $\mathcal{C}^\infty(\mathbb{R})$. In fact Colombeau's fundamental set can be readily identified with the set (or ultrapower) $(\mathcal{C}^\infty)^\Phi$ of all mappings $T$ which carry $\Phi$ into $\mathcal{C}^\infty(\mathbb{R})$. With an obvious modification of the convention adopted in the preceding discussion of operators and systems we shall use the notation $\mathcal{E}[\Phi, \mathcal{C}^\infty]$ for the associative, commutative differential algebra formed by the set of all operators $T : \Phi \to \mathcal{C}^\infty(\mathbb{R})$ under the natural componentwise definitions of addition and multiplication, and with derivation defined by $(DT)\varphi = D(T\varphi)$. We shall continue to denote by $T_{[f]}$ the operator in $\mathcal{E}[\Phi, \mathcal{C}^\infty]$ given by

$$T_{[f]}\varphi(x) = \int_{-\infty}^{+\infty} f(y)\varphi(y-x)dy = \int_{-\infty}^{+\infty} f(y+x)\varphi(y)dy$$

for all functions $\varphi \in \Phi$, and to refer to $T_{[f]}$ as the quasi-convolution operator generated by the continuous function $f$.

An embedding of $\mathcal{C}(\mathbb{R})$ into $\mathcal{E}[\Phi, \mathcal{C}^\infty]$ can now be obtained by defining an appropriate equivalence relation for a certain sub-algebra of $\mathcal{E}[\Phi, \mathcal{C}^\infty]$ which allows the identification of each function $f$ in $\mathcal{C}(\mathbb{R})$ with the equivalence class containing the operator $T_{[f]}$. Note that this is an embedding of linear spaces but not of algebras since multiplication in $\mathcal{E}[\Phi, \mathcal{C}^\infty]$ is not coherent with the pointwise product in $\mathcal{C}(\mathbb{R})$; in general the product operator $T_{[f]}T_{[g]}$ is not the same as the quasi-convolution operator $T_{[fg]}$.

### 5.2.4 The Colombeau Algebra $\mathcal{G}(\mathbb{R})$

Let $\varphi$ be some fixed function in $\Phi$ and $\varepsilon$ any positive real number. Then for any function $f \in \mathcal{C}^0(\mathbb{R})$ the function $T_{[f]}\varphi_{(\varepsilon)}$ is well defined as

$$T_{[f]}\varphi_{(\varepsilon)}(x) = \int_{-\infty}^{+\infty} f(y)\varepsilon^{-1}\varphi\left(\varepsilon^{-1}(y-x)\right)dy = \int_{-\infty}^{+\infty} f(x+\varepsilon y)\varphi(y)dy$$

and, for any $p \in \mathbb{N}$,

$$D^p T_{[f]}\varphi_{(\varepsilon)}(x) = (-\varepsilon)^{-p} \int_{-\infty}^{+\infty} f(x+\varepsilon y)\varphi^{(p)}(y)dy.$$

This shows that quasi-convolution operators satisfy a certain growth constraint. Accordingly Colombeau defines a sub-algebra $\mathcal{E}_M[\Phi, \mathcal{C}^\infty]$ of $\mathcal{E}[\Phi, \mathcal{C}^\infty]$ which consists of all the operators whose growth is similarly constrained.

**Definition 5.3 (Moderate Operators)** *An operator $T \in \mathcal{E}[\Phi, \mathcal{C}^\infty]$ is said to be* **moderate** *if and only if for every compact $K \subset \mathbb{R}$ and each derivation operator $D^p$, where $p \in \mathbb{N}$, there exists $m \in \mathbb{N}$ such that*

*if $\varphi \in \Phi_m$ then there are numbers $\eta > 0, c > 0$ for which*

$$|D^p T \varphi_{(\varepsilon)}(x)| \leq c/\varepsilon^m$$

*for all $x \in K$ and all $\varepsilon \in (0, \eta)$.*

*We denote by $\mathcal{E}_M[\Phi, \mathcal{C}^\infty]$ the sub-algebra of all moderate operators in $\mathcal{E}[\Phi, \mathcal{C}^\infty]$.*

To define a suitable equivalence relation for moderate operators we examine the relation between a typical function $f \in \mathcal{C}^\infty(\mathbb{R}) \subset \mathcal{C}^0(\mathbb{R})$ and the quasi-convolution operator $T_{[f]}$ generated by $f$. For any $\varepsilon > 0$ and any $\varphi \in \Phi_m, m \in \mathbb{N}$, we have

$$\begin{aligned} f(x) - T_{[f]}\varphi_{(\varepsilon)}(x) &= f(x) - \int_{-\infty}^{+\infty} f(x + \varepsilon y)\varphi(y)dy \\ &= -\int_{-\infty}^{+\infty} [f(x + \varepsilon y) - f(x)]\varphi(y)dy \\ &= -\left\{\varepsilon \int_{-\infty}^{+\infty} f'(x)y\varphi(y)dy + \ldots + \frac{\varepsilon^m}{m!}\int_{-\infty}^{+\infty} f^{(m)}(x)y^m\varphi(y)dy\right\} \\ &\quad - \frac{\varepsilon^{m+1}}{(m+1)!}\int_{-\infty}^{+\infty} f^{(m+1)}(x + \varepsilon\theta y)y^{m+1}\varphi(y)dy \end{aligned}$$

where $0 < \theta < 1$. Since $\varphi \in \Phi_m$ all terms within the braces will vanish; in particular this confirms that

$$\lim_{\varepsilon \downarrow 0} T_{[f]}\varphi_{(\varepsilon)}(x) = f(x)$$

for all $x \in \mathbb{R}$. More generally, for any $p \in \mathbb{N}$ a similar argument shows that

$$f^{(p)}(x) - D^p T_{[f]}\varphi_{(\varepsilon)}(x) = -\frac{\varepsilon^{m+1}}{(m+1)!}\int_{-\infty}^{+\infty} f^{(p+m+1)}(x + \varepsilon\theta y)y^{m+1}\varphi(y)dy.$$

Recalling that for $f \in \mathcal{C}^\infty(\mathbb{R})$, $T_f$ denotes the constant map

$$\varphi \rightsquigarrow T_f\varphi(x) = f(x),$$

the result obtained above can be interpreted as showing that, in some sense, the difference between the (moderate) operators $T_{[f]}$ and $T_f$ may be regarded as negligible. More precisely we can define the negligible operators in $\mathcal{E}_M[\Phi, \mathcal{C}^\infty]$ as those which satisfy the following criterion:

**Definition 5.4 (Negligible Operators)** *First let $\Gamma$ denote the set of all monotone increasing real sequences $\alpha = (\alpha_n)_{n \in \mathbb{N}}$ which are unbounded above. An operator $T \in \mathcal{E}_M[\Phi, \mathcal{C}^\infty]$ is then said to be negligible if, for every compact $K \subset \mathbb{R}$ and each derivation operator $D^p$, where $p \in \mathbb{N}$, there exists $q \in \mathbb{N}$ and $\alpha \in \Gamma$ such that*

*if $\varphi \in \Phi_m$ where $m \geq q$, then there exist numbers $\eta > 0, c > 0$ for which*

$$|D^p T \varphi_{(\varepsilon)}(x)| \leq c \varepsilon^{\alpha_m - q}$$

*for all $x \in K$ and all $\varepsilon \in (0, \eta)$.*

*We denote by $\mathcal{N}[\Phi, \mathcal{C}^\infty]$ the subset of all negligible operators in $\mathcal{E}_M[\Phi, \mathcal{C}^\infty]$.*

$\mathcal{N}[\Phi, \mathcal{C}^\infty]$, the sub-algebra of all negligible operators, is an ideal of the algebra $\mathcal{E}_M[\Phi, \mathcal{C}^\infty]$. The **Colombeau Algebra of New Generalised Functions** is now defined as the quotient

$$\mathcal{G}(\mathbb{R}) = \mathcal{E}_M[\Phi, \mathcal{C}^\infty] / \mathcal{N}[\Phi, \mathcal{C}^\infty].$$

Thus if $\gamma$ is any member of $\mathcal{G}(\mathbb{R})$ then $\gamma$ is an equivalence class $[T]$ of operators which are moderate (i.e. which belong to the algebra $\mathcal{E}_M[\Phi, \mathcal{C}^\infty]$) and which are such that if $T_1 \in \gamma$ and $T_2 \in \gamma$ then $T_1 - T_2 \in \mathcal{N}[\Phi, \mathcal{C}^\infty]$.

**Non-moderate operators and negligible operators** Colombeau gives the following example to show that the restriction to moderate operators is necessary if we are to obtain a suitable ideal of negligible operators.

For any function $\varphi$ in $\mathcal{D}$ we denote by $d(\varphi)$ the diameter of the support of $\varphi$:

$$d(\varphi) = \sup\{|x - y| \in \mathbb{R} : x, y \in \mathrm{supp}(\varphi)\}.$$

Then $d(\varphi_{(\varepsilon)}) = \varepsilon d(\varphi)$. We can now define an operator $T_d$ in $\mathcal{E}[\Phi, \mathcal{C}^\infty]$ by writing

$$T_d \varphi(x) = \exp\left(1/d(\varphi)\right).$$

Then $T_d$ certainly belongs to $\mathcal{E}[\Phi, \mathcal{C}^\infty]$ but it is not a moderate operator. Hence the set

$$\mathcal{E}[\Phi, \mathcal{C}^\infty] \setminus \mathcal{E}_M[\Phi, \mathcal{C}^\infty]$$

is not empty. Moreover $T_d$ has an algebraic inverse in $\mathcal{E}[\Phi, \mathcal{C}^\infty]$ which is given by

$$T_d^{-1} \varphi(x) = \exp\left(-1/d(\varphi)\right)$$

and it is clear that $T_d^{-1}$ belongs to $\mathcal{N}[\Phi, \mathcal{C}^\infty]$. We have $T_d \cdot T_d^{-1} = \mathbf{1}$, where $\mathbf{1}$ denotes the operator in $\mathcal{E}_M[\Phi, \mathcal{C}^\infty]$ which maps each function $\varphi \in \Phi$ into the constant function $f(x) = 1$. Since this operator is not negligible it follows that $\mathcal{N}[\Phi, \mathcal{C}^\infty]$ cannot be an ideal of $\mathcal{E}[\Phi, \mathcal{C}^\infty]$.

**Exercise 5.4 :**

**1.** Confirm the following facts about the Colombeau indexing sets $\Phi_m$:

(a) $\Phi_m \neq \emptyset$ for any $m \in \mathbb{N}$; that is, for each $m = 1, 2, \ldots$, there exists some $\varphi \in \mathcal{D}$ such that

$$\int_{-\infty}^{+\infty} \varphi(x)dx = 1 \text{ and } \int_{-\infty}^{+\infty} x^k \varphi(x)dx = 0, \ 1 \leq k \leq m,$$

(b) for each $m = 1, 2, \ldots$, there exists some $\varphi \in \Phi_m$ such that

$$\int_0^{+\infty} x\varphi(x)dx \neq 0,$$

(c) there exists no $\varphi \in \mathcal{D}$ such that $\varphi \in \Phi_m$ for all $m \in \mathbb{N}$.

**2.** Confirm that $\mathcal{N}[\Phi, \mathcal{C}^\infty]$ is an ideal of the algebra $\mathcal{E}[\Phi, \mathcal{C}^\infty]$.

**3.** Let $f(x) = x_+(x)$ and $g(x) = x_-(x)$. Show that although the pointwise product $h(x) = f(x)g(x)$ vanishes identically, the operator $T_{[fg]}$ is not negligible, and so the generalised function $\left[T_{[fg]}\right]$ is not zero in $\mathcal{G}(\mathbb{R})$.

### 5.2.5 Distributions viewed as Colombeau generalised functions

**The embedding $\mathcal{C}^\infty(\mathbb{R}) \subset \mathcal{G}(\mathbb{R})$** Let $f$ be any function in $\mathcal{C}^\infty(\mathbb{R})$. Then $f$ may be identified with the equivalence class $[T_f]$. This defines an embedding of $\mathcal{C}^\infty(\mathbb{R})$ into $\mathcal{G}(\mathbb{R})$, with the constant function $f(x) = 1$ in $\mathcal{C}^\infty(\mathbb{R})$ being identified with the unit in the algebra $\mathcal{G}(\mathbb{R})$. Furthermore, for each integer $p \in \mathbb{N}_0$, the classical derivative $f^{(p)}$ of the function $f$ will be identified in $\mathcal{G}(\mathbb{R})$ with the operator equivalence class

$$D^p[T_f] = [D^p T_f] = [T_{f^{(p)}}].$$

Hence the embedding $\mathcal{C}^\infty(\mathbb{R}) \subset \mathcal{G}(\mathbb{R})$ is actually an embedding of differential algebras. Note, however, that the element $f \equiv [T_f]$ of $\mathcal{G}(\mathbb{R})$ does represent a *generalisation* of the function $f$ as a member of $\mathcal{C}^0(\mathbb{R})$, and admits a whole family of operators as representatives. In particular recall that $T_{[f]} \in [T_f]$ since, as shown above, for any $\varepsilon > 0$ we can always find a non-negative integer $m$ such that for $\varphi \in \Phi_m$ we have

$$D^p \{T_f - T_{[f]}\} \varphi_{(\varepsilon)}(x) = \varepsilon^{m+1} \left\{ \frac{-1}{(m+1)!} \int_{-\infty}^{+\infty} f^{(p+m+1)}(x + \varepsilon\theta y) y^{m+1} \varphi(y) dy \right\}.$$

It is sometimes convenient to write $F \equiv [T_f]$ when we wish to distinguish the generalised function in $\mathcal{G}(\mathbb{R})$ from the ordinary function $f$ in $\mathcal{C}^\infty(\mathbb{R})$.

**The embedding $\mathcal{C}(\mathbb{R}) \subset \mathcal{G}(\mathbb{R})$** For $f \in \mathcal{C}^0(\mathbb{R})$ we make the identification with the equivalence class $[T_{[f]}]$ in $\mathcal{G}(\mathbb{R})$. If $f \in \mathcal{C}^\infty(\mathbb{R}) \subset \mathcal{C}^0(\mathbb{R})$ then this agrees with the embedding described above since in that case we would have $[T_{[f]}] = [T_f]$. However the embedding of $\mathcal{C}^0(\mathbb{R})$ in $\mathcal{G}(\mathbb{R})$ is an embedding of linear spaces but not of algebras; for $f$ and $g$ in $\mathcal{C}^0(\mathbb{R})$ in general it will not be the case that

$$T_{[f]} T_{[g]} - T_{[fg]}$$

is a negligible operator. For example, if $f(x) = x_-$ and $g(x) = x_+$ then the pointwise product $f(x)g(x)$ vanishes identically. But,

$$\{T_{[f]} T_{[g]}\} \varphi_{(\varepsilon)}(0) = \varepsilon^2 \left\{ \int_{-\infty}^{0} y\varphi(y) dy \right\} \left\{ \int_{0}^{+\infty} y\varphi(y) dy \right\}$$

and for each $m \in \mathbb{N}$ we can always find $\varphi \in \Phi_m$ such that neither of the integrals in braces on the right-hand side is zero. Hence although it is true that the difference $T_{[f]} T_{[g]} - T_{[fg]}$ is in some sense *small*, it is not an operator belonging to $\mathcal{N}[\Phi, \mathcal{C}^\infty]$ and so the generalised function $[T_{[f]} T_{[g]} - T_{[fg]}]$ is not zero in $\mathcal{G}(\mathbb{R})$.

In contrast, if $f$ and $g$ belong to $\mathcal{C}^\infty(\mathbb{R}) \subset \mathcal{C}^0(\mathbb{R})$ then we have

$$T_f T_g = T_{fg}$$

and

$$T_{[f]} T_{[g]} \in [T_{fg}] = [T_f T_g] = [T_{[fg]}].$$

**The embedding $\mathcal{D}' \subset \mathcal{G}(\mathbb{R})$** For $\mu \in \mathcal{D}'$ we can make the identification with the equivalence class $[T_{[\mu]}]$ in $\mathcal{G}(\mathbb{R})$ where we define

$$T_{[\mu]} \varphi(x) = <\mu(y), \varphi(y-x)>$$

for all $\varphi \in \Phi \subset \mathcal{D}$. The derivation operator $D^p$ on $\mathcal{G}(\mathbb{R})$ does agree with distributional differentiation when restricted to $\mathcal{D}'$. Nevertheless we have again an embedding of linear spaces but not of algebras since the incoherence of multiplication already remarked in the case of the embedding of $\mathcal{C}(\mathbb{R})$ remains (and is made worse by the fact that no comprehensive definition of multiplication in $\mathcal{D}'$ can be given).

It is customary to think of a distribution as a *functional* on $\mathcal{D}$. It is certainly true that the action of $\mu$ as a functional can be generated through its representative $T_{[\mu]}$ since we have

$$T_{[\mu]}\varphi(0) = <\mu(y), \varphi(y)>$$

for every $\varphi \in \mathcal{D}$. However just as we must consider the equivalence class $[T_f] \in \mathcal{G}(\mathbb{R})$ as the generalisation of the function $f \in \mathcal{C}^\infty(\mathbb{R})$ so we need to look at distributions in some more general sense when we regard them as members of $\mathcal{G}(\mathbb{R})$.

**Example 5.2.3** Consider $\delta \in \mathcal{D}' \subset \mathcal{G}(\mathbb{R})$. First we know that $\delta$ belongs to $\mathcal{G}(\mathbb{R})$ since there exists an equivalence class $[T_{[\delta]}]$ which contains the operator $T_{[\delta]}$ given by

$$T_{[\delta]}\varphi(x) = <\delta(y), \varphi(y-x)> = \varphi(-x).$$

We also have

$$\varphi \rightsquigarrow D^2 T_{[x_+]}\varphi(x) = D^2 \int_0^{+\infty} y\varphi(y-x)dy = \int_0^{+\infty} y\varphi''(y-x)dy$$
$$= -\int_0^{+\infty} \varphi'(y-x)dy = \varphi(-x).$$

Hence, when we write $\delta = D^2 x_+$, we may interpret the derivation operator either in the operational sense appropriate to representatives of elements of $\mathcal{G}(\mathbb{R})$ or in the more familiar distributional sense appropriate to $\mathcal{D}'$. Also the functional defined on $\mathcal{D}$ by $T_{[\delta]}\check{\varphi}(0)$ coincides precisely with the usual Dirac delta distribution in the classical Schwartz sense (strictly, with the restriction of that distribution to $\Phi \subset \mathcal{D}$).

**Other generalised functions in $\mathcal{G}(\mathbb{R})$**  So far we have only considered the embedding of $\mathcal{D}'$, as a well defined independent structure, in $\mathcal{G}(\mathbb{R})$. Colombeau [11] gives a formal *definition* of a distribution as a member of $\mathcal{G}(\mathbb{R})$ which satisfies certain conditions:

> **Definition 5.5 (Colombeau)** *An element $\mu$ of $\mathcal{G}(\mathbb{R})$ is a distribution on $\mathbb{R}$ if it is a finite order derivative of some continuous function on every relatively compact open subset $\Omega$ of $\mathbb{R}$. That is to say, $\mu \in \mathcal{G}(\mathbb{R})$ is defined to be a distribution if, for every relatively compact open subset $\Omega$ of $\mathbb{R}$ there exists some function $f \in \mathcal{C}^0(\mathbb{R})$, and an integer $p \in \mathbb{N}_0$, such that the restriction of $\mu$ to $\Omega$ is equal to $D^p f$.*

There exist many kinds of generalised functions in $\mathcal{G}(\mathbb{R})$ which cannot be identified with distributions. Consider, for example, the mapping $T_{[\delta^2]}$ in $\mathcal{E}_M[\mathcal{D}, \mathcal{C}^\infty]$ which is given by

$$T_{[\delta^2]}\varphi(x) = (<\delta(y), \varphi(y-x)>)^2 = \varphi^2(-x).$$

This is a representative of the product

$$[T_{[\delta]}][T_{[\delta]}] \equiv \left([T_{[\delta]}]\right)^2$$

in $\mathcal{G}(\mathbb{R})$, and generates the nonlinear functional

$$\varphi \rightsquigarrow T_{[\delta^2]}\check{\varphi}(0) = \varphi^2(0).$$

This mapping belong to an equivalence class which defines a new generalised function which we denote by $\delta^2$; $\delta^2$ is not the derivative, to any order, of a continuous function and hence cannot be a distribution.

Again, consider the product $x\delta$. In the classical context of Schwartz distributions and the S-product we have $x\delta(x) = 0$ so that this product is equal in $\mathcal{D}'$ to the null distribution. However the operator $T_x T_{[\delta]}$ is a representative of the Colombeau generalised function $[T_{x[\delta]}]$ where

$$T_{x[\delta]}\varphi(x) = \{T_x\varphi(x)\}\{T_{[\delta]}\varphi(x)\} = x\varphi(-x)$$

so that we have

$$T_{x[\delta]}\varphi_{(\varepsilon)}(x) = \frac{x}{\varepsilon}\varphi(-x/\varepsilon).$$

Thus $T_{x[\delta]}\varphi_{(\varepsilon)}(0) = 0$ for every $\varphi \in \Phi$. On the other hand we have

$$DT_{x[\delta]}\varphi_{(\varepsilon)}(x) = \frac{1}{\varepsilon}\varphi(-x/\varepsilon) - \frac{x}{\varepsilon^2}\varphi'(-x/\varepsilon)$$

so that $DT_{x[\delta]}\varphi_{(\varepsilon)}(0) = \frac{1}{\varepsilon}\varphi(0)$. For each $m \in \mathbb{N}$ there exists $\varphi \in \Phi_m$ with $\varphi(0) \neq 0$. Hence $T_{x[\delta]} \notin \mathcal{N}[\Phi, \mathcal{C}^\infty]$ and so $x\delta$ is a non-zero generalised function in $\mathcal{G}(\mathbb{R})$.

**Exercise 5.5 :**

1. Let $m \in \mathbb{N}$ and let $T_{x^m[\delta]}$ denote the product $T_{[x^m]}T_{[\delta]}$. Show that although $x^m\delta(x) = 0$ in the sense of $\mathcal{D}'$ the operator $T_{[x^m]}T_{[\delta]}$ is not negligible, and hence $\left[T_{x^m[\delta]}\right] \neq 0$ in $\mathcal{G}(\mathbb{R})$.

## 5.3 CALCULUS OF GENERALISED FUNCTIONS

### 5.3.1 Colombeau's generalised complex numbers

The classical concept of the *value of a function at a point* has presently no obvious meaning for arbitrary members of $\mathcal{G}(\mathbb{R})$. Consider however the case of an infinitely differentiable function $f$ which, under the embedding of $\mathcal{C}^\infty(\mathbb{R})$ into $\mathcal{G}(\mathbb{R})$, is identified with the generalised function $F \equiv [T_f] \in \mathcal{G}(\mathbb{R})$. Let $x$ be some arbitrarily fixed point of $\mathbb{R}$. If $T$ is any particular operator belonging to $[T_f]$ then for each $\varphi \in \Phi$ we surely have

$$\lim_{\varepsilon \to 0} \{T\varphi_{(\varepsilon)}(x) - T_f\varphi_{(\varepsilon)}(x)\} \equiv \lim_{\varepsilon \to 0} \{T\varphi_{(\varepsilon)}(x) - f(x)\} = 0.$$

Thus the value $f(x)$ has a particular significance with respect to the generalised function $F$ at the point $x$. So much so that it seems reasonable to speak (loosely) of the complex number $f(x)$ as the value assumed by the generalised function $F$ at the point $x$, and even to use the same symbol $f$ for both the ordinary function in $\mathcal{C}^\infty(\mathbb{R})$ and the generalised function in $\mathcal{G}(\mathbb{R})$. This idea of value at a point can be made rigorous, and extended to arbitrary members of $\mathcal{G}(\mathbb{R})$, by defining a certain enlargement of the complex number system $\mathbb{C}$ itself. Colombeau defines an enlargement of the standard (complex) number system $\mathbb{C}$ as follows:

Let $x$ be some fixed point in $\mathbb{R}$. Given a generalised function $\gamma = [T] \in \mathcal{G}(\mathbb{R})$ then to each function $\varphi \in \Phi$ there corresponds a family

$$\{T\varphi(x)\}_{T \in \gamma}$$

of complex numbers. Equivalently we can say that to each $T \in \gamma$ there corresponds a certain family $\{t_{(x)}\}_{x \in \mathbb{R}}$ of functionals on $\Phi$, where we have $t_{(x)}(\varphi) = T\varphi(x)$ for $\varphi \in \Phi$. More particularly we can associate $T \in \gamma$ with a family of complex numbers

$$\left\{t_{(x)}\left(\rho_{(\varepsilon)}\right)\right\}_{x \in \mathbb{R}, \rho \in \Phi} \equiv \left\{T\rho_{(\varepsilon)}(x)\right\}_{x \in \mathbb{R}, \rho \in \Phi}$$

where $\varepsilon \in (0,1)$, and $\rho$ is some specific function in $\Phi$.

Denote by $\mathcal{E}_0$ the set $\mathbb{C}^\Phi$ (that is, the set of all complex-valued functionals $t$ on $\Phi$) equipped with the natural, componentwise definitions of addition and multiplication. Then $\mathcal{E}_0$ is an associative, commutative algebra which contains a sub-algebra $\mathcal{E}_M$ comprising those functionals which are "*moderate*" in the following sense:

> The sub-algebra $\mathcal{E}_M$ is the set of all functionals $t \in \mathbb{C}^\Phi \equiv \mathcal{E}_0$ which are moderate in the sense that

there exists $m \in \mathbb{N}$ such that if $\rho \in \Phi_m$ then there exist numbers $\eta > 0, c > 0$ for which we have

$$|t(\rho_{(\varepsilon)})| \leq c/\varepsilon^m,$$

for all $\varepsilon \in (0, \eta)$.

There is a corresponding ideal of $\mathcal{E}_M$ defined as follows:

The ideal $\mathcal{N}_0$ of $\mathcal{E}_M$ is the sub-algebra of all functionals $t \in \mathcal{E}_0$ which are negligible in the sense that

there exists $q \in \mathbb{N}$ and $\alpha \in \Gamma$ such that if $\rho \in \Phi_m$ where $m \geq q$ then there exist numbers $\eta > 0, c > 0$ for which we have

$$|t(\rho_{(\varepsilon)})| \leq c\varepsilon^{\alpha(m)-q},$$

for all $\varepsilon \in (0, \eta)$.

The quotient $\overline{\overline{\mathbb{C}}} = \mathcal{E}_M/\mathcal{N}_0$ is an associative and commutative algebra of equivalence classes which Colombeau refers to as *generalised complex numbers*. As usual we use square brackets $[\cdot]$ to denote equivalence classes belonging to $\overline{\overline{\mathbb{C}}}$. Where convenient we write $\overline{\overline{z}}$ for an arbitrary typical member of $\overline{\overline{\mathbb{C}}}$, in preference to the form $\bar{z} \in \overline{\overline{\mathbb{C}}}$ used, for example, by Rosinger [92] since this conflicts with the orthodox notation for the complex conjugates.

### 5.3.2 Embedding and association

That $\overline{\overline{\mathbb{C}}}$ is an enlargement of $\mathbb{C}$ follows at once from the embedding of $\mathbb{C}$ into $\overline{\overline{\mathbb{C}}}$ defined by

$$z \rightsquigarrow [t_z] \in \overline{\overline{\mathbb{C}}}$$

where $t_z(\varphi) = z$ for all $\varphi \in \Phi$. It is clear that $\overline{\overline{\mathbb{C}}}$ will contain elements other than those which can be identified as (copies of) ordinary complex numbers $z \in \mathbb{C}$. Among such elements of $\overline{\overline{\mathbb{C}}}$ will nevertheless be those which are, in some sense, "*close to*" ordinary complex numbers. This can be made precise through the introduction of the idea of **association**.

**Definition 5.6** *A generalised complex number $\overline{\overline{z}} \in \overline{\overline{\mathbb{C}}}$ is said to be **associated** with the ordinary complex number $z \in \mathbb{C}$ if and only if we have $\overline{\overline{z}} = [t]$, where for some $m \in \mathbb{N}$,*

$$\lim_{\varepsilon \downarrow 0}\{t(\rho_{(\varepsilon)})\} = z$$

*for all functions $\rho \in \Phi_m$.*

We follow Rosinger [92] by writing $\overline{\overline{z}} \vdash z$ whenever the generalised complex number $\overline{\overline{z}}$ is associated with the classical complex number $z$ in the sense defined above. It is easily seen that there exist elements $\overline{\overline{z}}$ of $\overline{\overline{\mathbb{C}}}$ which are not associated with any classical complex number $z \in \mathbb{C}$. For example, take for $t$ the functional $t = \delta$ on $\Phi$ so that $t(\varphi) = \varphi(0)$ for all $\varphi \in \Phi$. Then we would get

$$t(\rho_{(\varepsilon)}) = \varepsilon^{-1} t(\rho) = \varepsilon^{-1} \rho(0)$$

for any $\rho \in \Phi$, and it follows at once that we cannot have

$$\lim_{\varepsilon \downarrow 0} \{t(\rho_{(\varepsilon)})\} = z$$

for any $z \in \mathbb{C}$. Hence the generalised complex number $\overline{\overline{z}} = [t]$ is not associated with any classical number $z \in \mathbb{C}$.

Following Rosinger again we shall denote by $\overline{\overline{\mathbb{C}}}_0$ the set of all generalised complex numbers $\overline{\overline{z}}$ such that $\overline{\overline{z}}$ has some associated complex number $z \in \mathbb{C}$:

$$\overline{\overline{\mathbb{C}}}_0 = \left\{ \overline{\overline{z}} \in \overline{\overline{\mathbb{C}}} : [\exists z \in \mathbb{C} : \overline{\overline{z}} \vdash z] \right\}.$$

The following properties of association are trivial consequences of the definition:

**(1)** $\overline{\overline{z}} \in \overline{\overline{\mathbb{C}}}_0$ is associated with a **unique** complex number $z \in \mathbb{C}$, i.e.

$$[\overline{\overline{z}} \vdash z_1 \wedge \overline{\overline{z}} \vdash z_2] \Rightarrow [z_1 = z_2],$$

**(2)** Each classical complex number $z$, as a member of $\overline{\overline{\mathbb{C}}}$, belongs to $\overline{\overline{\mathbb{C}}}_0$ and is associated with itself,

**(3)** If $\overline{\overline{z}}_1$ and $\overline{\overline{z}}_2$ belong to $\overline{\overline{\mathbb{C}}}_0$ and $\overline{\overline{z}}_1 \vdash z_1, \overline{\overline{z}}_2 \vdash z_2$, then we have

$$[\overline{\overline{z}}_1 + \overline{\overline{z}}_2 \vdash z_1 + z_2] \wedge [\overline{\overline{z}}_1 \overline{\overline{z}}_2 \vdash z_1 z_2].$$

### 5.3.3 The point values of generalised functions

With the definition of a generalised complex number system, $\overline{\overline{\mathbb{C}}}$, as given above we can now interpret the members of the Colombeau algebra $\mathcal{G}(\mathbb{R})$ as *functions*, defined on $\mathbb{R}$ and taking values in $\overline{\overline{\mathbb{C}}}$. If $T$ is any operator belonging to $\mathcal{E}[\Phi, \mathcal{C}^\infty]$ and $x$ any point in $\mathbb{R}$ then there exists a functional $t_{(x)}$ on $\Phi$ given by

$$t_{(x)}(\varphi) = T\varphi(x).$$

If $T \in \mathcal{E}_M[\Phi, \mathcal{C}^\infty]$ then $t_{(x)} \in \mathcal{E}_M$ for each $x \in \mathbb{R}$; similarly, if $T \in \mathcal{N}[\Phi, \mathcal{C}^\infty]$ then $t_{(x)} \in \mathcal{N}_0$ for each $x \in \mathbb{R}$.

Now let $\gamma \equiv [T]$ be a generalised function in $\mathcal{G}(\mathbb{R})$. Then for each $x \in \mathbb{R}$ the value $\gamma(x)$ of $\gamma$ at the point $x$ is defined to be the generalised complex number

$$\left[t_{(x)}\right] \in \overline{\mathbb{C}} \equiv \mathcal{E}_M/\mathcal{N}_0.$$

**Example 5.3.1** Let $\delta = [\Delta]$ where for convenience we use the symbol $\Delta$ to denote the operator $T_{[\delta]}$ which carries each function $\varphi$ into $\check{\varphi}(x) \equiv \varphi(-x)$. Then $\Delta$ corresponds to an equivalent family $\{\delta_{(x)}\}_{x \in \mathbb{R}}$ of functionals such that $\delta_{(x)}(\varphi) = \varphi(-x)$ for each $\varphi \in \Phi$. In particular we have

$$\delta_{(0)}(\varphi) = \varphi(0).$$

Hence if $\rho \in \Phi$, with $\rho(0) \neq 0$, and $\varepsilon \in (0,1)$ then we get

$$\delta_{(0)}\left(\rho_{(\varepsilon)}\right) = \varepsilon^{-1}\delta_{(0)}(\rho) = \varepsilon^{-1}\rho(0)$$

and we cannot have

$$\lim_{\varepsilon \downarrow 0}\{\delta_{(0)}\left(\rho_{(\varepsilon)}\right)\} = z$$

for any $z \in \mathbb{C}$. Therefore the value assumed by the generalised function $\delta$ at the point $x = 0$ is a generalised complex number which is not associated with any classical complex number.

On the other hand, since any function $\rho$ belonging to $\Phi$ has compact support, it follows that if $x \in \mathbb{R}$ is any non-zero point then for all sufficiently small values of $\varepsilon > 0$ we must have $\rho_{(\varepsilon)}(x) \equiv \varepsilon^{-1}\rho(x/\varepsilon) = 0$. Hence we certainly have

$$\lim_{\varepsilon \downarrow 0}\{\delta_{(x)}\left(\rho_{(\varepsilon)}\right)\} = 0$$

for every $x \neq 0$, so that at every point in $\mathbb{R}$ the value assumed by the generalised function $\delta$ is associated with the complex number 0. In fact we can say more than this. For each $x \neq 0$ we can find some $\eta > 0$ such that $\delta_{(x)}(\rho_{(\varepsilon)}) = 0$ for all $\varepsilon \in (0,\eta)$ and therefore $\delta_{(x)}$ belongs to $\mathcal{N}_0$.

Thus the generalised function $\delta$ actually takes the value 0 at every point $x \neq 0$, which corresponds precisely with the intuitive picture of the delta function.

**Example 5.3.2** The argument used for $\delta(x)$ above is enough to show also that for any point $x \neq 0$ the generalised function $x\delta(x)$ must take the value 0. In this case it is also obvious that $x\delta(x)$ takes the value 0 when $x = 0$. Thus $x\delta(x)$ is a generalised function which vanishes identically on $\mathbb{R}$ although, as we have seen above, $x\delta(x)$ is a non-zero member of the algebra $\mathcal{G}(\mathbb{R})$. This counter-example is quoted by Rosinger [92] with the comment that the point values (in $\overline{\mathbb{C}}$) assumed by a generalised function $\gamma(x)$ as $x$ ranges over $\mathbb{R}$ are not themselves enough to determine the function $\gamma$!

This may appear to be a less than satisfactory aspect of the Colombeau theory from a conceptual point of view. The difficulty can be resolved if we adopt a new perspective in which the members of $\mathcal{G}(\mathbb{R})$ are interpreted as functions defined on a suitable enlargement of $\mathbb{R}$, as well as taking values in an extended number system. We return to this point in the next chapter when we consider so-called nonstandard representations of distributions and other generalised functions.

**Example 5.3.3** For a function $f \in \mathcal{C}^\infty(\mathbb{R})$ there are no conceptual problems since $T_f$ is a representative of $f$ in $\mathcal{E}_M[\Phi, \mathcal{C}^\infty]$ and it is easy to confirm that for every $x \in \mathbb{R}$ the value assumed by the generalised function $f$ coincides with the ordinary number $f(x)$ assumed by the classical function $f$ at $x$. However, apparently paradoxical situations can arise in connection with non-smooth functions. If $T$ is a representative of the function $|x|$ in $\mathcal{G}(\mathbb{R})$, in particular if we take $T = T_{[|x|]}$, then for any $\varphi \in \Phi$ we have

$$T\varphi(x) = \int_{-\infty}^{+\infty} |x + y|\varphi(y)dy$$

so that

$$T\varphi(0) = \int_{-\infty}^{+\infty} |y|\varphi(y)dy.$$

Hence

$$T\varphi_{(\varepsilon)}(0) = \int_{-\infty}^{+\infty} |y||\varepsilon|\varphi(y)dy = \varepsilon T\varphi(0)$$

where $\varepsilon > 0$. This certainly shows that $T\varphi_{(\varepsilon)}(0) \vdash 0 \in \mathbf{C}$. But for each $m \in \mathbb{N}$ there exists $\varphi \in \Phi_m$ such that

$$\int_{-\infty}^{+\infty} |y|\varphi(y)dy \neq 0.$$

Therefore $T$ does not belong to $\mathcal{N}[\Phi, \mathcal{C}^\infty]$ and the corresponding functional $t_{(0)}$ does not belong to $\mathcal{N}_0$. The ordinary function $|x|$ vanishes at the origin, but we have

$$T\varphi_{(\varepsilon)}(0) \neq 0$$

so that the generalised function $|x|$ does not vanish at the origin. On the other hand the value in $\overline{\mathbb{C}}$ which it assumes when $x = 0$ has the ordinary number zero as its associated complex number.

### 5.3.4 Integrals of generalised functions

The introduction of the extended number system $\overline{\overline{\mathbb{C}}}$ allows a natural and readily comprehensible generalisation of the classical concept of integral. Suppose that $K$ is some fixed compact set in $\mathbb{R}$, and let $\gamma \equiv [T]$ be an arbitrary generalised function in $\mathcal{G}(\mathbb{R})$. For any function $\varphi \in \Phi$ we know that $T\varphi$ is an infinitely smooth function and therefore that the integral

$$\int_K T\varphi(x)dx$$

is certainly well defined in the classical Lebesgue (or, the elementary Riemann) sense. To each operator $T$ in $\gamma$, therefore, there corresponds a functional $g_T$ whose value $g_T(\varphi)$ at each function $\varphi \in \Phi$ is given by the above integral.

More particularly, to each $T \in \gamma$ there corresponds a certain family of complex numbers

$$\left\{g_T(\rho_{(\varepsilon)})\right\}_{\rho \in \Phi, 0 < \varepsilon < 1} \equiv \left\{\int_K T\rho_{(\varepsilon)}(x)dx\right\}_{\rho \in \Phi, 0 < \varepsilon < 1}.$$

Clearly, if $T$ belongs to $\mathcal{N}[\Phi, \mathcal{C}^\infty]$ then the functional $g_T$ will belong to $\mathcal{N}_0$. Hence we may unambiguously define the integral of the generalised function $\gamma$ over the set $K$ as the generalised complex number given by:

$$\oint_K \gamma(x)dx \equiv \left[\int_K T\rho_{(\varepsilon)}(x)dx\right] \in \overline{\overline{\mathbb{C}}}.$$

(To avoid confusion we shall always use a special symbol to distinguish the integral of a generalised function in the above sense.) As the following examples show this definition is certainly consistent with the classical integral over $K$ of an ordinary continuous function $f$, and extends to well known examples of generalised functions such as the delta function with familiar results. More generally the usual definition of the action of a distribution $\mu \in \mathcal{D}'$ on an arbitrary test function $\varphi \in \mathcal{D}$ can be represented as a generalised integration process in this sense.

**Example 5.3.4** Let $f$ be an infinitely smooth function. Then the generalised function $f$ is the equivalence class $[T_f]$ in $\mathcal{G}(\mathbb{R})$, where the operator $T_f$ maps every function $\varphi \in \Phi$ into the function $f \in \mathcal{C}^\infty(\mathbb{R})$. For any operator $T$ in $[T_f]$ we will have

$$T_f - T \in \mathcal{N}[\Phi, \mathcal{C}^\infty]$$

and so

$$g_{T_f} - g_T \in \mathcal{N}_0$$

where

$$g_{T_f}(\varphi) = \int_K f(x)dx$$

### Sec. 5.3] Calculus of Generalised Functions 245

for all $\varphi \in \Phi$, and certainly for all $\rho_{(\varepsilon)}$ where $\rho \in \Phi$ and $\varepsilon > 0$. It follows at once that the integral of the generalised function $f$ over the compact set $K$ coincides with the classical integral of the ordinary function $f$ over $K$.

**Example 5.3.5** Now let $f$ be any continuous function on $\mathbb{R}$. Then the generalised function $f$ is the equivalence class $[T_{[f]}]$ in $\mathcal{G}(\mathbb{R})$, where $T_{[f]}$ denotes as usual the quasi-convolution operator given by

$$T_{[f]}\varphi(x) = \int_{-\infty}^{+\infty} f(x+y)\varphi(y)dy$$

for all $\varphi \in \Phi$. Hence, for $\rho \in \Phi$ and $\varepsilon > 0$ we get

$$g_{T_{[f]}}\left(\rho_{(\varepsilon)}\right) = \int_K T_{[f]}\rho_{(\varepsilon)}(x)dx = \int_K \left\{\int_{-\infty}^{+\infty} f(x+\varepsilon y)\rho(y)dy\right\} dx.$$

Now,

$$\lim_{\varepsilon \downarrow 0} \int_{-\infty}^{+\infty} f(x+\varepsilon y)\rho(y)dy = f(x),$$

uniformly on $K$, and so

$$\lim_{\varepsilon \downarrow 0} g_{T_{[f]}}(\rho_{(\varepsilon)}) = \int_K f(x)dx.$$

Therefore if $f$ is an arbitrary function in $\mathcal{C}^0(\mathbb{R})$ then we always have

$$\oint_K f(x)dx \vdash \int_K f(x)dx.$$

That is, the value (in $\overline{\overline{\mathbb{C}}}$) of the integral of the generalised function $f$ over a compact set $K$ is associated with the ordinary complex number which is the value of the classical integral of $f$ over $K$. If $f$ happens to belong to $\mathcal{C}^\infty(\mathbb{R}) \subset \mathcal{C}^0(\mathbb{R})$ then we can replace association by equality, as shown in Example 5.3.4 above.

**Example 5.3.6** Consider again the continuous function $f \in \mathcal{C}^0(\mathbb{R})$, but this time as the generator of the (regular) distribution $\mu_{(f)} \in \mathcal{D}'$. Let $\psi$ be any function in $\mathcal{D}$; then $\psi$ determines a generalised function $[T_\psi]$ and the product $\psi f$ exists in $\mathcal{G}(\mathbb{R})$ as the generalised function $[T]$ where

$$T\varphi(x) = [T_\psi T_{[f]}]\varphi(x) = \psi(x)\int_{-\infty}^{+\infty} f(y)\varphi(y-x)dy.$$

If $K$ is any compact subset of $\mathbb{R}$ such that the support of $\psi$ is contained in the interior, $\Omega$, of $K$ then the integral of the generalised function $\psi f$ is independent of $K$ itself, and so we may define

$$\oint_{-\infty}^{+\infty}(\psi f)(x)dx = \oint_K (\psi f)(x)dx.$$

Now consider the functional $g_T$ where, for any $\varphi \in \Phi$,

$$g_T(\varphi) = \int_{-\infty}^{+\infty} \psi(x) \int_{-\infty}^{+\infty} f(y)\varphi(y-x) dy dx$$

$$= \int_{-\infty}^{+\infty} \int_{-\infty}^{+\infty} \psi(y-z) f(y) \varphi(z) dy dz.$$

Then for $\rho \in \Phi$ and $\varepsilon > 0$ we have

$$g_T\left(\rho_{(\varepsilon)}\right) = \int_{-\infty}^{+\infty} \int_{-\infty}^{+\infty} \psi(y - \varepsilon z) f(y) \rho(z) dy dz.$$

Expanding $\psi(y - \varepsilon z)$ as a Taylor series, and using the fact that we can choose $\rho \in \Phi_m$ for any given $m$, we can easily show that the functional $h$ given by

$$h(\varphi) = g_T(\varphi) - \int_{-\infty}^{+\infty} \psi(y) f(y) dy$$

belongs to $\mathcal{N}_0$. Hence we have the result,

$$< \mu_{(f)}(y), \psi(y) > = \fint_{-\infty}^{+\infty} (\psi f)(y) dy$$

for any test function $\psi \in \mathcal{D}$.

If $\mu$ is an arbitrary distribution in $\mathcal{D}'$ then for any compact $K$ we can represent $\mu$ as a finite-order derivative, $D^p f$, of a continuous function $f$ on $\Omega$, the interior of $K$. An extension of the argument used in Example 5.3.6 above then shows that for any $\psi \in \mathcal{D}$ we have

$$< \mu(y), \psi(y) > = \fint_{-\infty}^{+\infty} (\psi \mu)(y) dy$$

where $\psi\mu$ denotes the product in $\mathcal{G}(\mathbb{R})$ of the generalised function generated by $\psi \in \mathcal{D}$ and $\mu \in \mathcal{D}'$. In particular we get

$$< \delta(y), \psi(y) > = \fint_{-\infty}^{+\infty} (\psi \delta)(y) dy = \psi(0).$$

### 5.3.5 Association and distributional equivalence

It has been shown above that the product of the generalised functions $x$ and $\delta(x)$ is well defined in $\mathcal{G}(\mathbb{R})$ as a non-zero generalised function, despite the fact that it vanishes identically on $\mathbb{R}$. Now both $x$ and $\delta(x)$ are distributions and, as is well known, the Schwartz product $x\delta(x)$ is equal to the null distribution in $\mathcal{D}'$; that is to say, for any function $\varphi \in \mathcal{D}$ we have:

$$< x\delta(x), \varphi(x) > = < \delta(x), x\varphi(x) > = 0.$$

Colombeau resolves this inconsistency by the introduction of weaker senses of equality for generalised functions. Let $\gamma \equiv [T]$ be a generalised function in $\mathcal{G}(\mathbb{R})$. Then,

(a) In the sense of $\mathcal{G}(\mathbb{R})$ we say that $\gamma$ is **null** ($\gamma = 0$) if and only if for every $T \in \gamma$ we have $T \in \mathcal{N}[\Phi, \mathcal{C}^\infty]$.

(b) $\gamma$ is said to be **test null** ($\gamma \sim 0$) if and only if for every $\psi \in \mathcal{D}$ we have
$$\oint_{-\infty}^{+\infty} (\psi\gamma)(x)dx = 0$$
which belongs to $\overline{\overline{\mathbb{C}}}$. Then two generalised functions $\gamma_1, \gamma_2$ are said to be **test equal** ($\gamma_1 \sim \gamma_2$) if and only if $\gamma_1 - \gamma_2$ is test null.

(c) $\gamma$ is said to be **associated** with a distribution $\mu \in \mathcal{D}' \subset \mathcal{G}(\mathbb{R})$ if and only if, for any $\psi \in \mathcal{D}$ the generalised integral of the product $\psi\gamma$ is associated with the (ordinary) complex number $< \mu, \psi >$:
$$\oint_{-\infty}^{+\infty} (\psi\gamma)(x)dx \vdash < \mu(y), \psi(y) > .$$
Then two generalised functions $\gamma_1, \gamma_2$ are associated ($\gamma_1 \approx \gamma_2$) if and only if $\gamma_1 - \gamma_2$ is associated with the null distribution.

The following properties are immediate consequences of these definitions.

1. Test nullity is weaker than equality in $\mathcal{G}(\mathbb{R})$; that is we may have $\gamma \sim 0$ while $\gamma \neq 0$. For example the generalised function $x\delta(x)$ is non-null in $\mathcal{G}(\mathbb{R})$ but we have
$$\oint_{-\infty}^{+\infty} (\psi x \delta)(x) dx = 0 \ , \ \forall \psi \in \mathcal{D}.$$

2. If $\gamma \in \mathcal{G}(\mathbb{R})$ is test null then $\gamma$ is certainly associated with the null distribution. It follows that if $\gamma_1$ and $\gamma_2$ are test equal then we must also have $\gamma_1 \approx \gamma_2$.

3. For any $\mu \in \mathcal{D}' \subset \mathcal{G}(\mathbb{R})$, if $\mu \sim 0$ or if $\mu \approx 0$ then $\mu$ coincides with the null distribution.

The relations $\sim$ and $\approx$ are obviously equivalence relations on $\mathcal{G}(\mathbb{R})$. It is due to these (more particularly to $\approx$) that Colombeau is able to offer a resolution of the multiplication problem for distributions. First we recall the fundamental incoherence of the ordinary pointwise product of continuous functions with their product as distributions, or as generalised functions in $\mathcal{G}(\mathbb{R})$.

Let $f_1$ and $f_2$ belong to $\mathcal{C}^0(\mathbb{R})$. Then they each determine a generalised function in $\mathcal{G}(\mathbb{R})$ and, for the sake of clarity, we shall temporarily use upper case symbols to distinguish these generalised functions from their classical counterparts:

$f_i \in C^0(\mathbb{R})$ determines $F_i \in \mathcal{G}(\mathbb{R})$, where $F_i \equiv [T_{[f_i]}]$.

The pointwise product $f_1 f_2$ again belongs to $C^0(\mathbb{R})$ and itself determines a generalised function (distribution) which we shall write as $F \equiv [T_{[f_1 f_2]}]$. For the product in the sense of $\mathcal{G}(\mathbb{R})$ we have

$$F_1 F_2 = [T_{[f_1]}][T_{[f_2]}] \equiv [T_{[f_1]} T_{[f_2]}]$$

and in general this is not equal to $F$ in $\mathcal{G}(\mathbb{R})$.

Now let $\psi$ be any arbitrary function in $\mathcal{D}$. For the generalised integral

$$\oint_{-\infty}^{+\infty} (\psi(F_1 F_2))(x) dx$$

we need to consider a functional of the form

$$g(\varphi) = \int_{-\infty}^{+\infty} \psi(x) \left\{ \int_{-\infty}^{+\infty} f_1(x+y)\varphi(y) dy \int_{-\infty}^{+\infty} f_2(x+z)\varphi(z) dz \right\} dx,$$

while for the generalised integral

$$\oint_{-\infty}^{+\infty} (\psi F)(x) dx$$

we need to consider a functional of the form

$$h(\varphi) = \int_{-\infty}^{+\infty} \psi(x) \left\{ \int_{-\infty}^{+\infty} f_1(x+y) f_2(x+y)\varphi(y) dy \right\} dx.$$

Then, for $\rho \in \Phi$ and $\varepsilon > 0$ we get

$$g\left(\rho_{(\varepsilon)}\right) - h\left(\rho_{(\varepsilon)}\right) = \int_{-\infty}^{+\infty} \psi(x) \left\{ \int_{-\infty}^{+\infty} f_1(x+\varepsilon y)\rho(y) dy \cdot \int_{-\infty}^{+\infty} f_2(x+\varepsilon z)\rho(z) dz \right.$$
$$\left. - \int_{-\infty}^{+\infty} f_1(x+\varepsilon y) f_2(x+\varepsilon y)\rho(y) dy \right\} dx$$

and then the familiar Taylor expansion argument is enough to show that

$$\lim_{\varepsilon \downarrow 0} \left\{ g(\rho_{(\varepsilon)}) - h(\rho_{(\varepsilon)}) \right\} = 0 \ , \ \forall \rho \in \Phi_m,$$

for every $m \in \mathbb{N}$. Therefore, although $F_1 F_2 \neq F$ in $\mathcal{G}(\mathbb{R})$ in general, we will always have $F_1 F_2 \approx F$. In Rosinger [92] a counter-example is given to show that in this last result we cannot generally replace the weak equivalence relation $\approx$ with the slightly stronger one $\sim$. If we take

$$f_1(x) = x_- \equiv \begin{cases} x & \text{, when } x \leq 0 \\ 0 & \text{, when } x > 0 \ ; \end{cases}$$

and
$$f_2(x) = x_+ \equiv \begin{cases} 0 & \text{, when } x \leq 0 \\ x & \text{, when } x > 0 \end{cases};$$

then the pointwise product $f_1 f_2$ vanishes identically in $\mathcal{C}^0(\mathbb{R})$. However

$$T_{[f_1]} T_{[f_2]} \varphi(x) = \int_{-\infty}^{-x}(x+y)\varphi(y)dy \cdot \int_{-x}^{+\infty}(x+y)\varphi(y)dy$$

which gives, in particular,

$$T_{[f_1]} T_{[f_2]} \rho_{(\varepsilon)}(0) = \varepsilon^2 \left\{ \int_{-\infty}^{0} y\rho(y)dy \right\} \left\{ \int_{0}^{+\infty} y\rho(y)dy \right\}$$

and it follows that $F_1 F_2 = [T_{[f_1]} T_{[f_2]}]$ is not zero in $\mathcal{G}(\mathbb{R})$. If $\rho$ has support contained in the interval $[a, b]$ and if $\psi$ is an arbitrary function in $\mathcal{D}$ then,

$$g\left(\rho_{(\varepsilon)}\right) = \int_{-b\varepsilon}^{-a\varepsilon} \psi(x) \left\{ \int_{a}^{-x/\varepsilon}(x+\varepsilon z)\rho(z)dz \int_{-x/\varepsilon}^{b}(x+\varepsilon z)\rho(z)dz \right\} dx$$
$$= \varepsilon^2 \int_{a}^{b} \psi(-\varepsilon y) \left\{ \int_{a}^{y}(z-y)\rho(z)dz \int_{y}^{b}(z-y)\rho(z)dz \right\} dy$$

and it follows that
$$\oint_{-\infty}^{+\infty} (\psi(F_1 F_2))(x) dx \neq 0 \in \overline{\overline{\mathbb{C}}}.$$

Therefore we must have
$$F_1 F_2 \not\sim f_1 f_2.$$

## 5.4 THE SCHWARTZ PRODUCT AND MULTIPLICATION IN $\mathcal{G}(\mathbb{R})$

For $\theta \in \mathcal{C}^\infty(\mathbb{R})$ and a distribution $\mu \in \mathcal{D}'$ there exists a distribution $\gamma \in \mathcal{D}'$ defined by the orthodox Schwartz product as follows:

$$<\gamma(y), \psi(y)> = <\mu(y), \theta(y)\psi(y)>, \quad \forall \psi \in \mathcal{D}.$$

Now denote by $\Gamma$ the generalised function given by the product $\theta\mu$, in the sense of $\mathcal{G}(\mathbb{R})$, of the generalised functions defined in $\mathcal{G}(\mathbb{R})$ by $\theta$ and $\mu$ respectively. For an arbitrary test function $\psi \in \mathcal{D}$ we have

$$\oint_{-\infty}^{+\infty}(\psi\gamma)(x)dx = <\gamma(y), \psi(y)>$$
$$= <\mu(y), \theta(y)\psi(y)> = \oint_{-\infty}^{+\infty}((\psi\theta)\mu)(x)dx.$$

Since $\theta$ and $\psi$ belong to $C^\infty(\mathbb{R})$ there is agreement between the product $\psi\theta$ in the pointwise sense, in the sense of their product as (regular) distributions, and in the sense of multiplication as generalised functions in $\mathcal{G}(\mathbb{R})$. Hence, using the associativity of multiplication in $\mathcal{G}(\mathbb{R})$, we have

$$(\psi\theta)\mu = \psi(\theta\mu) \equiv \psi\Gamma$$

and so we can conclude that

$$\oint_{-\infty}^{+\infty} (\psi\Gamma)(x)dx = \oint_{-\infty}^{+\infty} (\psi\gamma)(x)dx = <\mu, \theta\psi> \in \mathbb{C}.$$

Therefore the generalised functions $\gamma$ and $\Gamma$, generated by the Schwartz product of $\theta \in C^\infty(\mathbb{R})$ and $\mu \in \mathcal{D}'$ on the one hand and by the product of the corresponding generalised functions in $\mathcal{G}(\mathbb{R})$ on the other, will be equal in the sense of test equality:

$$\Gamma \sim \gamma$$

and hence certainly in the sense of association,

$$\Gamma \approx \gamma.$$

The counter-example $x\delta(x)$ already discussed is enough to show that $\Gamma$ and $\gamma$ will not generally be equal as generalised functions.

### 5.4.1 Examples of products in $\mathcal{G}(\mathbb{R})$ and association

The Heaviside unit step function $H$ is defined in $\mathcal{G}(\mathbb{R})$ as the derivative of the function $x_+ \in C^0(\mathbb{R}) \subset \mathcal{G}(\mathbb{R})$, so that we have $H = [T]$ where

$$T\varphi(x) = D_x \int_{-x}^{+\infty} (x+y)\varphi(y)dy = \int_{-x}^{+\infty} \varphi(y)dy.$$

If we define

$$\sigma_\varphi(x) = \int_{-\infty}^{x} \varphi(y)dy,$$

then $\sigma'_\varphi(x) = \varphi(x)$ and $\sigma_\varphi(x)$ tends to 0 as $x \to -\infty$, and tends to 1 as $x \to +\infty$ for every function $\varphi \in \Phi$. Further, for $\rho \in \Phi$ and $\varepsilon > 0$,

$$T\rho_{(\varepsilon)}(x) = \int_{-x}^{+\infty} \varepsilon^{-1}\rho(y/\varepsilon)dy = \int_{-x/\varepsilon}^{+\infty} \rho(z)dz$$
$$= \int_{-\infty}^{+\infty} \rho(z)dz - \int_{-\infty}^{-x/\varepsilon} \rho(z)dz = 1 - \sigma_\rho(-x/\varepsilon).$$

The product $HH = H^2$ in $\mathcal{G}(\mathbb{R})$ is represented by the equivalence class $[T^2]$ where

$$T^2 \rho_{(\varepsilon)}(x) = [1 - \sigma_\rho(-x/\varepsilon)]^2$$

and it is clear that $H^2 \neq H$ in the sense of $\mathcal{G}(\mathbb{R})$. But now take $\psi \in \mathcal{D}$ and consider the generalised integral

$$\oint_{-\infty}^{+\infty} \left( \psi(H^2 - H) \right)(x) dx$$

which has a typical representative functional $g_T$ where

$$g_T\left(\rho_{(\varepsilon)}\right) = \int_{-\infty}^{+\infty} \psi(x) \left\{ (1 - \sigma_\rho(-x/\varepsilon))^2 - (1 - \sigma_\rho(-x/\varepsilon)) \right\} dx$$

It is easy to show that convergence is dominated as $\varepsilon \to 0$ and therefore that

$$\lim_{\varepsilon \downarrow 0} g_T(\rho_{(\varepsilon)}) = 0.$$

Therefore we have

$$\oint_{-\infty}^{+\infty} \left( \psi(H^2 - H) \right)(x) dx \vdash 0 \;,\; \forall \psi \in \mathcal{D}$$

so that although $H$ and $H^2$ are distinct generalised functions in $\mathcal{G}(\mathbb{R})$ (contrary to what the pointwise multiplication of $H(x)$ by $H(x)$ suggests) we certainly have that $H^2 \approx H$; the argument readily extends to show that $H^q \approx H$ for any $q \in \mathbb{N}$. It follows that, for any $q \in \mathbb{N}$, we have

$$DH^q = qH^{q-1}\delta \approx \delta$$

so that in particular when $q = 2$ we have the following version of a familiar (conjectural) result

$$H\delta \approx \frac{1}{2}\delta$$

in $\mathcal{G}(\mathbb{R})$.

# 6

# Nonstandard Treatments of Generalised Functions on $\mathbb{R}$

## 6.1 INTRODUCTION TO NONSTANDARD ANALYSIS

### 6.1.1 Hyperreal numbers: an elementary ultrapower model

Nonstandard Analysis (NSA) appeared explicitly on the scene with the publication in 1966 of the definitive text by Abraham Robinson [89]. In point of fact Robinson had been exploiting nonstandard methods for some years before this and had already explored the possibility of a rigorous development of the calculus using infinitely small and infinitely large numbers. Applications in his 1966 book cover a much wider field and include in particular a nonstandard treatment of the theory of distributions.

In standard real analysis we are concerned with the study of functions defined on the classical real number system $\mathbb{R}^1$ (or, more generally, on $\mathbb{R}^m$); $\mathbb{R}$ is an ordered field which is Archimedean and which is complete in the sense of Dedekind (that is, it satisfies the least upper bound property). In Nonstandard Analysis we deal instead with functions defined on a proper extension $^*\mathbb{R}$ of $\mathbb{R}$. Examples of classical field extensions of $\mathbb{R}$ are well known, but for NSA we need a structure rich enough to possess the same properties as $\mathbb{R}$ itself in a certain sense. Now any ordered field which contains $\mathbb{R}$ as a proper subordered field must be non-Archimedean and cannot be Dedekind complete. The apparent paradox which this suggests is resolved (as will be seen below) when we take into account the significance which must be given to the phrase "*the same properties*".

Robinson's original development of NSA is based on model theory (in particular on the Lowenheim-Skolem theorem) and inevitably demands some expertise in mathematical logic. Since then other, more readily approachable, treatments have appeared, most notably the so-called "*Ultrapower Construction*" introduced by Luxemburg [70] and the *Internal Set Theory* (IST) of E. Nelson [79]. A very clear and straightforward presentation of the former is given in the article by Tom Lindstrøm in the 1988 text edited by Nigel Cutland [14]. The book by Alain Robert [88] gives a correspondingly simple introduction to the IST approach. In what follows we give a brief elementary account of the ultrapower construction which is self-contained and which is adequate for a nonstandard theory of distributions and other generalised functions. It is based on the fact that an extension *$\mathbb{R}$ of the real number system which is sufficiently rich to form the basis of a nonstandard treatment of analysis can be obtained in terms of appropriately defined equivalence classes of infinite real sequences. This lends itself rather easily to theories of generalised functions which are closely related to sequential treatments of distributions such as those of Mikusinski or of Temple. Reference should be made to the Lindstrøm article for such supplementary information as may be necessary.

### 6.1.1.1 Definition of a hyperreal number system

The set $\mathbb{R}^{\mathbb{N}}$ of all infinite real sequences $(x_n)_{n \in \mathbb{N}}$ forms a ring under the natural, componentwise definitions of addition and multiplication; there is a canonical embedding of $\mathbb{R}$ into $\mathbb{R}^{\mathbb{N}}$ if we identify each $x \in \mathbb{R}$ with the constant sequence $x_n = x$. Unfortunately $\mathbb{R}^{\mathbb{N}}$ is not a field but only a ring, and contains divisors of zero. For example if $(x_n)_{n \in \mathbb{N}} = (0, 1, 0, 1, \ldots)$ while $(y_n)_{n \in \mathbb{N}} = (1, 0, 1, 0, \ldots)$ then $(x_n)_{n \in \mathbb{N}} \cdot (y_n)_{n \in \mathbb{N}}$ is the null sequence although neither $(x_n)_{n \in \mathbb{N}}$ nor $(y_n)_{n \in \mathbb{N}}$ is null. Accordingly we seek to define an equivalence relation on $\mathbb{R}^{\mathbb{N}}$ which is just strong enough to eliminate such divisors of zero. To do this we fix a finitely additive measure $m$ on the subsets of the set $\mathbb{N}$ of natural numbers which is such that

**M1** for each $A \subset \mathbb{N}$ we have either $m(A) = 0$ or else $m(A) = 1$,

**M2** for any disjoint subsets $A, B$ of $\mathbb{N}$ we have

$$m(A \cup B) = m(A) + m(B), \quad \text{(finite additivity)},$$

**M3** $m(A) = 0$ for every finite subset $A$ of $\mathbb{N}$.

Then two infinite real sequences $(x_n)_{n\in\mathbb{N}}$ and $(y_n)_{n\in\mathbb{N}}$ are said to be equivalent, $(x_n)_{n\in\mathbb{N}} \sim (y_n)_{n\in\mathbb{N}}$, if and only if the set $\{n \in \mathbb{N} : x_n = y_n\}$ has measure 1. We shall often say for brevity that this criterion holds if $x_n = y_n$ for **nearly all** $n$. If, in particular, $x_n = y_n$ for all but finitely many values of $n$ then it is certainly true that $x_n = y_n$ for nearly all $n$ and therefore that $(x_n)_{n\in\mathbb{N}} \sim (y_n)_{n\in\mathbb{N}}$; the converse is of course not generally true. Then $\sim$ is an equivalence relation on $\mathbb{R}^\mathbb{N}$, and we denote by $^*\mathbb{R} \equiv \mathbb{R}^\mathbb{N}/\sim$ the corresponding set of equivalence classes. The equivalence class containing a sequence $(x_n)_{n\in\mathbb{N}}$ will be written as $[(x_n)_{n\in\mathbb{N}}]$ and will be called a **hyperreal number**. $^*\mathbb{R}$ is equipped with algebraic operations and an ordering relation in the natural way:

(a) **addition**
$$[(z_n)_{n\in\mathbb{N}}] = [(x_n)_{n\in\mathbb{N}}] + [(y_n)_{n\in\mathbb{N}}] \text{ if } z_n = x_n + y_n \text{ for nearly all } n,$$

(b) **multiplication**
$$[(z_n)_{n\in\mathbb{N}}] = [(x_n)_{n\in\mathbb{N}}] \cdot [(y_n)_{n\in\mathbb{N}}] \text{ if } z_n = x_n \cdot y_n \text{ for nearly all } n,$$

(c) **order**
$$[(x_n)_{n\in\mathbb{N}}] < [(y_n)_{n\in\mathbb{N}}] \text{ if } x_n < y_n \text{ for nearly all } n.$$

The hyperreal number system $^*\mathbb{R}$ is an ordered field, and $\mathbb{R}$ is embedded in $^*\mathbb{R}$ as a sub-ordered field by identifying each $x \in \mathbb{R}$ with the equivalence class $[(x)_{n\in\mathbb{N}}]$ in $^*\mathbb{R}$ which contains the constant sequence $x_n = x$ for all $n$. The map $^* : \mathbb{R} \to {^*\mathbb{R}}$ defined by
$$\mathbb{R} \ni x \rightsquigarrow {^*x} \equiv [(x, x, x, \ldots)] \in {^*\mathbb{R}}$$
is an onto, order-preserving field homomorphism which embeds $\mathbb{R}$ into $^*\mathbb{R}$. Identifying $\mathbb{R}$ with the set $\{^*x \in {^*\mathbb{R}} : x \in \mathbb{R}\}$ we consider all real numbers as elements of $^*\mathbb{R}$. The real numbers, as members of $^*\mathbb{R}$, are often referred to as the standard elements of $^*\mathbb{R}$. We have the following classification of the general (standard and nonstandard) elements of $^*\mathbb{R}$,

$\xi \in {^*\mathbb{R}}$ *is* **finite** *if* $\xi = [(x_n)_{n\in\mathbb{N}}]$ *and* $|x_n| < r$ *for nearly all $n$, where $r$ is some positive real number,*

$\xi \in {^*\mathbb{R}}$ *is* **infinitesimal** *if* $\xi = [(x_n)_{n\in\mathbb{N}}]$ *and* $|x_n| < r$ *for nearly all $n$, where $r$ is any positive real number,*

$\xi \in {^*\mathbb{R}}$ *is* **infinite** *if* $\xi = [(x_n)_{n\in\mathbb{N}}]$ *and* $|x_n| > r$ *for nearly all $n$, where $r$ is any positive real number,*

For example the numbers $0 = [(0)_{n \in \mathbb{N}}]$, $\varepsilon = [(1/n)_{n \in \mathbb{N}}]$, and $\varepsilon^2 = [(1/n^2)_{n \in \mathbb{N}}]$ are distinct infinitesimal hyperreals, and we have $0 < \varepsilon^2 < \varepsilon$. Similarly, the reciprocals $\varepsilon^{-1} \equiv \omega = [(n)_{n \in \mathbb{N}}]$ and $\varepsilon^{-2} \equiv \omega^2 = [(n^2)_{n \in \mathbb{N}}]$ are (positive) infinite hyperreals such that $\omega^2 > \omega$. It is convenient to denote the set of all finite hyperreals by $^*\mathbb{R}_{bd}$, and the set of all infinite hyperreals by $^*\mathbb{R}_\infty$; thus $^*\mathbb{R} = {^*\mathbb{R}_{bd}} \cup {^*\mathbb{R}_\infty}$. $\mathbb{R}$ is a proper subset of $^*\mathbb{R}_{bd}$ and each finite hyperreal number $\xi$ can be written uniquely as a sum

$$\xi = x + \varepsilon$$

where $x$ is a standard (i.e. real) number and $\varepsilon$ is an infinitesimal; we call $x$ the **standard part** of $\xi$ and write

$$x = \text{st}(\xi), \quad \text{or} \quad x = {}^\circ\xi.$$

**Definition 6.1** *Given any two finite hyperreal numbers $\xi$ and $\eta$ we say that they are **infinitely close**, and write $\xi \approx \eta$, if and only if $\xi - \eta$ is infinitesimal.*

The relation $\approx$ is an equivalence relation on $^*\mathbb{R}_{bd}$, and it can be shown that $^*\mathbb{R}_{bd}/\approx$ is isomorphic to $\mathbb{R}$.

**Definition 6.2** *For each real number $x \in \mathbb{R} \subset {}^*\mathbb{R}$ the set of hyperreal numbers $\{\xi \in {}^*\mathbb{R} : \text{st}(\xi) = x\}$ is called the **monad** of $x$ and is written as $\text{mon}(x)$.*

### 6.1.1.2 Canonical extensions of sets and functions

For any standard function $f$ mapping $\mathbb{R}$ into $\mathbb{R}$ (more generally, mapping $A \subset \mathbb{R}$ into $\mathbb{R}$) there is a canonical mode of extension to a function $^*f$ mapping $^*\mathbb{R}$ into $^*\mathbb{R}$. First, if $A$ is any standard subset of $\mathbb{R}$ we define the **nonstandard extension** of $A$ to be the set $^*A$ of hyperreals $\xi = [(x_n)_{n \in \mathbb{N}}]$ such that $x_n \in A$ for nearly all $n$,

$$\xi = [(x_n)_{n \in \mathbb{N}}] \in {}^*A \Leftrightarrow m\{n \in \mathbb{N} : x_n \in A\} = 1.$$

For example, the set $\mathbb{N}$ of all natural numbers is a subset of $\mathbb{R}$ which has a nonstandard extension $^*\mathbb{N}$ which we call the set of all **hypernatural numbers**. $^*\mathbb{N}$ contains infinite elements, such as $\omega = [(n)_{n \in \mathbb{N}}]$, as well as (copies of) the natural numbers themselves. It will be convenient to denote by $^*\mathbb{N}_\infty$ the set $^*\mathbb{N}\setminus\mathbb{N}$ of all infinite hypernatural numbers. Similarly if $[a,b]$ is any real interval in $\mathbb{R}$ then its extension $^*[a,b]$ is a hyperreal interval in $^*\mathbb{R}$ containing all hyperreal numbers $x = [(x_n)_{n \in \mathbb{N}}]$ such that $a \leq x_n \leq b$ for (nearly) all $n$. The set $^*\mathbb{R}$ of all hyperreal numbers is, of course, itself the nonstandard extension of $\mathbb{R}$.

We can now define the canonical extension of an arbitrary standard function $f$:

**Definition 6.3** *Let $f$ be a standard function defined on a set $A \subset \mathbb{R}$. The **nonstandard extension** of $f$ is the function $^*f$ defined on $^*A$ by*

$$^*f(\xi) = [(f(x_n))_{n \in \mathbb{N}}],$$

*for each $\xi = [(x_n)_{n \in \mathbb{N}}] \in {^*A}$.*

The restriction of $^*f$ to $A$ coincides with the original standard function $f$. For example, the standard function $f(x) = x^{-1}$ is defined on $A = \mathbb{R} \backslash \{0\}$. Its nonstandard extension $^*f(\xi) = \xi^{-1}$ is defined on $^*\mathbb{R} \backslash \{0\}$; it takes infinite hyperreal values when $\xi$ is a non-zero infinitesimal, infinitesimal values when $\xi$ is infinite and coincides with $f(x) = x^{-1}$ for all real, non-zero, values of $x$.

**Remark 6.1** *The hypercomplex number system $^*\mathbb{C}$ is defined in the obvious way as $^*\mathbb{C} = {^*\mathbb{R}} + i{^*\mathbb{R}}$. Any standard function $f$ mapping a set $A \subset \mathbb{R}$ into $\mathbb{C}$ has a canonical extension $^*f$ mapping $^*A$ into $^*\mathbb{C}$.*

**Exercise 6.1 :**

1. Prove that any finite hyperreal number $x \in {^*\mathbb{R}}$ can be written *uniquely* as a sum $x = a + \xi$, where $a \in \mathbb{R}$ and $\xi$ is infinitesimal.

2. Prove that if $^*A \subseteq {^*\mathbb{R}}$ is the nonstandard extension of a set $A \subset \mathbb{R}$ then $A \subseteq {^*A}$ with equality if and only if $A$ is a finite set.

### 6.1.2 Elementary applications of NSA

The value of NSA at an elementary level is usually illustrated by its applications to calculus and real analysis. We do so here, very briefly, to indicate some of the essential ideas.

In the usual topology for $\mathbb{R}$ a set $U$ of real numbers is a neighbourhood of a point $a \in \mathbb{R}$ if and only if $U$ contains all real $x$ sufficiently close to $a$; that is, if and only if there exists some $r > 0$ such that $\{x \in \mathbb{R} : |a - x| < r\} \subset U$. It is easy to restate this criterion in nonstandard terms. If $x = [(x_n)_{n \in \mathbb{N}}]$ is any hyperreal which is infinitely close to $^*a \equiv a$ then for any real $r > 0$ we must have $|a - x_n| < r$ for nearly all $n$; hence if $U$ is a neighbourhood of $a$ then $^*U \supset \mathrm{mon}(a)$. Suppose on the other hand that $U$ is any set of reals such that $^*U \supset \mathrm{mon}(a)$. If $U$ were not a neighbourhood of $a$ then for each $n \in \mathbb{N}$ we could find a point $x_n \in \mathbb{R}$ such that $|a - x_n| < 1/n$ and $x_n \in \mathbb{R} \backslash U$. But then $x = [(x_n)_{n \in \mathbb{N}}]$ would be a hyperreal which belongs to $\mathrm{mon}(a)$ but not to $^*U$, contrary to hypothesis. Thus we have the result

*A set $U \subset \mathbb{R}$ is a neighbourhood of $a \in \mathbb{R}$ if and only if $\mathrm{mon}(a) \subset {^*U}$.*

This nonstandard characterisation of neighbourhood allows us to frame simple nonstandard definitions of open, closed and compact sets in $\mathbb{R}$. Thus $G \subset \mathbb{R}$ is open if and only if for every real $x \in G$ we have $\text{mon}(x) \subset {}^*G$. Similarly $F \subset \mathbb{R}$ is closed if and only if for every finite hyperreal $x \in {}^*F$ we have $\text{st}(x) \in F$. And since the compact sets of $\mathbb{R}$ are just the bounded and closed subsets we have that $K \subset \mathbb{R}$ is compact if and only if every hyperreal $x \in {}^*K$ is a finite point such that $\text{st}(x) \in K$. Continuity and uniform continuity are standard concepts which can be re-expressed with advantage in nonstandard terms. Thus a standard function $f$ is continuous at a point $a \in \mathbb{R}$ if and only if ${}^*f(x)$ is infinitely close to $f(a)$ for every $x \in {}^*\mathbb{R}$ which is infinitely close to $a$. That is, $f$ is continuous at $a \in \mathbb{R}$ if and only if

$$\forall x \in {}^*\mathbb{R} : x \approx a \Rightarrow {}^*f(x) \approx f(a).$$

Similarly $f : \mathbb{R} \to \mathbb{R}$ is uniformly continuous on $A \subset \mathbb{R}$ if and only if for every $x, y \in {}^*A$ such that $x \approx y$ we have ${}^*f(x) \approx {}^*f(y)$.

A nice example of the simplifying power of nonstandard methods can be seen in a proof of the standard result that continuity on a compact set implies uniform continuity there: Suppose that $f$ is continuous on a compact set $K \subset \mathbb{R}$. Let $x$ and $y$ be any points of ${}^*K$ such that $x \approx y$. Since $K$ is compact there must exist some point $a \in K$ such that $\text{st}(x) = \text{st}(y) = a$. But by the continuity of $f$ at $a$ we must have ${}^*f(x) \approx f(a)$ and ${}^*f(y) \approx f(a)$. Hence ${}^*f(x) \approx {}^*f(y)$ and the uniform continuity of $f$ follows.

We can give the following nonstandard definition for the differentiability of a standard function $f$:

**Definition 6.4** *A function $f : \mathbb{R} \to \mathbb{R}$ is differentiable at the point $a \in \mathbb{R}$ if and only if there is a number $b \in \mathbb{R}$ such that*

$$b \approx \frac{{}^*f(x) - f(a)}{x - a}$$

*for all $x \in {}^*\mathbb{R}$ such that $x \approx a, x \neq a$.*

This allows, for example, a particularly simple derivation of the chain formula: if $g$ is differentiable at $a \in \mathbb{R}$ and $f$ is differentiable at $g(a) \in \mathbb{R}$ then $f \circ g$ is differentiable at $a$ and

$$(f \circ g)'(a) = f'(g(a)) g'(a).$$

Let $x \in \text{mon}(a), x \neq a$; all we have to prove is that

$$\text{st}\left(\frac{{}^*f({}^*g(x)) - f(g(a))}{x - a}\right) = f'(g(a)) g'(a).$$

If $^*g(x) = g(a)$ then both sides are zero; if $^*g(x) \neq g(a)$ then

$$\frac{^*f(^*g(x)) - f(g(a))}{x-a} = \frac{^*f(^*g(x)) - f(g(a))}{^*g(x) - g(a)} \cdot \frac{^*g(x) - g(a)}{x-a}$$

and the result follows at once.

Finally note the following nonstandard criterion for sequential convergence. A sequence $(a_n)_{n \in \mathbb{N}}$ of real numbers is a mapping $a : \mathbb{N} \to \mathbb{R}$ such that $a(n) = a_n$. As such it extends to a mapping $^*a$ of $^*\mathbb{N}$ into $^*\mathbb{R}$ which agrees with the original sequence on $\mathbb{N}$. If $\mu = [(m_n)_{n \in \mathbb{N}}]$ is an infinite hypernatural number then

$$^*a(\mu) \equiv {}^*a_\mu = [(a_{m_n})_{n \in \mathbb{N}}].$$

The sequence converges to $\alpha \in \mathbb{R}$ as its limit if and only if we have $^*a_\mu \approx \alpha$ for every infinite hypernatural number $\mu$:

$$\lim_{n \to \infty} a_n = \alpha \Leftrightarrow {}^*a_\mu \in \mathrm{mon}(\alpha), \, \forall \mu \in {}^*\mathbb{N}_\infty.$$

**Exercise 6.2 :**

1. (a) Use the nonstandard definition of derivative to prove that every polynomial is infinitely differentiable;

    (b) Let $f(x) = x\sin(1/x)$ for $x \neq 0$ and set $f(0) = 0$; show that $^*f(x) \approx 0$ for all infinitesimal $x \in {}^*\mathbb{R}$, but that $f$ is not differentiable at the origin.

2. Give a nonstandard proof of the following version of the Brouwer fixed point theorem: if $f$ is a continuous map of $[0,1]$ onto $[0,1]$ then there exists $x \in [0,1]$ such that $f(x) = x$.

3. Prove that a necessary and sufficient condition for an infinite series $\sum_{k=1}^{\infty} a_k$ to be convergent is that

$$\sum_{k=\nu+1}^{\eta} a_k \approx 0$$

for every pair $\nu, \eta$ of infinite hypernatural numbers such that $\eta > \nu$.

### 6.1.3 Internal and external sets and functions

It is time to qualify the remark in 6.1 that $^*\mathbb{R}$ has "*the same properties*" as $\mathbb{R}$. It is certainly true that $^*\mathbb{R}$ inherits all the first order properties of $\mathbb{R}$; for example, $^*\mathbb{R}$ is an ordered field just as $\mathbb{R}$ is itself an ordered field. But $^*\mathbb{R}$ does not inherit *all* the properties of $\mathbb{R}$, at least in a direct sense. For example the least upper bound property of $\mathbb{R}$ does not carry over directly to $^*\mathbb{R}$; $\mathbb{R}$ is itself a non-empty subset of

*ℝ which is bounded above (by any positive infinite hyperreal) but which has no least upper bound. Nevertheless there is a sense in which a form of the least upper bound property remains true in *ℝ in that it applies to a certain distinguished class of subsets of *ℝ.

> **Definition 6.5** *Each sequence $(A_n)_{n \in \mathbb{N}}$ of subsets of ℝ defines a subset $\mathcal{A}$ of *ℝ according to the rule that $x = [(x_n)_{n \in \mathbb{N}}]$ is a member of $\mathcal{A}$ if and only if $x_n \in A_n$ for nearly all $n \in \mathbb{N}$; that is,*
>
> $$x \in \mathcal{A} \equiv [(A_n)_{n \in \mathbb{N}}] \Leftrightarrow m\{n \in \mathbb{N} : x_n \in A_n\} = 1. \qquad (6.1)$$
>
> *Any such subset of *ℝ is said to be an **internal** set; all other subsets of *ℝ are called **external**.*

It is the internal subsets of *ℝ which inherit the properties of the standard subsets of ℝ. Thus it is easy to see that the least upper bound property of ℝ carries over to *ℝ only in the following modified sense:

> *Every nonempty internal subset of *ℝ which is bounded above has a least upper bound in *ℝ.*

By contrast the set mon(0) of all infinitesimal elements of *ℝ is nonempty and surely bounded above (by any positive real number), but it has no least upper bound; hence it must be an external subset of *ℝ.

If $A$ is any standard subset of ℝ then its nonstandard extension *$A$ is an internal set since it can be defined by the sequence $(A_n)_{n \in \mathbb{N}}$ in which $A_n = A$ for (nearly) all $n \in \mathbb{N}$. In particular the nonstandard extension *ℕ of the set ℕ of all natural numbers is internal. It inherits properties peculiar to ℕ in the sense that

(1) - *ℕ is a discrete subset of *ℝ (which properly contains ℕ);

(2) - *ℕ is closed under addition and multiplication;

(3) - Each member $\nu$ of *ℕ has an immediate successor, $\nu + 1$, which also belongs to *ℕ;

(4) - Each non-zero member $\nu$ of *ℕ has an immediate predecessor, $\nu - 1$, which is also in *ℕ;

(5) - Every internal subset of *ℕ which is nonempty has a first element.

On the other hand not every nonempty subset of *ℕ has a first element. The set *ℕ\ℕ ≡ *ℕ$_\infty$ of all infinite hypernatural numbers is a nonempty subset of *ℕ which has no first element (since any infinite hypernatural number has a predecessor which also is an infinite hypernatural number). Hence $^*\mathbb{N}_\infty$ cannot be an internal set; it is another example of an external set.

In a similar fashion we distinguish between internal and external functions on $^*\mathbb{R}$.

**Definition 6.6** *If $(f_n)_{n\in\mathbb{N}}$ is a sequence of ordinary (standard) functions defined on $\mathbb{R}$, then the function $F = [(f_n)_{n\in\mathbb{N}}] : {}^*\mathbb{R} \to {}^*\mathbb{R}$ defined for all $x = [(x_n)_{n\in\mathbb{N}}] \in {}^*\mathbb{R}$ by*

$$F(x) \equiv [(f_n)_{n\in\mathbb{N}}]([(x_n)_{n\in\mathbb{N}}]) = [(f_n(x_n))_{n\in\mathbb{N}}]$$

*is said to be an **internal function** on $^*\mathbb{R}$.*

More generally let $(A_n)_{n\in\mathbb{N}}$ be a sequence of nonempty subsets of $\mathbb{R}$ and $(f_n)_{n\in\mathbb{N}}$ a sequence of functions such that, for each $n$, $f_n : A_n \to \mathbb{R}$. Then the internal function $F = [(f_n)_{n\in\mathbb{N}}]$ has as its domain the internal subset $\mathcal{A} = [(A_n)_{n\in\mathbb{N}}]$ of $^*\mathbb{R}$. In particular the nonstandard extension $^*f$ of any standard function $f$ is internal since it can be defined by the constant sequence $(f_n)_{n\in\mathbb{N}}$ in which $f_n = f$ for (nearly) all $n \in \mathbb{N}$. Properties of, and concepts appropriate to, standard functions can be transferred to internal ones. Thus if each one of the functions $f_n$ is differentiable in the standard sense then we can define the nonstandard derivative $^*DF$ of the internal function $F = [(f_n)_{n\in\mathbb{N}}]$ by setting

$$^*DF(x) = [(f'_n(x_n))_{n\in\mathbb{N}}]. \qquad (6.2)$$

Then the nonstandard differential operator $^*D$ exhibits the same formal properties as the standard differential operator $D$. Thus we have, for example

$$^*D(F+G) = {}^*DF + {}^*DG$$
$$^*D(FG) = ({}^*DF)G + F({}^*DG)$$

and so on. Similarly if, for each $n$, the set $A_n$ is compact and the function $f_n$ is integrable over $A_n$ then we can define the nonstandard integral of the internal function $F$ over the internal set $\mathcal{A} \equiv [(A_n)_{n\in\mathbb{N}}]$ by

$$^*\!\!\int_\mathcal{A} F(x)dx = \left[\left(\int_{A_n} f_n(x)dx\right)_{n\in\mathbb{N}}\right] \qquad (6.3)$$

The usual elementary properties of the standard integral (linearity, mean value theorems, etc.) transfer to this nonstandard integral. If $A_n = [a_n, b_n] \subset \mathbb{R}$ and we write $\alpha = [(a_n)_{n\in\mathbb{N}}]$, $\beta = [(b_n)_{n\in\mathbb{N}}]$ then the integral of the internal function $F$ over the (possibly infinite) hyperreal interval $[\alpha, \beta]$ is given by

$$*\!\!\int_\alpha^\beta F(x)dx \equiv \left[\left(\int_{a_n}^{b_n} f_n(t)dt\right)_{n\in\mathbb{N}}\right]. \qquad (6.4)$$

Finally if in particular all the $f_n$ vanish outside the same finite interval $[a, b]$ then without ambiguity we can use the conventional $\infty$ signs as the limits in the nonstandard integral and write

$$*\!\!\int_{-\infty}^{+\infty} F(x)dx \equiv \left[\left(\int_{-\infty}^{+\infty} f_n(t)dt\right)_{n\in\mathbb{N}}\right] = \left[\left(\int_a^b f_n(t)dt\right)_{n\in\mathbb{N}}\right]. \qquad (6.5)$$

**Example 6.1.1** Let $\omega$ be the infinite number in $^*\mathbb{R}$ defined by the sequence $(n)_{n\in\mathbb{N}}$. The function $F_1 : {^*\mathbb{R}} \to {^*\mathbb{R}}$ defined by

$$F_1(\xi) = \frac{1}{2} + \frac{1}{\pi}\arctan(\omega\xi), \quad \xi \in {^*\mathbb{R}}$$

is an internal function since we have $F_1 = [(f_{1n})_{n\in\mathbb{N}}]$ where

$$f_{1n}(x) = \frac{1}{2} + \frac{1}{\pi}\arctan(nx), \quad n = 1, 2, \ldots, \quad x \in \mathbb{R}.$$

The restriction of $F_1$ to $\mathbb{R}$ is the Heaviside unit step function $H_{1/2}(x)$. Note, however, that the nonstandard extension of $H_{1/2}(x)$ is the function defined by

$$^*H_{1/2}(\xi) = \begin{cases} 1 & \text{for all } \xi > 0, \\ 1/2 & \text{for } \xi = 0, \\ 0 & \text{for all } \xi < 0, \end{cases}$$

and this does not coincide with $F_1(\xi)$ on $^*\mathbb{R}$; $F_1(\xi)$ is a nonstandard internal function which is not the extension of any standard function on $\mathbb{R}$. The (nonstandard) derivative $F_2(\xi)$ of $F_1(\xi)$ is given by

$$^*DF_1(\xi) = F_2(\xi) = \frac{1}{\pi}\frac{\omega}{1+\omega^2\xi^2}$$

and is a nonstandard internal function which can be defined as the equivalence class $[(f_{2n})_{n\in\mathbb{N}}]$ where

$$f_{2n}(x) = \frac{1}{\pi}\frac{n}{1+n^2x^2}, \quad n = 1, 2, \ldots.$$

As will be seen below $F_2$ is an example of an internal function which behaves as a nonstandard representative of a delta function. The restriction of $F_2(\xi)$ to $\mathbb{R}$ is a function which is infinitesimal for all non-zero reals and which takes infinite values at points within $\text{mon}(0)$.

**Exercise 6.3 :**

1. Prove that every nonempty internal subset of *$\mathbb{R}$ which is bounded above in *$\mathbb{R}$ has a least upper bound, and derive the so-called *overflow* and *underflow* principles:

   **overflow:** if an internal set $\mathcal{A}$ contains arbitrarily large finite elements then $\mathcal{A}$ contains at least one infinite element,

   **underflow:** if an internal set $\mathcal{A}$ contains arbitrarily small (positive) infinite elements then $\mathcal{A}$ contains at least one finite element.

2. An internal set $\mathcal{A} = [(A_n)_{n \in \mathbb{N}}]$ is said to be **hyperfinite** if and only if the sets $A_n \subset \mathbb{R}$ are finite for (nearly) all $n \in \mathbb{N}$. The hypernatural number $\mathrm{card}(\mathcal{A}) = [(\mathrm{card}(A_n))_{n \in \mathbb{N}}]$ is called the **internal cardinality** of $\mathcal{A}$. Prove that

   (a) an internal set $\mathcal{A} \subset {}^*\mathbb{R}$ is hyperfinite, with internal cardinality $\nu \in {}^*\mathbb{N}$, if and only if there exists a bijection
   $$\psi : \{1, 2, \ldots, \nu\} \to \mathcal{A},$$

   (b) every hyperfinite set has a smallest and a largest member.

### 6.1.4  NSA and distributions

Examples of internal functions such as those given above suggests the possibility of a nonstandard approach to distributions in general. If $\mu$ is any distribution and $\varphi$ any function in $\mathcal{D}$ then the convolution $\varphi * \mu$ is well defined as an infinitely smooth function by
$$(\varphi * \mu)(x) = <\mu(y), \varphi(x-y)>.$$
For $n = 1, 2, \ldots$, define $\mu_n \equiv \delta_n * \mu$, where $(\delta_n)_{n \in \mathbb{N}}$ is a delta sequence of infinitely differentiable functions. Then the sequence $(\mu_n)_{n \in \mathbb{N}}$ of regular distributions (each generated by a $\mathcal{C}^\infty$ function) converges in $\mathcal{D}'$ to $\mu$:
$$<\mu, \psi> = \lim_{n \to \infty} <\mu_n, \psi>$$
for all $\psi \in \mathcal{D}$. The sequence $(\mu_n)_{n \in \mathbb{N}}$ also defines an internal function $[(\mu_n)_{n \in \mathbb{N}}]$ which is infinitely differentiable in the nonstandard sense.

**Definition 6.7** *We denote by* $^*\mathcal{C}^\infty(\mathbb{R})$ *the set of all internal functions* $F = [(f_n)_{n \in \mathbb{N}}]$ *where* $f_n \in \mathcal{C}^\infty(\mathbb{R})$ *for (nearly) all* $n \in \mathbb{N}$.

The *differential operator *$D$ is well defined for all $F$ in $^*\mathcal{C}^\infty(\mathbb{R})$ by *$DF = [(f_n')_{n \in \mathbb{N}}]$, and each $F \in {}^*\mathcal{C}^\infty(\mathbb{R})$ is infinitely *differentiable. We have immediately that,

(1) $F \in {}^*\mathcal{C}^\infty(\mathbb{R}) \Rightarrow {}^*DF \in {}^*\mathcal{C}^\infty(\mathbb{R})$,

(2) $F, G \in {}^*\mathcal{C}^\infty(\mathbb{R}) \Rightarrow FG \in {}^*\mathcal{C}^\infty(\mathbb{R})$, and
$${}^*D(FG) = ({}^*DF)G + F({}^*DG),$$

so that ${}^*\mathcal{C}^\infty(\mathbb{R})$ is a differential algebra.

If $F = [(f_n)_{n \in \mathbb{N}}]$ is any function in ${}^*\mathcal{C}^\infty(\mathbb{R})$ then, for any $\psi \in \mathcal{D}$ with support in $[a, b]$, the nonstandard integral ${}^*\int_{-\infty}^{+\infty} F(x){}^*\psi(x)dx$ is well defined as

$${}^*\int_{-\infty}^{+\infty} F(x){}^*\psi(x)dx = \left[\left(\int_{-\infty}^{+\infty} f_n(t)\psi(t)dt\right)_{n \in \mathbb{N}}\right] \equiv \left[\left(\int_a^b f_n(t)\psi(t)dt\right)_{n \in \mathbb{N}}\right]$$

since, for each $n \in \mathbb{N}$, the standard function $f_n(t)\psi(t)$ has compact support in $[a, b]$. Therefore we can define ${}^*<F, {}^*\psi>$ by

$${}^*<F, {}^*\psi> = {}^*\int_{-\infty}^{+\infty} F(x){}^*\psi(x)dx \equiv [(<f_n, \psi>)_{n \in \mathbb{N}}].$$

If $f_n = \mu_n \equiv \delta_n * \mu$, then $\mu_n \to \mu$ in $\mathcal{D}'$ and it follows that the hyperreal number ${}^*<F, {}^*\psi>$ differs from the standard number $<\mu, \psi>$ by at most an infinitesimal. Hence, every standard distribution can be represented (non-uniquely) as a function in ${}^*\mathcal{C}^\infty(\mathbb{R})$.

**Remark 6.2** *A nonstandard treatment of distributions was already considered by Abraham Robinson in his original 1966 text [89]. The essential ideas are in fact to be found even earlier, in the classic 1958 paper by Schmieden and Laugwitz [95]. In the latter the theory is based on an extension of $\mathbb{R}$ to a partially ordered ring, ${}^\times\mathbb{R}$, rather than to a totally ordered field ${}^*\mathbb{R}$. This lacks the power of NSA proper, but has the advantage that the treatment is genuinely elementary. ${}^\times\mathbb{R}$ can be constructed explicitly by defining a weak equivalence relation $\sim_\times$ on $\mathbb{R}^\mathbb{N}$; sequences $(x_n)_{n \in \mathbb{N}}$ and $(y_n)_{n \in \mathbb{N}}$ are weakly equivalent, $(x_n)_{n \in \mathbb{N}} \sim_\times (y_n)_{n \in \mathbb{N}}$, if and only if the set $\{n \in \mathbb{N} : x_n = y_n\}$ is cofinite (that is, has finite complement) in $\mathbb{N}$. The resulting quotient ${}^\times\mathbb{R} = \mathbb{R}^\mathbb{N}/\sim_\times$ has many of the desirable properties of ${}^*\mathbb{R}$, though it is not a field and still contains divisors of zero. Nevertheless standard sets and functions can be extended from $\mathbb{R}$ to ${}^\times\mathbb{R}$ in much the same way as for the hyperreal case, and nonstandard representatives of $H(x)$ and $\delta(x)$, etc. can be defined. The definition of a true hyperreal number system ${}^*\mathbb{R}$ requires a strong equivalence relation $\sim$ and this depends on the assumption that a measure $m$ on $\mathbb{N}$ exists which satisfies the conditions $M(1)$ - $M(3)$ of §6.1.1.1. The existence of such a measure can be inferred from Zorn's Lemma. We cannot give an explicit characterization of $m$ since this would involve infinitely many special definitions. Somewhat surprisingly this turns* out not to matter. For example, since *ℝ is a field (and therefore has no divisors of zero) we know that either

$$[(x_n)_{n\in\mathbb{N}}] \equiv [(0,1,0,1,0,\ldots)] = 0 \ \text{and} \ [(y_n)_{n\in\mathbb{N}}] \equiv [(1,0,1,0,1,\ldots)] = 1,$$

or else that

$$[(x_n)_{n\in\mathbb{N}}] \equiv [(0,1,0,1,0,\ldots)] = 1 \ \text{and} \ [(y_n)_{n\in\mathbb{N}}] \equiv [(1,0,1,0,1,\ldots)] = 0,$$

but we do not need to know in practice which alternative actually obtains.

**Exercise 6.4 :**

1. For each $n \in \mathbb{N}$ let $f_n$ be a continuous function of the real interval $[a_n, b_n]$. If $\alpha = [(a_n)_{n\in\mathbb{N}}]$ and $\beta = [(b_n)_{n\in\mathbb{N}}]$ then the internal function $F = [(f_n)_{n\in\mathbb{N}}]$ is well defined on the hyperreal interval $[\alpha, \beta]$. Show that there exists a hyperreal $\xi \in (\alpha, \beta)$ such that
$$^*\!\!\int_{[\alpha,\beta]} F(x)dx = (\beta - \alpha)F(\xi).$$

2. Show that there is a similar "*transfer*" of the generalised Mean Value Theorem of the Integral Calculus to *ℝ.

3. Let $F(x) = \omega/\pi(1 + \omega^2 x^2)$ for all $x \in {}^*\mathbb{R}$, where $\omega = [(n)_{n\in\mathbb{N}}]$. If $\varphi$ is any continuous function with compact support $K$ show that
$$^*\!\!\int_{^*K} F(x)^*\varphi(x)dx \approx \varphi(0).$$

4. Establish a corresponding "*sampling property*" for the nonstandard derivative of the internal function $F(x)$ defined in question 3.

## 6.2 NONSTANDARD REPRESENTATIONS OF DISTRIBUTIONS

### 6.2.1 Internal functions and pre-distributions

The differential algebra $^*\mathcal{C}^\infty(\mathbb{R})$ of internal functions on $^*\mathbb{R}$ contains all internal functions $^*f$ which are the nonstandard extensions of standard infinitely differentiable functions in $\mathcal{C}^\infty(\mathbb{R})$. It also contains internal functions which are not extensions of any standard function. Among these there will be representatives of distributions. We cannot identify a distribution $\mu$ with a particular internal function $F$ in $^*\mathcal{C}^\infty(\mathbb{R})$; there are infinitely many nonstandard representatives of $\mu$, each of which will be referred to as a **pre-distribution**. What is more there are internal functions in $^*C^\infty(\mathbb{R})$ which do not represent distributions at all. The case of the delta function itself illustrates both these points.

Suppose that $\theta$ is a function in $\mathcal{D} \subset C^\infty(\mathbb{R})$ for which we have $\int_{-\infty}^{+\infty} \theta(x)dx = 1$ and that we define a sequence $(\theta_n)_{n \in \mathbb{N}}$ by setting $\theta_n(x) = n\theta(nx)$ for $n = 1, 2, \ldots$. Then the internal function $\delta_{(\theta)}(\xi) = \omega^*\theta(\omega\xi)$ belongs to $^*C^\infty(\mathbb{R})$. Since $\theta$ has compact support it follows that $\delta_{(\theta)}$ must vanish outside $\mathrm{mon}(0)$ and therefore that $\delta_{(\theta)}(x) = 0$ at every non-zero point $x \in \mathbb{R} \subset {}^*\mathbb{R}$. If $\theta(0) = 0$ then $\delta_{(\theta)}(0) = 0$ also; if $\theta(0) \neq 0$ then $\delta_{(\theta)}(0)$ will be an infinite hyperreal number. Hence in neither case can $\delta_{(\theta)}$ be the nonstandard extension of some standard function. $\delta_{(\theta)}$ is of course a nonstandard representative of the Dirac delta distribution. There are infinitely many such representatives, corresponding to arbitrary choices of $\theta$ in $\mathcal{D}$, and each of them is a *function* in the proper sense of the term, albeit a function which maps $^*\mathbb{R}$ into $^*\mathbb{R}$. There is therefore some justification for the use of the term *delta function* in this context. Moreover the properties usually attributed to the delta function in the conventional sense of the engineering or physics literature can be established quite rigorously in nonstandard terms. Thus, following for example the treatment given by Laugwitz [61], let $\eta$ be an arbitrary infinite hypernatural number, say $\eta = [(m_n)_{n \in \mathbb{N}}] \in {}^*\mathbb{N}_\infty$. Then the nonstandard integral of $^*\theta$ from $-\eta$ to $+\eta$ is well defined as

$$^*\!\!\int_{-\eta}^{+\eta} {}^*\theta(x)dx = \left[\left(\int_{-m_n}^{+m_n} \theta(t)dt\right)_{n \in \mathbb{N}}\right] \approx 1,$$

and it follows readily that, for any hyperreal $\alpha$ which is such that $\alpha \geq \eta/\omega$, we have

$$1 \approx {}^*\!\!\int_{-\eta}^{+\eta} {}^*\theta(x)dx$$
$$= {}^*\!\!\int_{-\eta/\omega}^{+\eta/\omega} {}^*\theta(\omega\tau)d(\omega\tau) = {}^*\!\!\int_{-\eta/\omega}^{+\eta/\omega} \delta_{(\theta)}(\tau)d\tau, \qquad (6.6)$$

and

$$^*\!\!\int_{-\alpha}^{-\eta/\omega} \delta_{(\theta)}(\tau)d\tau \approx 0 \approx {}^*\!\!\int_{+\eta/\omega}^{+\alpha} \delta_{(\theta)}(\tau)d\tau. \qquad (6.7)$$

Note that (6.6) and (6.7) remain true for any positive infinite hyperreal $\eta$, whether or not $\eta$ is an integer and whether or not $\eta > \omega$. We could, for example, take $\eta = \sqrt{\omega}$ so that $\eta/\omega \approx 0$.

Suppose in particular that $\theta(x) \geq 0$ everywhere, so that we have $\delta_{(\theta)}(\xi) \geq 0$ for all $\xi \in {}^*\mathbb{R}$. Then for any standard function $f$ which is bounded and continuous on $\mathbb{R}$ and for any positive infinite hyperreal $\eta$ and any positive infinitesimal $\varepsilon$ such that $\varepsilon\omega$ is infinite,

$$^*\!\!\int_{-\eta}^{+\eta} \delta_{(\theta)}(\tau) {}^*f(\tau)d\tau \approx \left\{{}^*\!\!\int_{-\eta}^{-\varepsilon} + {}^*\!\!\int_{+\varepsilon}^{+\eta} + {}^*\!\!\int_{-\varepsilon}^{+\varepsilon}\right\} \delta_{(\theta)}(\tau) {}^*f(\tau)d\tau.$$

Since $f$ is bounded and continuous it follows at once that the first two integrals are infinitesimal. For the remaining integral we can use the fact that the (generalised) mean value theorem of the integral calculus transfers to nonstandard integrals to get

$$\int_{-\varepsilon}^{*+\varepsilon} \delta_{(\theta)}(\tau)\, {}^*f(\tau)d\tau \approx {}^*f(\xi)\int_{-\varepsilon}^{*+\varepsilon} \delta_{(\theta)}(\tau)d\tau \approx {}^*f(\xi),$$

where $|\xi| < \varepsilon$. The continuity of $f$ ensures that ${}^*f(\xi) \approx f(0)$ and so we can write

$$\int_{-\eta}^{*+\eta} \delta_{(\theta)}(\tau)\, {}^*f(\tau)d\tau \approx {}^*f(\xi)\int_{-\varepsilon}^{*+\varepsilon} \delta_{(\theta)}(\tau)d\tau \approx f(0).$$

The delta function therefore appears quite naturally in an elementary context as a genuine function $\delta_{(\theta)}$ (although one which is defined on, and takes values in, a hyperreal number system). Each such internal function $\delta_{(\theta)}$ is a pre-distribution. On the other hand, given any such function $\theta \in \mathcal{D}$, there will exist an internal function $\delta^2_{(\theta)} \in {}^*\mathcal{C}^\infty(\mathbb{R})$ determined by the pointwise product

$$\delta_{(\theta)}(x) \cdot \delta_{(\theta)}(x) = \delta^2_{(\theta)}(x), \; \forall x \in {}^*\mathbb{R}.$$

This time if $\varphi \in \mathcal{D}$ then for some positive infinitesimal we have

$$\int_{-\eta}^{*+\eta} \delta^2_{(\theta)}(\tau)\, {}^*\varphi(\tau)d\tau \approx \int_{-\varepsilon}^{*+\varepsilon} \delta^2_{(\theta)}(\tau)\, {}^*\varphi(\tau)d\tau$$
$$= \delta_{(\theta)}(\xi)\int_{-\eta}^{*+\eta} \delta_{(\theta)}(\tau)\, {}^*\varphi(\tau)d\tau \approx \delta_{(\theta)}(\xi)\varphi(0)$$

where $|\xi| < \varepsilon$. In general $\delta_{(\theta)}(\xi)$ may be infinite and it is clear that $\delta^2_{(\theta)}$ cannot a pre-distribution.

To develop a systematic nonstandard theory of distributions we need some criterion to distinguish those internal functions in ${}^*\mathcal{C}^\infty(\mathbb{R})$ which are pre-distributions. There are several ways in which such a criterion might be defined. Among the most straightforward is to use the fact that every distribution is locally the finite order derivative of a continuous function. To interpret this in a nonstandard context it is necessary to look a little more closely at the concept of continuity as it applies to nonstandard internal functions.

### Exercise 6.5 :

1. Let $\theta$ a function in $\mathcal{D}$ such that $\int_{-\infty}^{+\infty} \theta(x)dx = 1$, but which is not necessarily everywhere non-negative. For $n = 1, 2, \ldots$, define

$$h_n(x) = \int_{-\infty}^{x} n\theta(nt)dt \equiv \int_{-\infty}^{nx} \theta(\tau)d\tau.$$

Show that the internal function $H_{(\theta)} = [(h_n)_{n\in\mathbb{N}}]$ is a nonstandard representative of the Heaviside step function in the sense that

$$H_{(\theta)}(x) = \begin{cases} 1 & \text{for } x \geq +1/\sqrt{\omega}, \\ 0 & \text{for } x \leq -1/\sqrt{\omega}. \end{cases}$$

2. Let $H_{(\theta)}$, as defined in question 1, satisfy the condition that there exists a real constant $M$ such that $|H_{(\theta)}(x)| \leq M$ for all $x \in {}^*\mathbb{R}$. Prove that if $f$ is any continuously differentiable function which vanishes identically outside some finite real interval $[-a, +a]$ then

$${}^*\!\!\int_{-\infty}^t \delta_{(\theta)}(\tau)\, {}^*\!f(\tau)\,d\tau \approx {}^*\!f(1/\omega) \approx f(0),$$

for all $t \geq 1/\sqrt{\omega}$.

### 6.2.2 *continuity and S-continuity

The transfer to ${}^*\mathbb{R}$ of the standard criterion for continuity on $\mathbb{R}$ gives rise to the concept of *continuity:

**Definition 6.8** *The internal function $F : {}^*\mathbb{R} \to {}^*\mathbb{R}$ is said to be *continuous at a point $a \in {}^*\mathbb{R}$ (standard or nonstandard) if and only if given $\xi \in {}^*\mathbb{R}^+$ there exists a corresponding $\eta \in {}^*\mathbb{R}^+$ such that, for all $x \in {}^*\mathbb{R}$,*

$$|x - a| < \eta \Rightarrow |F(x) - F(a)| < \xi.$$

A necessary and sufficient condition for an internal function $F = [(f_n)_{n\in\mathbb{N}}]$ to be *continuous on ${}^*\mathbb{R}$ is that, for (nearly) all $n \in \mathbb{N}$ each function $f_n$ is continuous in the standard sense on $\mathbb{R}$. Hence, in particular, every function in ${}^*\mathcal{C}^\infty(\mathbb{R})$ is *continuous.

**Definition 6.9** *If an internal function $F$ is such that $F(x)$ is finite for all hyperreal $x$, then the map*

$$°F \equiv \text{st}(F) : \mathbb{R} \to \mathbb{R}$$

*defined for all $x \in {}^*\mathbb{R}_{\text{bd}}$ by $°F(x) = \text{st}(F(x))$, is called the* **standard part** *or the* **shadow** *of $F$.*

The shadow of an internal function $F = [(f_n)_{n\in\mathbb{N}}]$, whenever it does exist, is the pointwise limit of the representative sequence $(f_n)_{n\in\mathbb{N}}$. The pointwise limit of a sequence of continuous functions is not necessarily itself continuous. Hence the fact that an internal function $F = [(f_n)_{n \in \mathbb{N}}]$ is *continuous at a standard point of its domain does not generally imply that its shadow, $°F$, will be continuous in the standard sense at that point. However, another concept of continuity results if we modify Definition 6.8 slightly so as to apply the standard criterion for continuity *directly* in *$\mathbb{R}$:

> **Definition 6.10** *The internal function $F : {}^*\mathbb{R} \to {}^*\mathbb{R}$ is said to be* **S-continuous** *at a point $a \in {}^*\mathbb{R}$ (standard or nonstandard) if and only if, given $r \in \mathbb{R}^+$ there exists a corresponding $s \in \mathbb{R}^+$ such that, for all $x \in {}^*\mathbb{R}$*
> $$|x - a| < s \;\Rightarrow\; |F(x) - F(a)| < r.$$

A necessary and sufficient condition for an internal function $F$ to be S-continuous at $a$ is that for all $x \in {}^*\mathbb{R}$, we have

$$x \approx a \;\Rightarrow\; F(x) \approx F(a).$$

S-continuity and *continuity are independent properties of internal functions; an internal function $F$ may be *continuous but not S-continuous, or S-continuous but not *continuous. However the situation when $F$ is the nonstandard extension of a standard function is comparatively straightforward:

If $I$ is any real interval then from the nonstandard criteria for (standard) continuity and uniform continuity we have immediately that

> *(1) A standard function $f : I \to \mathbb{R}$ is continuous at a (standard) point $a \in I$ if and only if its nonstandard extension ${}^*f : {}^*I \to {}^*\mathbb{R}$ is S-continuous at $a$.*

> *(2) If ${}^*f$ is S-continuous at every point $x \in {}^*I$ (standard or nonstandard) then $f$ is uniformly continuous on $I$.*

For a general internal function $F$ the above statements must be modified as follows:

> *(1) If the internal function $F : {}^*I \to {}^*\mathbb{R}$ is finite-valued and S-continuous at every (standard) point then the shadow of $F$ exists and is continuous in the standard sense on $I$.*

> *(2) If $F : {}^*I \to {}^*\mathbb{R}$ is finite-valued and S-continuous at every point (standard or nonstandard) of ${}^*I \subset {}^*\mathbb{R}$ then the shadow of $F$ exists and is uniformly continuous on $I$.*

Every S-continuous function in $^*\mathcal{C}^\infty(\mathbb{R})$ is a pre-distribution. However, not every function in $^*\mathcal{C}^\infty(\mathbb{R})$ is S-continuous; consider for example the internal delta function $\delta_{(\theta)}(\xi) = \omega^*\theta(\omega\xi)$, $\xi \in {}^*\mathbb{R}$, where $\theta(x) = \frac{1}{\pi(1+x^2)}$, $x \in \mathbb{R}$. This belongs to $^*\mathcal{C}^\infty(\mathbb{R})$ and is a pre-distribution but it is not S-continuous. On the other hand $\delta_{(\theta)}$ is the second-order nonstandard derivative of an internal function in $^*\mathcal{C}^\infty(\mathbb{R})$ which is itself S-continuous. For we have

$$\delta_{(\theta)}(x) = {}^*D\left\{\frac{1}{\pi}\arctan(\omega x)\right\} = {}^*D^2\Phi(x)$$

where $\Phi(x) = \frac{x}{\pi}\arctan(\omega x) - \frac{1}{2\pi\omega}\log(1+\omega^2 x^2)$.

In general we know that every distribution can be expressed locally as a finite-order derivative of a continuous function. Hence we should be able to identify the pre-distributions as those internal functions in $^*\mathcal{C}^\infty(\mathbb{R})$ which are either S-continuous or which can be expressed (locally) as finite-order nonstandard derivatives of S-continuous functions. This is equivalent to showing that $\mathcal{D}'$ is in some sense contained in the algebra $^*\mathcal{C}^\infty(\mathbb{R})$. This inclusion is not canonical since to each distribution there will correspond an infinite family of pre-distributions in $^*\mathcal{C}^\infty(\mathbb{R})$. Hence it will also be necessary to define an appropriate equivalence relation on the set of pre-distributions. We consider first the relatively simple case of pre-distributions of finite order.

**Exercise 6.6 :**

1. The internal function $F = [(f_n)_{n \in \mathbb{N}}]$ is *differentiable at the point $x = [(x_n)_{n \in \mathbb{N}}] \in {}^*\mathbb{R}$ if and only if for (nearly) all $n \in \mathbb{N}$ each standard function $f_n$ is differentiable in the standard sense at $x_n \in \mathbb{R}$. Prove that a necessary and sufficient condition for $F$ to be *differentiable at $x$ is that there exist a number $b$ in $^*\mathbb{R}$ such that given $\xi \in {}^*\mathbb{R}^+$ there exists a corresponding $\eta \in {}^*\mathbb{R}^+$ such that

$$\forall \tau \in {}^*\mathbb{R}\ [0 < |\tau| < \eta \Rightarrow |\{F(x+\tau) - F(x)\}/\tau| < \xi].$$

2. Prove that an internal function $F$ is S-continuous at a point $a \in {}^*\mathbb{R}$ if and only if for all $x \in {}^*\mathbb{R}$,

$$x \approx a \Rightarrow F(x) \approx F(a).$$

3. Verify that the internal function

$$\Phi(x) = \frac{x}{\pi}\arctan(\omega x) - \frac{1}{2\omega}\log(1+\omega^2 x^2)$$

considered above is S-continuous on $^*\mathbb{R}_b$.

4. Give an example of an internal S-continuous function on $^*\mathbb{R}$ which is not *continuous.

### 6.2.3 Finite-order pre-distributions on an interval

For simplicity we have worked so far in terms of standard functions defined on $\mathbb{R}$, and correspondingly with internal functions defined on $^*\mathbb{R}$. It is more convenient to continue with a mildly more general approach. Let $I$ be an interval of $\mathbb{R}$ with more than one point. Then we denote by $^*\mathcal{C}^\infty(I)$ the algebra of all infinitely $^*$differentiable internal functions on $^*I$. The restriction of $F$ in $^*\mathcal{C}^\infty(\mathbb{R})$ to $^*I$, denoted by $F|_{^*I}$, is determined by the canonical restriction of the representative sequences to $I$; that is, if $F = [(f_n)_{n \in \mathbb{N}}] \in {^*\mathcal{C}^\infty}(\mathbb{R})$ then $F|_{^*I} = [(f_n|_I)_{n \in \mathbb{N}}]$, where $f_n|_I$ is the restriction of $f_n$ to $I$. Clearly we have that $F|_{^*I}$ belongs to $^*\mathcal{C}^\infty(I)$; conversely, we can show that $^*\mathcal{C}^\infty(I)$ is the restriction of $^*\mathcal{C}^\infty(\mathbb{R})$ to $^*I$. (More generally we may define the algebra $^*\mathcal{C}^\infty(\Omega)$ for any open subset $\Omega$ of $\mathbb{R}$.)

> **Definition 6.11** *Given any interval $I \subset \mathbb{R}$ (with more than one point), we denote by $^s\mathcal{C}(I)$ the $^*\mathbb{R}_{bd}$-submodule of all functions in $^*\mathcal{C}^\infty(I)$ which are finite-valued and S-continuous at every point of $I$.*

It follows immediately from the definition that the restriction to $I$ of any member of $^s\mathcal{C}(I)$ is a well defined continuous standard function on $I$. Every function in $^s\mathcal{C}(I)$ is therefore certainly a pre-distribution.

We define a $^*$indefinite integral operator $^*\mathfrak{I}_a$, where $a$ is a given point in $^*I$, as follows:

> **Definition 6.12** *Given any function $F = [(f_n)_{n \in \mathbb{N}}] \in {^*\mathcal{C}^\infty}(\mathbb{R})$, and any point $a = [(a_n)_{n \in \mathbb{N}}] \in {^*\mathbb{R}_{bd}}$, we define the internal function $^*\mathfrak{I}_a F$ by*
> $$^*\mathfrak{I}_a F(x) = \left[ \left( \int_{a_n}^{x_n} f_n(t) dt \right)_{n \in \mathbb{N}} \right]
> *where $x = [(x_n)_{n \in \mathbb{N}}] \in {^*\mathbb{R}}$.*

Once again it follows immediately from the definition of $^*\mathfrak{I}_a F$ that

(1) $F \in {^*\mathcal{C}^\infty}(\mathbb{R}) \Rightarrow {^*\mathfrak{I}_a F} \in {^*\mathcal{C}^\infty}(\mathbb{R})$,

(2) $^*\mathfrak{I}_a \{^*DF\} = F - F(a)$ and $^*D\{^*\mathfrak{I}_a F\} = F$.

Moreover we can show that if $F$ is S-continuous then so also is every $^*$indefinite integral $^*\mathfrak{I}_a F$ of $F$. On the other hand for an arbitrary function $F$ belonging to $^s\mathcal{C}(I)$ we cannot generally assert that $^*DF$ belongs to $^s\mathcal{C}(I)$. If we denote by $^*D\{^s\mathcal{C}(I)\}$ the set of all $^*$derivatives of the functions in $^s\mathcal{C}(I)$ then the most we can say in general is that
$$^*D\{^s\mathcal{C}(I)\} \subset {^*\mathcal{C}^\infty}(I).$$

More generally, for any $j \in \mathbb{N}_0$, let $^*D^j\{{}^s\mathcal{C}(I)\}$ denote the set of all $j$th $^*$derivatives of functions in ${}^s\mathcal{C}(I)$. Then for $F \in {}^*D^j\{{}^s\mathcal{C}(I)\}$ there exists some function $\Phi \in {}^s\mathcal{C}(I)$ such that $F = {}^*D^j\Phi$ and therefore such that $F = {}^*D^{j+1}\{{}^*\mathfrak{J}_a\Phi\}$, where ${}^*\mathfrak{J}_a\Phi$ is a function in ${}^s\mathcal{C}(I)$. Hence we have the result,

$$\forall j \in \mathbb{N}_0 : {}^*D^j\{{}^s\mathcal{C}(I)\} \subset {}^*D^{j+1}\{{}^s\mathcal{C}(I)\},$$

(the inclusion being strict).

**Definition 6.13** *We denote by ${}^s\mathbf{D}(I) \equiv {}^*D^\infty\{{}^s\mathcal{C}(I)\}$ the space of all internal functions $F \in {}^*\mathcal{C}^\infty(I)$ such that $F$ is a finite-order $^*$derivative of some S-continuous function:*

$$\begin{aligned}{}^s\mathbf{D}(I) &= \bigcup_{j=0}^\infty {}^*D^j\{{}^s\mathcal{C}(I)\} \\ &= \left\{F \in {}^*\mathcal{C}^\infty(I) : [\exists j \in \mathbb{N}_0, \Phi \in {}^s\mathcal{C}(I) : F = {}^*D^j\Phi]\right\}.\end{aligned}$$

The $^*\mathbb{R}_{bd}$-submodule ${}^s\mathbf{D}(I)$ contains all those functions in $^*\mathcal{C}^\infty(I)$ which are nonstandard representatives of (finite-order) distributions on $I$; that is, the members of ${}^s\mathbf{D}(I)$ are precisely the pre-distributions of finite order.

**Remark 6.3** *It is sometimes convenient to describe a function $F$ belonging to ${}^s\mathcal{C}(I)$ as S-differentiable if it does have a derivative $^*DF$ which is S-continuous itself. More generally if ${}^*D^jF \in {}^s\mathcal{C}(I)$ for some $j \in \mathbb{N}_0$ then $F$ will be said to be $j$-times S-differentiable. We write ${}^s\mathcal{C}^j(I)$ for the space of all functions which are $j$-times S-differentiable, and ${}^s\mathcal{C}^\infty(I)$ for the space of all internal functions in ${}^s\mathcal{C}(I)$ which are infinitely S-differentiable:*

$$^s\mathcal{C}^\infty(I) = \{F \in {}^s\mathcal{C}(I) : [{}^*DF \in {}^s\mathcal{C}(I), \forall j \in \mathbb{N}_0]\}.$$

*Then we have the following (strict) inclusions*

$$^s\mathcal{C}^\infty(I) \subset {}^s\mathcal{C}(I) \subset {}^s\mathbf{D}(I) \subset {}^*\mathcal{C}^\infty(I).$$

### 6.2.4 Distributional equivalence

In the approach to nonstandard distribution theory proposed by Laugwitz [61] distributional equivalence is defined in the following terms:

> *Two internal functions $F, G$ will be called distributionally equivalent if there exist an integer $k \in \mathbb{N}_0$ and two internal functions $\Phi, \Psi$ such that*

(1) $\Phi(\xi) \approx \Psi(\xi)$ *for all finite* $\xi \in {}^*\mathbb{R}$, *and*

(2) $F = {}^*D^k\Phi$, *and* $G = {}^*D^k\Psi$.

In the context of the nonstandard distribution theory developed above we define an equivalence relation on ${}^s\mathbf{D}(I)$ which is consistent with the Laugwitz definition as follows:

If $F_1$ and $F_2$ belong to ${}^s\mathbf{D}(I)$ then there exist functions $\Phi_1$ and $\Phi_2$ in ${}^s\mathcal{C}(I)$, and integers $r, s \in \mathbb{N}_0$ such that

$$F_1 = {}^*D^r\Phi_1 \text{ and } F_2 = {}^*D^s\Phi_2.$$

Taking $m \in \mathbb{N}_0$ such that $m \geq \max\{r, s\}$ and defining recursively ${}^*\mathfrak{J}_a^m = {}^*\mathfrak{J}_a \circ {}^*\mathfrak{J}_a^{m-1}$, with ${}^*\mathfrak{J}_a^0$ the identity, we get

$$ {}^*\mathfrak{J}_a^m F_1 = {}^*\mathfrak{J}_a^{m-r}\Phi_1 - \sum_{j=0}^{r-1} \frac{{}^*D^j\Phi_1(a)}{(m-r+j)!} x^{m-r+j} \equiv {}^*\mathfrak{J}_a^{m-r}\Phi_1 + p_{1m},$$

and,

$$ {}^*\mathfrak{J}_a^m F_2 = {}^*\mathfrak{J}_a^{m-s}\Phi_2 - \sum_{j=0}^{s-1} \frac{{}^*D^j\Phi_2(a)}{(m-s+j)!} x^{m-s+j} \equiv {}^*\mathfrak{J}_a^{m-s}\Phi_1 + p_{2m}$$

where ${}^*\mathfrak{J}_a^{m-r}\Phi_1$ and ${}^*\mathfrak{J}_a^{m-s}\Phi_2$ belong to ${}^s\mathcal{C}(I)$ and $p_{1m}, p_{2m}$ are polynomials of degree smaller than $m$ (with coefficients, possibly infinite, belonging to ${}^*\mathbb{C}$). Hence $F_1$ and $F_2$ are the $m$-th *derivatives of the functions $\Psi_1 = {}^*\mathfrak{J}_a^{m-r}\Phi_1$ and $\Psi_2 = {}^*\mathfrak{J}_a^{m-s}\Phi_2$ in ${}^s\mathcal{C}(I)$. If we have $\Psi_1 - \Psi_2 \approx q_m$, where $q_m$ is a polynomial function in ${}^s\mathcal{C}(I)$ of degree $< m$, then ${}^*D^m(\Psi_1 - \Psi_2) \approx 0$; also we will have

$$ {}^*\mathfrak{J}_a^m (F_1 - F_2) = \left({}^*\mathfrak{J}_a^{m-r}\Phi_1 - {}^*\mathfrak{J}_a^{m-s}\Phi_2\right) + (p_{1m} - p_{2m}) \approx p_m,$$

where $p_m$ is a polynomial of degree $< m$ with coefficients in ${}^*\mathbb{C}$. Accordingly we make the following definition of distributional equivalence:

**Definition 6.14** *Two internal functions* $F_1, F_2 \in {}^s\mathbf{D}(I)$ *are said to be distributionally equivalent if there is an integer* $m \in \mathbb{N}_0$ *and a polynomial* $p_m$ *of degree* $< m$, *(with coefficients in* ${}^*\mathbb{C}$*) such that*

$$\forall x \in {}^*I_{\text{bd}} : {}^*\mathfrak{J}_a^m (F_1 - F_2)(x) \approx p_m(x)$$

*where* ${}^*I_{\text{bd}} \equiv \bigcup_{a,b \in I, a<b} {}^*[a,b]$ *is the bounded part of* ${}^*I$.

For consistency in the Definition 6.14 a polynomial of degree $< 0$ is taken to mean an infinitesimal constant. When $F_1$ and $F_2$ are distributionally equivalent in this sense we write $F_1 \Xi F_2$. Clearly Definition 6.14 agrees with the Laugwitz definition.

**Definition 6.15** *If $F \in {}^s\mathbf{D}(I)$ we shall denote by ${}^\Xi[F]$ the equivalence class which contains $F$, and by ${}^\Xi\mathcal{C}_\infty(I)$ the set of all such equivalence classes:*

$$ {}^\Xi\mathcal{C}_\infty(I) = {}^s\mathbf{D}(I)/\Xi \equiv {}^*D^\infty\{{}^s\mathcal{C}(I)\}/\Xi. $$

${}^\Xi\mathcal{C}_\infty(I)$ will be called the set of all finite-order ${}^\Xi$distributions on $I$.

### 6.2.4.1 ${}^\Xi$derivatives and differential order

Let $\mu \equiv {}^\Xi[F]$ be a finite order ${}^\Xi$distribution and suppose that $F_1, F_2$ are any two ${}^*\mathcal{C}^\infty(I)$ internal functions belonging to ${}^\Xi[F]$. Then there exists $m \in \mathbb{N}_0$ and a polynomial $p_m$ of degree $< m$ such that

$$ {}^*\mathfrak{J}_a^m(F_1 - F_2) \approx p_m $$

on ${}^*I_{\text{bd}}$. For any $r \in \mathbb{N}_0$ we have

$$ {}^*\mathfrak{J}_a^{m+r}({}^*D^r F_1 - {}^*D^r F_2) = {}^*\mathfrak{J}_a^m(F_1 - F_2) + p_r $$

where $p_r$ is a polynomial of degree $< r$ and with coefficients in ${}^*\mathbb{C}$. Hence

$$ {}^*\mathfrak{J}_a^{m+r}({}^*DF_1 - {}^*D^r F_2) \approx p_{m+r} $$

where $p_{m+r} = p_m + p_r$ is a polynomial of degree $< \max\{m, r\} \leq m + r$. This shows that the $r$th derivatives of all functions belonging to the equivalence class ${}^\Xi[F]$ are again distributionally equivalent. Accordingly we may define the equivalence class to which all such $r$th derivatives belong to be the $r$th ${}^\Xi$derivative of the finite order ${}^\Xi$distribution $\mu$. Denoting this derivative by ${}^\Xi D^r \mu$ we have

$$ \forall r \in \mathbb{N}_0 : \mu = {}^\Xi[F] \Rightarrow {}^\Xi D^r \mu = {}^\Xi[{}^*D^r F]. $$

Further, if $\mu = {}^\Xi[F]$ then $F \in {}^s\mathbf{D}(I) \equiv {}^*D^\infty\{{}^s\mathcal{C}(I)\}$ and so there exists an internal function $\Phi \in {}^s\mathcal{C}(I)$ and an integer $j \in \mathbb{N}_0$ such that $F = {}^*D^j \Phi$. Therefore we have

$$ \mu \equiv {}^\Xi[F] = {}^\Xi[{}^*D^j \Phi] = {}^\Xi D\left({}^\Xi[\Phi]\right) = {}^\Xi D^j \nu, $$

where $\nu = {}^\Xi[\Phi] \in {}^s\mathcal{C}(I)/\Xi$.

**Definition 6.16** *The (differential) order of a ${}^\Xi$distribution $\mu$ in ${}^\Xi\mathcal{C}_\infty(I)$ is defined as the smallest number $r \in \mathbb{N}_0$ for which there exists a ${}^\Xi$distribution $\nu = {}^\Xi[G] \in {}^\Xi\mathcal{C}_\infty(I)$ with $G \in {}^s\mathcal{C}(I)$ such that $\mu = {}^\Xi D^r \nu$.*

It is easy to see that $\Xi\mathcal{C}_\infty(I)$ is a nonstandard model for the axiomatic definition of finite-order distributions on $I$ as proposed by J. S. Silva [102] and outlined in §1.4. In this model the concept of standard continuous function on $I$ has been replaced by that of $\Xi$-equivalence class of nonstandard ${}^s\mathcal{C}(I)$ internal functions, and the concept of distributional derivative by that of $\Xi$-derivative. The addition of $\Xi$-distributions and multiplication by scalars can be readily defined within the model as follows: Given $\Xi$-distributions $\mu, \nu \in \Xi\mathcal{C}_\infty(I)$ there must exist internal functions $\Phi, \Psi$ in ${}^s\mathcal{C}(I)$ and an integer $j \in \mathbb{N}_0$ such that

$$\mu = {}^\Xi D^j\left({}^\Xi[\Phi]\right) \;,\;\; \nu = {}^\Xi D^j\left({}^\Xi[\Psi]\right).$$

Accordingly we can define the sum $\mu + \nu$ to be the $\Xi$-distribution on $I$ such that

$$\mu + \nu = {}^\Xi D^j\left({}^\Xi[\Phi + \Psi]\right),$$

while for any $a \in {}^*\mathbb{R}_{bd}$ we similarly define the (scalar) multiple $a\mu$ of $\mu$ to be the $\Xi$-distribution

$$a\mu = {}^\Xi D^j\left({}^\Xi[a\Phi]\right).$$

These definitions are independent of the particular internal functions taken to represent each equivalence class concerned. Equipped with the operations of addition and multiplication by scalars $\Xi\mathcal{C}_\infty(I)$ constitutes a ${}^*\mathbb{C}_{bd}$-module, which may be identified with $\mathcal{D}'_{\text{fin}}(I)$.

### 6.2.5 Global $\Xi$-distributions

There is a straightforward extension of the theory to the general case of distributions of arbitrary order. Let $\{K_\alpha\}_{\alpha \in A}$ be the family of all compact subintervals of $I$ with more than one point and let ${}^\pi\mathbf{D}(I)$ denote the ${}^*\mathbb{R}_{bd}$-submodule of ${}^*\mathcal{C}^\infty(I)$ such that for each $\alpha \in A$ there exist $r_\alpha \in \mathbb{N}_0$ and $\Phi_\alpha \in {}^s\mathcal{C}(K_\alpha)$ for which we have

$$\forall x \in {}^*K_\alpha : F(x) = {}^*D^{r_\alpha}\Phi_\alpha(x).$$

We can extend the equivalence relation $\Xi$ to ${}^\pi\mathbf{D}(I)$ as follows

$$\forall F_1, F_2 \in {}^\pi\mathbf{D}(I) : \{F_1 \Xi F_2 \Leftrightarrow \forall \alpha \in A, \exists m_\alpha \in \mathbb{N}_0,$$
$$\forall x \in {}^*K_\alpha : [{}^*\mathfrak{I}^{m_\alpha}_{a_\alpha}(F_1 - F_2) \approx p_{m_\alpha}]\}$$

where, for each $\alpha \in A$, $a_\alpha$ is an arbitrarily fixed point in ${}^*K_\alpha$ and $p_{m_\alpha}$ is a polynomial of degree $< m_\alpha$. Then we write

$$\Xi\mathcal{C}_\pi(I) = {}^\pi\mathbf{D}(I)/\Xi$$

and again denote by $\Xi[F]$ the equivalence class which contains $F \in {}^\pi\mathbf{D}(I)$. Each such class $\mu \equiv \Xi[F] \in \Xi\mathcal{C}_\pi(I)$ will be called a **global** $\Xi$**distribution** on I.

It is clear that $\Xi\mathcal{C}_\pi(I)$ is non-empty since any equivalence class containing an S-continuous function in ${}^s\mathcal{C}(I)$ belongs to $\Xi\mathcal{C}_\pi(I)$. Moreover, for any $\mu \in \Xi\mathcal{C}_\infty(I)$ there exists $r \in \mathbb{N}_0$ such that $\mu = \Xi D^r(\Xi[\Phi])$ for some $\Phi \in {}^s\mathcal{C}(I)$. Hence if we let $\Phi_\alpha$ denote the restriction of $\Phi$ to $K_\alpha$, it follows immediately that $\mu$ belongs to $\Xi\mathcal{C}_\pi(I)$. That is to say

$$\Xi\mathcal{C}_\infty(I) \subseteq \Xi\mathcal{C}_\pi(I)$$

where equality holds if and only if $I$ is itself compact. Whenever $I$ is not compact $\Xi\mathcal{C}_\pi(I) - \Xi\mathcal{C}_\infty(I)$ is not empty; its elements will be called $\Xi$distributions of **infinite order**.

Consider for example the particular case when $I = \mathbb{R}$. Take the family of compacts $\{K_n\}_{n\in\mathbb{N}}$ with $K_n = [-n, n]$ for $n \in \mathbb{N}$ so that $\mathbb{R} = \bigcup_{n\in\mathbb{N}} K_n$. Let $\mu$ be a $\Xi$distribution in $\Xi\mathcal{C}_\pi(\mathbb{R})$, and for each $n \in \mathbb{N}$ let $r_n \in \mathbb{N}_0$ be the differential order of the restriction of $\mu$ to $K_n$. Then we can assign to $\mu$ the hypernatural number $r(\mu) = [(r_n)_{n\in\mathbb{N}}] \in {}^*\mathbb{N}_0$. If $\mu$ is in $\Xi\mathcal{C}_\infty(\mathbb{R})$ then $r(\mu) \in \mathbb{N}_0$; on the other hand if $\mu$ is a $\Xi$distribution of infinite order then $r(\mu) \in {}^*\mathbb{N}_\infty$. The hypernatural number $r(\mu) \in {}^*\mathbb{N}_0$ is called the (differential) order of $\mu \in \Xi\mathcal{C}_\pi(\mathbb{R})$.

To each standard Schwartz distribution $\mu \in \mathcal{D}'$ there corresponds a unique $\Xi$distribution $\Xi[F]$ and we have

$$<\mu, \varphi> = \mathrm{st}\left(\int_{{}^*\mathbb{R}}^* F(x)^*\varphi(x) dx\right)$$

for any $\varphi \in \mathcal{D}$ and any internal function $F$ in the class $\Xi[F]$. In fact if $F_1$ and $F_2$ are any two internal functions in $\Xi[F]$ then for each compact $K \sqsubset \mathbb{R}$ there exists $m \in \mathbb{N}_0$ such that ${}^*\mathfrak{I}_a^m(F_1 - F_2) \approx p_m$ where $a \in {}^*\mathbb{R}_{\mathrm{bd}}$ is fixed and $p_m$ is a polynomial of degree $< m$. For any $\varphi \in \mathcal{D}_K$ we have

$$\begin{aligned} {}^*<F_1 - F_2, {}^*\varphi> &= {}^*<{}^*D^m\{{}^*\mathfrak{I}_a(F_1 - F_2)\}, {}^*\varphi> \\ &= {}^*<{}^*\mathfrak{I}_a^m(F_1 - F_2), (-1)^m {}^*\varphi^{(m)}> \\ &\approx {}^*<p_m, (-1)^m {}^*\varphi^{(m)}> = 0. \end{aligned}$$

Since every function in $\mathcal{D}$ has compact support it follows that, for every pair of internal functions $F_1, F_2 \in {}^\pi\mathbf{D}(I)$ such that $F_1 \Xi F_2$ we have

$$\forall \varphi \in \mathcal{D} : {}^*<F_1, {}^*\varphi> \approx {}^*<F_2, {}^*\varphi>.$$

Hence we may identify $\mathcal{D}'$ with $\Xi\mathcal{C}_\pi(\mathbb{R})$.

**Remark 6.4** *The nonstandard theory of $\Xi$ distributions developed above is relatively concise and could be presented at a genuinely elementary level. It presupposes only a very basic acquaintance with a simple model of a hyperreal number system, together with ad hoc definitions of internal sets and functions; no explicit discussion of the general transfer principle is needed. It has the conceptual appeal common to all nonstandard theories; the representation of distributions as functions offers a straightforward explanation of some of the apparent paradoxes which standard theories seem to involve. For example, a student can accept fairly readily the usual definition of a null function as a standard real function $f : \mathbb{R} \to \mathbb{R}$ which vanishes almost everywhere. The statement that the delta distribution vanishes on $\mathbb{R} \backslash \{0\}$ but is nevertheless something other than a null function is often found to be less convincing. On the other hand the fact that the restriction to $\mathbb{R}$ of a nonstandard delta function such as $\omega/\pi(1 + \omega^2 x^2)$ is zero for all real $x \neq 0$ raises no conceptual difficulties; the "spike" which, so to speak, is responsible for the action of $\delta$ exists in a hyperreal context. This illustrates the fact that, so far as distributions are concerned, the archimedean property of $\mathbb{R}$ and the resulting absence of infinitely large and infinitely small elements is a disadvantage. There is a strong case for the preferred use of nonstandard methods in introductory courses on the theory of distributions, particularly for engineers and physicists. See for example the pioneering work of Detlef Laugwitz in this respect [61].*

**Exercise 6.7** :

1. If $F$ is any function in $^s\mathcal{C}(I)$ and if $a$ is any point in $^*I$ prove that $^*\Im_a F$ also belongs to $^s\mathcal{C}(I)$.

2. Confirm that $\Xi\mathcal{C}_\infty(\mathbb{R})$ is a nonstandard model for the J. S. Silva axioms for finite-order distributions.

3. Which of the following internal functions belong to $^s\mathbf{D}(\mathbb{R})$:

    (a) $\exp(-\omega x^2)$;
    (b) $\omega^2/\pi(1 + \omega^2 x^2)$;
    (c) $\pi^{-1} x^{-2} \{\omega x \cos(\omega x) - \sin(\omega x)\}$.

4. Show that the internal functions $F_1(x) = \omega^3 \sin(\omega x)$ and $F_2(x) \equiv 0$ are distributionally equivalent (although they are clearly not infinitely close).

## 6.3 PRODUCTS OF $^\Xi$DISTRIBUTIONS

### 6.3.1 Nonstandard Schwartz products

The nonstandard model $^\Xi\mathcal{C}_\pi(\mathbb{R})$ makes the problems associated with the definition of a multiplicative product for distributions particularly obvious. For linear algebraic operations, translation and differentiation we can work with arbitrary representatives (pre-distributions) $F, G$ in $^\pi\mathbf{D}(\mathbb{R})$ of $^\Xi$distributions $\mu = {}^\Xi[F]$ and $\nu = {}^\Xi[G]$ in $^\Xi\mathcal{C}_\pi(\mathbb{R})$. However if $F_1 \in {}^\Xi[F]$ and $G_1 \in {}^\Xi[G]$ are representatives of $\mu$ and $\nu$ respectively then although the pointwise product $F_1 G_1$ always exists as an internal function belonging to $^*\mathcal{C}^\infty(\mathbb{R})$ it will not generally be a representative of a unique equivalence class in $^\Xi\mathcal{C}_\pi(\mathbb{R})$. For example, let $^\Xi[\Delta]$ be the $^\Xi$distribution corresponding to the delta distribution $\delta \in \mathcal{D}'$. Given any number $c \in \mathbb{R}$ whatsoever we can choose internal functions $\Delta_1, \Delta_2$ in $^\Xi[\Delta]$ such that the pointwise product $\Delta_1 \Delta_2$ is a member of the equivalence class $^\Xi[c\Delta]$. Hence we cannot define a $^\Xi$distribution corresponding to a distributional product $\delta^2$ as an equivalence class containing all pointwise products of the form $\Delta_1 \Delta_2$. Nevertheless, as with standard theories, it is possible to define restricted forms of distributional product for special classes of $^\Xi$distributions. In particular we can establish a nonstandard analogue of the Schwartz product between an arbitrary $^\Xi$distribution and a $^\Xi$distribution represented by an infinitely S-differentiable function. It is convenient to define the subspace

$$^\Xi\mathcal{E}(\mathbb{R}) = {}^s\mathcal{C}^\infty(\mathbb{R})/\Xi = \left\{ {}^\Xi[F] : \text{st}(F) \in \mathcal{C}^\infty(\mathbb{R}) \right\}.$$

Then we have the following theorem:

**Theorem 6.1** *Let $^\Xi[F]$ be an arbitrary $^\Xi$distribution in $^\Xi\mathcal{C}_\pi(\mathbb{R})$ and $^\Xi[\Theta]$ a $^\Xi$distribution belonging to $^\Xi\mathcal{E}(\mathbb{R})$. Then for any $F \in {}^\Xi[F]$ and $\Theta \in {}^\Xi[\Theta]$ the product $F\Theta$ is a member of a well defined equivalence class $^\Xi[F\Theta]$ in $^\Xi\mathcal{C}_\pi(\mathbb{R})$, which is independent of the choice of $F$ and $\Theta$.*

**Proof:** For simplicity we prove the result for the case when $^\Xi[F]$ is a $^\Xi$distribution of finite order. If $F \in {}^s\mathbf{D}(\mathbb{R})$ then there exists $r \in \mathbb{N}_0$ and $\Phi \in {}^s\mathcal{C}(\mathbb{R})$ such that $F = {}^*D^r \Phi$. For any internal function $\Theta \in {}^s\mathcal{C}^\infty(\mathbb{R})$ we have

$$\begin{aligned}
\Theta F &= \Theta \cdot {}^*D^r \Phi \\
&= \sum_{j=0}^{r} {}^*D^{r-j} \left\{ (-1)^j \binom{r}{j} \left( {}^*D^j \Theta \right) \cdot F \right\} = {}^*D^r \sum_{j=0}^{r} {}^*\vartheta_0^j F_j,
\end{aligned}$$

where, since $F_j \equiv (-1)^j \binom{r}{j}({}^*D^j \Theta) F$ is a function in $^s\mathcal{C}(\mathbb{R})$ for every $j = 0, 1, 2, \ldots, r$, it follows that $\Theta F$ belongs to $^s\mathbf{D}(\mathbb{R})$.

Next we show that if $F_1, F_2 \in {}^s\mathbf{D}(\mathbb{R})$ are two distributionally equivalent functions then for any $\Theta \in {}^s\mathcal{C}^\infty(\mathbb{R})$ the functions $\Theta F_1$ and $\Theta F_2$ are also distributionally equivalent. For if $F_1 \Xi F_2$ then there exists $m \in \mathbb{N}_0$ and functions $\Phi_1, \Phi_2$ in ${}^s\mathcal{C}(\mathbb{R})$ such that

$$\Phi_1 \approx \Phi_2 \text{ on } {}^*\mathbb{R}_{bd},$$

and

$$F_1 = {}^*D^m \Phi_1 \, , \, F_2 = {}^*D^m \Phi_2.$$

Hence from

$$\Theta(F_1 - F_2) = {}^*D^m \sum_{j=0}^{m} (-1)^j \binom{m}{j} {}^*\mathfrak{S}_0^j \left[ \left({}^*D^j \Theta\right)(\Phi_1 - \Phi_2) \right]$$

we get

$$ {}^*\mathfrak{S}_0^m \{\Theta(F_1 - F_2)\} = \sum_{j=0}^{m} (-1)^j \binom{m}{j} {}^*\mathfrak{S}_0^j \left[ \left({}^*D^j \Theta\right)(\Phi_1 - \Phi_2) \right] + p_m \approx p_m,$$

where $p_m$ is a polynomial of degree $< m$.

This is easily extended to the case of two functions $\Theta_1, \Theta_2$ in ${}^s\mathcal{C}^\infty(\mathbb{R})$ such that $\Theta_1 \Xi \Theta_2$. For we would get

$$\Theta_1 F_1 - \Theta_2 F_2 = {}^*D^m \sum_{j=0}^{m} (-1)^j \binom{m}{j} {}^*\mathfrak{S}_0^j \left[ \left({}^*D^j \Theta_1\right) \Phi_1 - \left({}^*D^j \Theta_2\right) \Phi_2 \right]$$

and since ${}^*D^j \Theta_1 \approx {}^*D^j \Theta_2, \forall j \in \mathbb{N}_0$, then

$${}^*\mathfrak{S}_0^m (\Theta_1 F_1 - \Theta_2 F_2) \approx p_m.$$

Therefore it follows that for any $\Xi[F] \in {}^\Xi \mathcal{C}_\pi(\mathbb{R})$ and any $\Xi[\Theta] \in {}^\Xi \mathcal{E}(\mathbb{R})$ we must have $\Theta F \in \Xi[\Theta F]$ where $\Xi[\Theta F]$ is a uniquely defined $\Xi$distribution in ${}^\Xi \mathcal{C}_\pi(\mathbb{R})$. □

It is convenient to write this result in the form

$$\left(\Xi[\Theta]\right) \cdot \left(\Xi[F]\right) = \Xi[\Theta F].$$

**Corollary 6.3.1** *If $\theta \in \mathcal{C}^\infty(\mathbb{R})$ and if $\Xi[F] \in {}^\Xi \mathcal{C}_\pi(\mathbb{R})$ corresponds to the Schwartz distribution $\mu \in \mathcal{D}'$ then the product $\Xi[{}^*\theta] \cdot \Xi[[F]]$ in ${}^\Xi \mathcal{C}_\pi(\mathbb{R})$ corresponds to the Schwartz product $\theta \mu$.*

## 6.4 THE DIFFERENTIAL ALGEBRA ${}^\omega \mathcal{G}(\mathbb{R})$

### 6.4.1 The Biagioni simplified algebra

The algebra of all internal functions in ${}^*\mathcal{C}^\infty(\mathbb{R})$ is, in a sense, too big. It contains pre-distributions and products of pre-distributions, and also other elements which are not immediately recognisable as representatives of (currently familiar) standard objects. On the other hand the use of the equivalence relation $\Xi$ and the restriction of $^*\mathcal{C}^\infty(\mathbb{R})$ to $^\pi \mathbf{D}(\mathbb{R})$ is too severe; the space

$$^\Xi \mathcal{C}_\pi(\mathbb{R}) \equiv {}^\pi \mathbf{D}(\mathbb{R})/\Xi$$

is not closed under multiplication and is therefore not an algebra. For a theory of generalised functions which will include (representatives of) distributions and arbitrary distributional products we need a subspace of $^*\mathcal{C}^\infty(\mathbb{R})$ which is also a *subalgebra*. This can be obtained via a nonstandard interpretation of a simplified form of the Colombeau theory of generalised functions due to H. Biagioni [3]:

In the Colombeau theory each member of the space $\mathcal{E}[\Phi, \mathcal{C}^\infty]$ can be associated with (and characterised by) an indexed family of functions $\{T\varphi_{(\varepsilon)}\}_{\varphi \in \Phi, 0 < \varepsilon < 1}$. Biagioni uses a subspace $\mathcal{E}_s[\Phi, \mathcal{C}^\infty]$ of $\mathcal{E}[\Phi, \mathcal{C}^\infty]$ consisting of operators $T$ which may be considered as independent of $\varphi \in \Phi$, and so characterised by an indexed family $\{T\varphi_{(\varepsilon)}\}_{0 < \varepsilon < 1}$, where $\varphi$ may be any function in $\Phi$. The members of $\mathcal{E}_s[\Phi, \mathcal{C}^\infty]$ can therefore be treated as mappings of the form

$$\begin{aligned} T : (0,1) \times \mathbb{R} &\to \mathbb{C} \\ (\varepsilon, x) &\to T(\varepsilon, x), \end{aligned}$$

which are infinitely differentiable in $x$ for each $\varepsilon \in (0,1)$. The definitions of moderate and negligible operators in this space are as follows:

**Definition 6.17 (Biagioni Moderate Operators)** *The operator $T$ belongs to $\mathcal{E}_{s,M}[\Phi, \mathcal{C}^\infty]$ if and only if, given a compact $K \sqsubset \mathbb{R}$ and $k \in \mathbb{N}_0$, there exists $m \in \mathbb{N}$ and $c > 0, \eta > 0$, such that*

$$|D^k T(\varepsilon, x)| \leq c \cdot \varepsilon^{-m}$$

*for all $x \in K$ and every $\varepsilon$ such that $0 < \varepsilon < \eta$.*

**Definition 6.18 (Biagioni Negligible Operators)** *The operator $T$ belongs to $\mathcal{N}_s[\Phi, \mathcal{C}^\infty]$ if and only if, given a compact $K \sqsubset \mathbb{R}$ and $k \in \mathbb{N}_0$, there exists $m \in \mathbb{N}$ and a sequence $\gamma \in \Gamma$ such that for all $q \geq m$ there exist $c > 0, \eta > 0$, such that*

$$|D^k T(\varepsilon, x)| \leq c \cdot \varepsilon^{\gamma(q) - m}$$

*for all $x \in K$ and every $\varepsilon$ such that $0 < \varepsilon < \eta$.*

Biagioni shows that the criterion for negligibility can be replaced in this context by the following simpler version:

**Theorem 6.2** *$T$ belongs to $\mathcal{N}_s[\Phi, \mathcal{C}^\infty]$ if and only if, given a compact $K \sqsubset \mathbb{R}$ and $k \in \mathbb{N}_0$, and given an integer $q \in \mathbb{N}_0$, there exist $c > 0$ and $\eta > 0$ such that*
$$|D^k T(\varepsilon, x)| \leq c \cdot \varepsilon^q$$
*for all $x \in K$ and every $\varepsilon$ such that $0 < \varepsilon < \eta$.*

**Proof:** The simplified version clearly implies the original form.
Suppose then that the original form holds and that a compact $K \sqsubset \mathbb{R}$ and integers $k, q$ are given. By hypothesis there will be $m \in \mathbb{N}$ and $\gamma \in \Gamma$ for which the original criterion works. Since $\gamma(n) \to \infty$ with $n$ there must exist $q' \in \mathbb{N}$ (which, without loss of generality, we assume $\geq m$) such that
$$r \in \mathbb{N} \text{ and } r \geq q' \Rightarrow \gamma(r) \geq m + q.$$
Since $r \geq m$ it follows that there exist $c > 0, \eta > 0$, such that
$$|D^k T(\varepsilon, x)| \leq c \cdot \varepsilon^{\gamma(r)-m} \leq c \cdot \varepsilon^q$$
for all $x \in K$ and for every $\varepsilon$ such that $0 < \varepsilon < \eta' = \min\{1, \eta\}$. □

The Biagioni algebra of generalised functions is now given by the quotient
$$\mathcal{G}_s(\mathbb{R}) \equiv \mathcal{E}_{s,M}[\Phi, \mathcal{C}^\infty]/\mathcal{N}_s[\Phi, \mathcal{C}^\infty].$$

As Biagioni points out, $\mathcal{G}_s(\mathbb{R})$ is a subspace of $\mathcal{G}(\mathbb{R})$ which is in some respects more natural for physical applications. In developing a nonstandard version of $\mathcal{G}_s(\mathbb{R})$ we replace the continuous parameter $\varepsilon$ by the discrete integral parameter $n \ (\equiv \varepsilon^{-1})$ in $\mathbb{N}$. All statements of the form *for every $\varepsilon$ such that $0 < \varepsilon < \eta$* will therefore be replaced by *for all sufficiently large $n \in \mathbb{N}$*, and will certainly be subsumed under *for nearly all $n \in \mathbb{N}$*.

### 6.4.2 $^\omega$Moderate functions

We begin by imposing growth conditions on members of $^*\mathcal{C}^\infty(\mathbb{R})$ which are compatible with known behaviour of pre-distributions and of products of pre-distributions.

**Definition 6.19** *An internal function $F \in {}^*\mathcal{C}^\infty(\mathbb{R})$ is said to be of $^\omega$growth on $^*\mathbb{R}_{bd}$ if, for each compact $K \sqsubset \mathbb{R}$, there exists an integer $m = m(K) \in \mathbb{N}_0$ such that $|F(x)| \leq C\omega^m$ for every $x \in {}^*K$, where $C$ is some hyperreal in $^*\mathbb{R}_{bd}$.*

For $k \in \mathbb{N}_0$, let $^\omega\mathcal{C}^k(\mathbb{R})$ denote the set of all functions $F \in {}^*\mathcal{C}^\infty(\mathbb{R})$ which are such that $^*D^k F$ is of $^\omega$growth on $^*\mathbb{R}_{bd}$. Then we have

$$^\omega\mathcal{C}^k(\mathbb{R}) \subset {}^\omega\mathcal{C}^{k-1}(\mathbb{R}) \subset \ldots \subset {}^\omega\mathcal{C}^0(\mathbb{R}) \subset {}^*\mathcal{C}^\infty(\mathbb{R})$$

and we can go on to define the intersection of all the $^\omega\mathcal{C}^k(\mathbb{R})$:

**Definition 6.20** *A function $F \in {}^*\mathcal{C}^\infty(\mathbb{R})$ is said to be an $^\omega$moderate function if $F \in {}^\omega\mathcal{C}^k(\mathbb{R})$ for all $k \in \mathbb{N}_0$. We denote by $^\omega\mathcal{C}^\infty(\mathbb{R})$ the $^*\mathbb{R}_{bd}$ submodule of all $^\omega$moderate internal functions in $^*\mathcal{C}^\infty(\mathbb{R})$.*

Then $^\omega\mathcal{C}^\infty(\mathbb{R})$ is a differential sub-algebra of $^*\mathcal{C}^\infty(\mathbb{R})$.

Next we define the following two subsets of $^*\mathbb{R}$:

$$^\omega\mathbf{fin}(^*\mathbb{R}) = \{x \in {}^*\mathbb{R} : [\exists m \in \mathbb{N}_0 : |x| < \omega^{+m}]\}$$
$$^\omega\mathbf{inf}(^*\mathbb{R}) = \{x \in {}^*\mathbb{R} : [\forall m \in \mathbb{N}_0 : |x| < \omega^{-m}]\}.$$

Both $^\omega\mathbf{fin}(^*\mathbb{R})$ and $^\omega\mathbf{inf}(^*\mathbb{R})$ are rings with respect to the algebraic operations of $^*\mathbb{R}$, and $^\omega\mathbf{inf}(^*\mathbb{R})$ is a maximal ideal of $^\omega\mathbf{fin}(^*\mathbb{R})$. The relation $\simeq$ defined by

$$\forall x, y \in {}^\omega\mathbf{fin}(^*\mathbb{R}) : [x \simeq y \Leftrightarrow (x - y) \in {}^\omega\mathbf{inf}(^*\mathbb{R})]$$

is an equivalence relation on $^\omega\mathbf{fin}(^*\mathbb{R})$. The quotient

$$^\omega\mathbb{R} = {}^\omega\mathbf{fin}(^*\mathbb{R})/{}^\omega\mathbf{inf}(^*\mathbb{R})$$

is a field which we call the **Robinson field**. It was introduced by Abraham Robinson in [90] and is studied intensively by Lightstone and Robinson [66]. We write $^\omega[x]$ for the equivalence class in $^\omega\mathbb{R}$ which contains the element $x \in {}^\omega\mathbf{fin}(^*\mathbb{R})$. $^\omega[x]$ is positive in $^\omega\mathbb{R}$ whenever $x$ is a positive hyperreal not in $^\omega\mathbf{inf}(^*\mathbb{R})$; this defines an ordering relation on $^\omega\mathbb{R}$ which extends it to an ordered field. $^\omega\mathbb{R}$ is an extension of $\mathbb{R}$ which includes all "*sufficiently large*" infinitesimals and all "*sufficiently small*" infinite numbers, but which excludes those infinitesimals (other than zero) which are too small and those infinite numbers which are too large. Each (non-zero) member $\iota$ of $^\omega\mathbf{inf}(^*\mathbb{R})$ is an infinitesimal which is too small in this sense, and is called an **iota**; the inverse $\iota^{-1}$ is an infinite number which is too large to belong to $^\omega\mathbb{R}$, and is called a **mega**.

**Definition 6.21** *An $^\omega$moderate function $F$ in $^\omega\mathcal{C}^\infty(\mathbb{R})$ is said to be **iota-null** on $^\omega\mathbb{R}$ if and only if $F(x) \simeq 0$ for all $x \in {}^\omega\mathbb{R}$.*

The set of all iota-null functions on $^\omega\mathbb{R}$, denoted by $^\omega\mathcal{N}_0(\mathbb{R})$, is an ideal of $^\omega\mathcal{C}^\infty(\mathbb{R})$ and so $^\omega\mathcal{C}^\infty(\mathbb{R})/^\omega\mathcal{N}_0(\mathbb{R})$ is an algebra. In fact it is the algebra formed by the restrictions of the members of $^\omega\mathcal{C}^\infty(\mathbb{R})$ to the field $^\omega\mathbb{R}$. However $^\omega\mathcal{N}_0(\mathbb{R})$ is not closed under the $^\star D$-operator and therefore $^\omega\mathcal{C}^\infty(\mathbb{R})/^\omega\mathcal{N}_0(\mathbb{R})$ is not a *differential* algebra. We consider instead the set $^\omega\mathcal{N}(\mathbb{R})$ of all functions $F$ in $^\omega\mathcal{C}^\infty(\mathbb{R})$ such that $^\star D^k F$ is iota-null on $^\omega\mathbb{R}$ for each $k \in \mathbb{N}_0$. Then $^\omega\mathcal{N}(\mathbb{R})$ is an ideal of $^\omega\mathcal{C}^\infty(\mathbb{R})$ and moreover we have

$$^\star D\{^\omega\mathcal{N}(\mathbb{R})\} \subset {^\omega\mathcal{N}(\mathbb{R})}.$$

Hence we use $^\omega\mathcal{N}(\mathbb{R})$ instead of $^\omega\mathcal{N}_0(\mathbb{R})$ to make the definition:

**Definition 6.22** *The quotient*

$$^\omega\mathcal{G}(\mathbb{R}) = {^\omega\mathcal{C}^\infty(\mathbb{R})}/{^\omega\mathcal{N}(\mathbb{R})}$$

*is called the algebra of $^\omega$***generalised functions***. We denote by $^\omega[F]$ the equivalence class containing $F \in {^\omega\mathcal{C}^\infty(\mathbb{R})}$.*

$^\omega\mathcal{G}(\mathbb{R})$ is a differential algebra with respect to the operations:

$$^\omega[F] + {^\omega[G]} = {^\omega[F+G]}$$
$$^\omega[F] \cdot {^\omega[G]} = {^\omega[FG]} \quad ; \quad a \cdot {^\omega[F]} = {^\omega[\alpha F]}$$
$$^\omega D\{^\omega[F]\} = {^\omega[^\star DF]},$$

where $a = {^\omega[\alpha]}$ is any number in $^\omega\mathbb{R}$.

Now let $T$ be an element of the Biagioni subspace $\mathcal{E}_{s,M}[\Phi, \mathcal{C}^\infty]$, and for each $n \in \mathbb{N}$ define a function $T_n : \mathbb{R} \to \mathbb{C}$ by writing

$$T_n(x) = T(1/n, x).$$

Then $T_n \in \mathcal{C}^\infty(\mathbb{R})$ and, for any compact $K \sqsubset \mathbb{R}$ and $r \in \mathbb{N}_0$, there exist $q \in \mathbb{N}_0$ and $c > 0$ such that for all sufficiently large $n \in \mathbb{N}$, the inequality

$$|T_n^{(r)}(x)| \leq c\, n^q$$

holds for all $x \in K$. Hence the internal function $F_T \equiv [(T_n)_{n \in \mathbb{N}}]$ belongs to $^\omega\mathcal{C}^\infty(\mathbb{R})$. We can define an injective mapping

$$\Psi : \mathcal{E}_{s,M}[\Phi, \mathcal{C}^\infty] \to {^\omega\mathcal{C}^\infty(\mathbb{R})}$$

by setting $\Psi(T) = F_T$ for each $T$ in $\mathcal{E}_{s,M}[\Phi, \mathcal{C}^\infty]$. Further if $T$ belongs to $\mathcal{N}_s[\Phi, \mathcal{C}^\infty]$ then, given a compact $K \sqsubset \mathbb{R}$ and $r \in \mathbb{N}_0$, there exists $p \in \mathbb{N}$ such that for all sufficiently large $n$

$$|T_n^{(r)}(x)| \leq c\, n^{-q}$$

holds for all $x \in K$. Hence for every $r \in \mathbb{N}$ it follows that $^*D^r F_T = [(T_n^{(r)})_{n \in \mathbb{N}}]$ is an iota, and so that

$$\Psi(\mathcal{N}_s[\Phi, \mathcal{C}^\infty]) \subset {}^\omega\mathcal{N}(\mathbb{R}).$$

Thus there is a homomorphism of algebras between $\mathcal{G}_s(\mathbb{R})$ and ${}^\omega\mathcal{G}(\mathbb{R})$.

**Remark 6.5** $F \in {}^*\mathcal{C}^\infty(\mathbb{R})$ *is said to be of $^\omega$growth of finite order if and only if there exists $m \in \mathbb{N}_0$ and $c \in {}^*\mathbb{R}_{bd}^+$ such that for each compact $K \sqsubset \mathbb{R}$ we have $|F(x)| \leq c \cdot \omega^m$ for all $x \in {}^*K$ (the constant $c$ and the integer $m$ being independent of $K$). If $^*D^k F$ is of $^\omega$growth of finite order for each $k \in \mathbb{N}_0$ then $F$ is said to be $^\omega$moderate of finite order. The set ${}^\omega\mathcal{C}_{fin}^\infty(\mathbb{R})$ of all internal functions in $^*\mathcal{C}^\infty(\mathbb{R})$ which are $^\omega$moderate of finite order is a proper subset of ${}^\omega\mathcal{C}^\infty(\mathbb{R})$, and it is clear that ${}^\omega\mathcal{N}(\mathbb{R})$ is an ideal of ${}^\omega\mathcal{C}_{fin}^\infty(\mathbb{R})$. Hence*

$$^\omega\mathcal{G}_{fin}(\mathbb{R}) \equiv {}^\omega\mathcal{C}_{fin}^\infty/{}^\omega\mathcal{N}(\mathbb{R})$$

*is a subalgebra of ${}^\omega\mathcal{G}(\mathbb{R})$; its elements may be called $^\omega$generalised functions of finite order.*

### 6.4.3  Embedding of $\mathcal{D}'$ into ${}^\omega\mathcal{G}(\mathbb{R})$

Let $f$ be any function defined and continuous on $\mathbb{R}$. If $\theta \in \Phi$ and we write $\theta_n(x) = n\theta(nx)$ for $n = 1, 2, \ldots$, then the internal function $\delta_{(\theta)} \equiv [(\theta_n)_{n \in \mathbb{N}}]$ belongs to $^*\mathcal{C}^\infty(\mathbb{R})$ and is a nonstandard representative of a delta function. Now define

$$f_{(\theta_n)}(x) \equiv f * \theta_n(x) = \int_{-\infty}^{+\infty} f(t)\theta_n(x-t)dt, \quad n = 1, 2, \ldots.$$

Then the sequence $(f_{(\theta_n)})_{n \in \mathbb{N}}$ converges uniformly on compacts in $\mathbb{R}$ to $f$. Also if $\theta$ has support contained in $[-a, +a]$ and if $r \in \mathbb{N}_0$ then we have

$$|f_{(\theta_n)}^{(r)}(x)| \leq \left\{ \sup_{x-a/n \leq t \leq x+a/n} |f(t)| \right\} \cdot \int_{x-a/n}^{x+a/n} n^{r+1} |\theta^{(r)}(n(x-t))| dt$$

$$= C_{(\theta,r)} \cdot \left\{ \sup_{x-a/n \leq t \leq x+a/n} |f(t)| \right\} \cdot n^r,$$

where

$$C_{(\theta,r)} \equiv \int_{-a}^{+a} |\theta^{(r)}(\tau)| d\tau.$$

Thus the internal function $F_{(\theta)} = [(f_{(\theta_n)})_{n \in \mathbb{N}}]$ is a member of $^*\mathcal{C}^\infty(\mathbb{R})$ such that for any $r \in \mathbb{N}_0$ and any compact $K \sqsubset \mathbb{R}$ there exists an integer $m = m(r) \leq r$ such that $|^*D^r F_{(\theta)}(x)| \leq c \cdot \omega^m$ for all $x \in {}^*K$, where $c$ is a positive constant in $^*\mathbb{R}_{bd}$; that is to say we have $F_{(\theta)} \in {}^\omega\mathcal{C}^\infty(\mathbb{R})$.

Note that if $\psi$ is some other function in $\Phi$ then the equivalence classes $^\omega[F_{(\theta)}]$ and $^\omega[F_{(\psi)}]$ will not generally coincide in $^\omega\mathcal{G}(\mathbb{R})$; each continuous function $f$ on $\mathbb{R}$ (and more generally each distribution in $\mathcal{D}'$) will have (infinitely) many distinct representatives in $^\omega\mathcal{G}(\mathbb{R})$. Nevertheless such representatives will satisfy a weaker form of equality which we call $\mathcal{D}$-association:

Let $^\omega f \equiv {}^\omega[F] \in {}^\omega\mathcal{G}(\mathbb{R})$, where $F = [(f_n)_{n\in\mathbb{N}}]$ is a function in $^\omega\mathcal{C}^\infty(\mathbb{R})$. For any $\varphi \in \mathcal{D}$ the number

$$^\star<F,{}^\star\varphi> \;=\; {}^\star\!\!\int_{-\infty}^{+\infty} F(t){}^\star\varphi(t)dt \;=\; \left[\left(\int_{-\infty}^{+\infty} f_n(\tau)\varphi(\tau)d\tau\right)_{n\in\mathbb{N}}\right]$$

certainly belongs to $^\omega\mathbf{fin}(^\star\mathbb{R})$, since on every compact $F$ is bounded by a power of $\omega$ and $\varphi$ has compact support. Moreover if $F \in {}^\omega\mathcal{N}(\mathbb{R})$ then $^\star<F,{}^\star\varphi>$ is an iota since the integration is only over a finite range. Hence it makes sense to define

$$^\omega<{}^\omega f,\varphi> \;\equiv\; {}^\omega[{}^\star<F,{}^\star\varphi>]$$

where $F$ is an arbitrary representative of $^\omega f \in {}^\omega\mathcal{G}(\mathbb{R})$.

Now denote by $^\omega\mathrm{mon}(0)$ the $^\omega$infinitesimals in $^\omega\mathbb{R}$, that is the set

$$^\omega\mathrm{mon}(0) \;=\; \mathrm{mon}(0)/{}^\omega\mathbf{inf}(^\star\mathbb{R}).$$

Then we have

**Definition 6.23** *Two $^\omega$generalised functions $^\omega f, {}^\omega g \in {}^\omega\mathcal{G}(\mathbb{R})$ are said to be $\mathcal{D}$-associated if and only if, for every $\varphi \in \mathcal{D}$,*

$$^\omega<{}^\omega f - {}^\omega g,\varphi> \;\in\; {}^\omega\mathrm{mon}(0).$$

Finally we can show that $\mathcal{D}$-association agrees with the concept of distributional equivalence defined in §6.2.4.

**Theorem 6.3** *Two functions $F, G \in {}^s\mathbf{D}(\mathbb{R}) \cap {}^\omega\mathcal{C}^\infty(\mathbb{R})$ are distributionally equivalent if and only if the equivalence classes $^\omega f \equiv {}^\omega[F], {}^\omega g \equiv {}^\omega[G] \in {}^\omega\mathcal{G}(\mathbb{R})$ are $\mathcal{D}$-associated.*

**Proof:** (a) If $F \equiv G$ then there exists $m \in \mathbb{N}_0$ such that

$$\begin{aligned}
^\star<F-G,{}^\star\varphi> \;&=\; {}^\star<{}^\star\mathfrak{I}_0^m(F-G),(-1)^m\,{}^\star\varphi^{(m)}> \\
&\approx\; {}^\star<p_m,(-1)^m\,{}^\star\varphi^{(m)}> \;=\; 0
\end{aligned}$$

and it follows that $^\omega f$ and $^\omega g$ are $\mathcal{D}$-associated.

(b) Now suppose that $^*<F-G,{}^*\varphi> \approx 0$ for every $\varphi \in \mathcal{D}$. Since

$$^*<F-G,{}^*\varphi> = {}^*<{}^*\mathfrak{I}_0^m(F-G),(-1)^m \, {}^*\varphi^{(m)}>$$

for any $m \in \mathbb{N}_0$ and $F - G \in {}^s\mathbf{D}(\mathbb{R})$, there exists $m_0 \in \mathbb{N}_0$ such that, for all $m \geq m_0$, ${}^*\mathfrak{I}_0^m(F-G)$ is equal to the sum of some function $\Phi \in {}^s\mathcal{C}(\mathbb{R})$ and a polynomial $p_m$ of degree $< m$. Hence

$$^*<F-G,{}^*\varphi> = {}^*<\Phi + p_m, (-1)^m \, {}^*\varphi^{(m)}> = <\Phi, (-1)^m \, {}^*\varphi^{(m)}>$$

and so we must have $\Phi \approx 0$. This means that $^*\mathfrak{I}_0^m(F-G) \approx p_m$, and therefore that $F \Xi G$. □

## 6.5 CANONICAL EMBEDDING OF $\Xi\mathcal{C}_\pi(\mathbb{R})$ INTO AN ALGEBRA

### 6.5.1 Nonstandard analysis and superstructures

The Biagioni algebra $\mathcal{G}_s(\mathbb{R})$ is a simplification of the Colombeau theory which is perfectly adequate for most applications. Nevertheless it does suffer from the disadvantage that there is no privileged inclusion of $\mathcal{C}(\mathbb{R})$, and hence of $\mathcal{D}'$, into $\mathcal{G}_s(\mathbb{R})$. To each distribution $\mu \in \mathcal{D}'$ there will correspond not one, but infinitely many elements of $\mathcal{G}_s(\mathbb{R})$ or, equivalently, infinitely many equivalence classes of internal functions in $^\omega\mathcal{G}(\mathbb{R})$. It is not possible to construct a nonstandard interpretation of the full Colombeau algebra $\mathcal{G}(\mathbb{R})$ using the simple model for $^*\mathbb{R}$ described above. A richer structure is needed, together with a more fully developed account of the concepts and techniques of Nonstandard Analysis proper. Limitations of space allow only the following brief sketch of the essential ideas.

A typical mathematical theory may be described formally as a collection of properly constructed statements about a specific universe of objects, most of which are sets. At the lowest level there will be some basic set $X$ whose elements are to be regarded as non-sets within the context of the theory; primitive elements such as these are variously referred to as the *atoms, individuals* or *urelements* of the theory. Higher order objects will consist of sets of atoms and various supersets of these. The most comprehensive collection of such objects is the so-called **superstructure** $\mathcal{V}(X)$ with individuals in $X$, defined as follows:

**Definition 6.24** *Let $X_0 = X$, and for $n = 1, 2, \ldots$, define $X_n$ recursively by*

$$X_n = X_{n-1} \bigcup \mathcal{P}(X_{n-1})$$

where $\mathcal{P}(Y)$ denotes, as usual, the power set of $Y$. The superstructure is the union of all the $X_n$,

$$\mathcal{V}(X) = \bigcup_{n \in \mathbb{N}_0} X_n.$$

If we call $\mathcal{V}(X)$ the **standard universe** for the theory in question then the essential assumption of Nonstandard Analysis is the existence of another, larger structure $^\star\mathcal{V}(X)$ which we call the **nonstandard universe** for the theory. This is a proper extension of the standard universe $\mathcal{V}(X)$ in which to each standard object $\alpha \in \mathcal{V}(X)$ there corresponds a specific member $^\star\alpha$ of $^\star\mathcal{V}(X)$ which is an object of the same type as $\alpha$ but which is generally richer in content. Thus if $\alpha \in \mathcal{V}(X)$ is a set then $^\star\alpha \in {}^\star\mathcal{V}(X)$ will also be a set, with $\alpha \subset {}^\star\alpha$ and, generally, $^\star\alpha \backslash \alpha \neq \emptyset$. The (standard) mathematical theory of $\mathcal{V}(X)$ will consist of well formed statements about the standard objects; more precisely, statements of the form

$$\forall x \in y : (\ldots) \quad \text{or} \quad \exists x \in y : (\ldots)$$

in which the scope of the quantifiers is restricted to sets belonging to $\mathcal{V}(X)$. For brevity we refer to these as bounded quantifier statements, or b.q.s.

Finally the nonstandard universe $^\star\mathcal{V}(X)$ is to be such that a **Transfer Principle** is satisfied. That is to say, if $\alpha_1, \ldots, \alpha_n$ are standard objects in $\mathcal{V}(X)$ and if $\Phi(x_1, \ldots, x_n)$ is any b.q.s., then $\Phi(\alpha_1, \ldots, \alpha_n)$ holds in $\mathcal{V}(X)$ if and only if $\Phi(^\star\alpha_1, \ldots, {}^\star\alpha_n)$ holds in $^\star\mathcal{V}(X)$.

For example, taking $X = \mathbb{R}$ we obtain the standard real universe $\mathcal{V}(\mathbb{R})$ whose individuals are the real numbers and whose other standard objects comprise sets of reals, functions and sets of functions, functionals and so on. In the earlier part of this chapter a model for an extension $^\star\mathbb{R}$ of $\mathbb{R}$ was defined as a set of equivalence classes of infinite real sequences (that is of members of $\mathbb{R}^\mathbb{N}$). Treating these hyperreal numbers as the members of a new (enlarged) set of individuals we can construct a new superstructure $\mathcal{V}(^\star\mathbb{R})$. The nonstandard real universe $^\star\mathcal{V}(\mathbb{R})$ is not identical with this new superstructure but is a certain proper part of it. It comprises just those members of $\mathcal{V}(^\star\mathbb{R})$ which can be defined in terms of equivalence classes of standard objects; these are the so-called **internal** objects of the nonstandard superstructure. With such a construction the Transfer Principle can be established as a theorem.

In contexts where sequences are inadequate (for example, in the theory of topological spaces with uncountable bases) a generalisation of this approach is necessary. Instead of defining an extended set of individuals, $^\star X$, in terms of equivalence classes of simple sequences (that is, members of $X^{\mathbb{N}}$) we may need to work with equivalence classes of *generalised sequences*; that is, with members of $X^J$, where $J$ is some (uncountably) infinite indexing set. Generalising the account given in §6.1.1.1 we fix a finitely additive Boolean measure $m_J$ defined on all subsets of $J$ which is such that $m_J(A) = 0$ if $A$ is any finite subset of $J$, and $m_J(J) = 1$. We can then define an equivalence relation on $X^J$ by saying that (generalised) sequences $\{x_j\}_{j \in J}$ and $\{y_j\}_{j \in J}$ are equivalent if and only if the set $\{j \in J : x_j = y_j\}$ has $m_J$-measure 1. The set $^\star X$ of individuals for the nonstandard superstructure $\mathcal{V}(^\star X)$ is then the set of all equivalence classes of (generalised) sequences belonging to $X^J$, modulo this equivalence relation. The definition of the general internal objects of this superstructure, and ultimately of the nonstandard universe itself, then proceeds by considering like equivalence classes of standard objects in $\mathcal{V}(X)$.

The existence of the measure $m_J$ on $J$ calls for some comment. First recall that a **filter** on $J$ is a non-empty family $\mathcal{F} \subset \mathcal{P}(J)$ such that

**F1** $A, B \in \mathcal{F} \Rightarrow A \cap B \in \mathcal{F}$,

**F2** $A \in \mathcal{F}$ and $A \subset B \Rightarrow B \in \mathcal{F}$,

**F3** $\emptyset \notin \mathcal{F}$.

A filter is said to be free if it contains no finite subset of $J$. Any filter $\mathcal{U}$ on $J$ which is such that for each $A \subset J$ either $A \in \mathcal{U}$ or else $J \setminus A \in \mathcal{U}$ is called an **ultrafilter** on $J$. Then if $m_J$ is a finitely additive Boolean measure of the required type on $J$, the family of subsets of $J$ defined by

$$\mathcal{U} = \{A \subset J : m_J(A) = 1\}$$

is a (free) ultrafilter on $J$. Conversely, to each free ultrafilter $\mathcal{U}$ on $J$ there corresponds a measure $m_J$ defined on $J$ such that $m_J(A) = 1$ if $A \in \mathcal{U}$ and $m_J(A) = 0$ if $A \notin \mathcal{U}$. Finally an appeal to Zorn's Lemma shows that any free filter on $J$ can be extended to a free ultrafilter on $J$ and therefore defines an appropriate Boolean measure.

It is easily confirmed that the set $\mathcal{F}_\mathcal{D} = \{\Phi_m : m \in \mathbb{N}\}$ satisfies the conditions

(1) $\emptyset \notin \mathcal{F}_\mathcal{D}$ and $\mathcal{F}_\mathcal{D} \neq \emptyset$,

(2) $\Phi_m \cap \Phi_n = \Phi_n$, for all $n \geq m$,

and is therefore the basis of a filter on $\Phi \subset \mathcal{D}$. With some slight modifications this fact has been used by Todor Todorov [114] to construct a nonstandard version of the Colombeau algebra $\mathcal{G}(\mathbb{R})$. We give below a brief sketch of Todorov's treatment and then an alternative approach which uses the algebra $^\omega\mathcal{G}(\mathbb{R})$ of $^\omega$generalised functions already established.

## 6.5.2 Todorov's treatment

Todorov uses an index set $\mathcal{D}_+ \equiv \mathcal{D} \times \mathbb{R}_+$ with elements $(\varphi, \varepsilon)$ denoted by $\varphi \otimes \varepsilon$, together with an ultrafilter $\mathcal{U}$ on $\mathcal{D}_+$ which contains a slightly modified form of the Colombeau sets $\Phi_n$. To be precise, $\mathcal{U}$ is to contain a sequence $(\mathcal{D}_n)_{n \in \mathbb{N}}$ of subsets of $\mathcal{D}$ which are such that

(1) $\mathcal{D}_n \times (0, 1/n) \in \mathcal{U}$ for all $n \in \mathbb{N}$,

(2) $\mathcal{D}_1 \supset \mathcal{D}_2 \supset \ldots \supset \mathcal{D}_n \supset \ldots$,

(3) for any $\varphi \in \mathcal{D}_1$ we have

   (a) $\operatorname{supp}(\varphi) \subset (-1, +1)$,

   (b) $\int_{-\infty}^{+\infty} |\varphi(x)| dx \leq M$ for some real constant $M > 1$, and

   (c) $\int_{-\infty}^{+\infty} \varphi(x) dx = 1$,

(4) for any $\varphi \in \mathcal{D}_n$,
$$\int_{-\infty}^{+\infty} x^k \varphi(x) dx = 0$$
for $1 \leq k \leq n$.

Taking $X = \mathbb{C}$ as the basic set of individuals for the standard universe we define $^*X \equiv {^*\mathbb{C}}$ to be the set of all equivalence classes of members of $\mathbb{C}^{\mathcal{D}_+}$, modulo the ultrafilter $\mathcal{U}$. Thus $(A_{\varphi \otimes \varepsilon})_{\varphi \otimes \varepsilon \in \mathcal{D}_+}$ and $(B_{\varphi \otimes \varepsilon})_{\varphi \otimes \varepsilon \in \mathcal{D}_+}$ in $\mathbb{C}^{\mathcal{D}_+}$ are equivalent if and only if
$$\{\varphi \otimes \varepsilon \in \mathcal{D}_+ : A_{\varphi \otimes \varepsilon} = B_{\varphi \otimes \varepsilon}\} \in \mathcal{U}.$$

$^*\mathbb{R}$ is the set of all $\alpha \equiv [(A_{\varphi \otimes \varepsilon})_{\varphi \otimes \varepsilon \in \mathcal{D}_+}]$ such that $\{\varphi \otimes \varepsilon \in \mathcal{D}_+ : A_{\varphi \otimes \varepsilon} \in \mathbb{R}\} \in \mathcal{U}$. Then we have the inclusions $\mathbb{R} \subset {^*\mathbb{R}}$ and $\mathbb{C} \subset {^*\mathbb{C}}$ defined by the constant elements of $\mathbb{C}^{\mathcal{D}_+}$; with the natural componentwise definitions of algebraic operations $^*\mathbb{R}$ and $^*\mathbb{C}$ are non-archimedean field extensions of $\mathbb{R}$ and $\mathbb{C}$ respectively. Now let $\rho$ denote the (non-zero) positive infinitesimal in $^*\mathbb{R}$ defined by $(A_{\varphi \otimes \varepsilon})_{\varphi \otimes \varepsilon \in \mathcal{D}_+} \in \mathbb{C}^{\mathcal{D}_+}$ for which $A_{\varphi \otimes \varepsilon} = \varepsilon$. Then $^*\mathbb{C}_M$ denotes the subring of $^*\mathbb{C}$ defined by
$$^*\mathbb{C}_M = \{c \in {^*\mathbb{C}} : |c| < \rho^{-m} \text{ for some } m \in \mathbb{N}\}$$
and the elements of $^*\mathbb{C}_M$ are referred to as **moderate hypercomplex numbers**. The set
$$^*\mathbb{C}_0 = \{c \in {^*\mathbb{C}} : |c| < \rho^m \text{ for all } m \in \mathbb{N}\}$$
is a maximal ideal of $^*\mathbb{C}_M$ and Todorov refers to the quotient
$$\hat{\mathbb{C}} = {^*\mathbb{C}_M}/{^*\mathbb{C}_0}$$

as the system of **meta-complex numbers**.

Todorov now defines a class $^*\mathcal{E}$ of nonstandard functions of the type $f : {}^*\mathbb{R} \to {}^*\mathbb{C}$ as the set of all equivalence classes of members of $\mathcal{E}^{\mathcal{D}_+}$, modulo the ultrafilter $\mathcal{U}$, where $\mathcal{E} \equiv C^\infty(\mathbb{R})$. The **moderate** $^*\mathcal{E}$**-functions** are the members of the subalgebra $^*\mathcal{E}_M$ of $^*\mathcal{E}$ defined by

$$^*\mathcal{E}_M = \{f \in {}^*\mathcal{E} : [\forall k \in \mathbb{N}_0, \exists m \in \mathbb{N} : |D^k f(x)| < \rho^{-m}, \forall x \in {}^*\mathbb{R}_{bd}]\}.$$

Similarly the **negligible** $^*\mathcal{E}$**-functions** are defined by

$$^*\mathcal{E}_0 = \{f \in {}^*\mathcal{E} : [\forall k \in \mathbb{N}_0 : |D^k f(x)| < \rho^{+m}, \forall m \in \mathbb{N}, \forall x \in {}^*\mathbb{R}_{bd}]\}.$$

Then the quotient

$$\hat{\mathcal{D}} = {}^*\mathcal{E}_M / {}^*\mathcal{E}_0$$

is a differential algebra over $\hat{\mathbb{C}}$.

### 6.5.3 Superhyperreal numbers

Let $\Phi \equiv \Phi_0$ denote, as before, the set of all functions $\varphi \in \mathcal{D}$ such that

$$\int_{-\infty}^{+\infty} \varphi(t) dt = 1,$$

and for $q = 1, 2, \ldots$, again let

$$\Phi_q \equiv \left\{\varphi \in \Phi_0 : \int_{-\infty}^{+\infty} t^k \varphi(t) dt = 0,\ 1 \leq k \leq q\right\}.$$

In addition we shall write, for $p = 0, 1, 2, \ldots$,

$$\Psi_p = \left\{\varphi \in \Phi_0 : \int_{-\infty}^{+\infty} |\varphi^{(j)}(t)| dt \leq c_p,\ 0 \leq j \leq p\right\}$$

where $(c_p)_{p \in \mathbb{N}}$ is some given sequence of positive real numbers. Now let $\mathcal{U}_\Phi$ be a free ultrafilter on $\Phi$ which is such that for all $q \in \mathbb{N}$ we have $\Phi_q \in \mathcal{U}_\Phi$ and for all $p \in \mathbb{N}_0$ we have $\Psi_p \in \mathcal{U}_\Phi$. Taking the Robinson field $^\omega\mathbb{R}$ as the basic set $X$ of individuals we then define a further enlargement of the number system

$$^\flat\mathbb{R} = ({}^\omega\mathbb{R})^\Phi / \sim_\Phi,$$

where $\sim_\Phi$ is the equivalence relation generated on the ultrapower $({}^\omega\mathbb{R})^\Phi$ by the ultrafilter $\mathcal{U}_\Phi$. Equipped with algebraic operations and with an ordering relation defined in the natural pointwise sense $^\flat\mathbb{R}$ is an ordered field, which we shall call the field of **superhyperreal numbers**.

The Robinson field $^\omega\mathbb{R}$ is isomorphically embedded in $^\natural\mathbb{R}$, but $^\natural\mathbb{R}$ includes elements other than those belonging to $^\omega\mathbb{R}$ just as $^\omega\mathbb{R}$ itself includes elements other than those belonging to $\mathbb{R}$. For example let $\Omega$ denote the equivalence class $[(\omega)_{\varphi \in \Phi}] \in {^\omega\mathbb{R}} \subset {^\natural\mathbb{R}}$; then the superhyperreal number $[(a_\varphi)_{\varphi \in \Phi}] \in {^\natural\mathbb{R}}$ such that

$$\forall q \in \mathbb{N}_0 \left[ \varphi \in \Phi_q \Rightarrow a_\varphi = \Omega^{-q} \right]$$

is a (non-zero) infinitesimal in $^\natural\mathbb{R}$ which is smaller than any positive number in $^\omega\mathbb{R}$. We call such infinitesimal an $^\Omega$infinitesimal.

We say that a superhyperreal number $\xi \in {^\natural\mathbb{R}}$ is bounded in $^\omega\mathbb{R}$ if and only if there exists $k \in \mathbb{N}$ such that $|\xi| \le \Omega^k$. The set of all superhyperreal numbers bounded in $^\omega\mathbb{R}$ will be denoted by $^\natural\mathbb{R}_{b\Omega}$, and its elements referred to as $^\Omega$**bounded numbers**. Every $^\Omega$bounded number $\xi$ may be expressed uniquely as a sum of a number in $^\omega\mathbb{R}$, called its $^\Omega$**standard part** and written as $^\Omega\mathbf{st}(\xi)$, and an $^\Omega$infinitesimal. For any $\alpha \in {^\omega\mathbb{R}}$ we define the $^\Omega$**monad** of $\alpha$ to be the set

$$^\Omega\mathbf{mon}(\alpha) = \{\xi \in {^\natural\mathbb{R}} : {^\Omega\mathbf{st}(\xi)} = \alpha\}.$$

Finally, any number $\xi \in {^\natural\mathbb{R}}$ is said to be **simply bounded** if and only if $^\Omega\mathbf{st}(\xi)$ is bounded in $\mathbb{R}$; the set

$$^\natural\mathbb{R}_{bd} = \{\xi \in {^\natural\mathbb{R}} : [\exists a \in \mathbb{R}^+ : |\xi| \le a]\}$$

is the set of all simply bounded superhyperreal numbers.

### 6.5.4 The Colombeau-Todorov algebra $^\natural\mathcal{G}_C(\mathbb{R})$

Now consider the $^\natural\mathbb{R}_{b\Omega}$-algebra $^\natural\mathcal{G}(\mathbb{R}) = ({^\omega\mathcal{G}(\mathbb{R})})^\Phi / \sim_\Omega$ comprising all $^\natural$internal functions $\Gamma = [(g_\varphi)_{\varphi \in \Phi}]$ such that $g_\varphi$ belongs to $^\omega\mathcal{G}(\mathbb{R})$ for $\mathcal{U}_\Phi$-almost all $\varphi \in \Phi$. We define differentiation in this algebra in the natural pointwise sense, so that if $\partial^k$ denotes the $k$th derivative of $\Gamma$ then we have

$$\partial^k \Gamma = [(^\star D g_\varphi)_{\varphi \in \Phi}].$$

An internal function $\Gamma \in {^\natural\mathcal{G}(\mathbb{R})}$ is $^\Omega$moderate if and only if $\partial^k \Gamma(x) \in {^\natural\mathbb{R}_{b\Omega}}$ for all $x \in {^\natural\mathbb{R}_{bd}}$ and for each $k \in \mathbb{N}_0$. We denote by $^\natural\mathcal{G}_\mathcal{M}(\mathbb{R})$ the differential algebra of all $^\Omega$moderate functions. A function $\Gamma$ in $^\natural\mathcal{G}_\mathcal{M}(\mathbb{R})$ is $^\Omega$null if and only if $\partial^k \Gamma(x) \in {^\Omega\mathbf{mon}(0)}$ for all $x \in {^\natural\mathbb{R}_{bd}}$ and each $k \in \mathbb{N}_0$. We denote by $^\natural\mathcal{N}(\mathbb{R})$ the set of all $^\Omega$null functions. Then $^\natural\mathcal{N}(\mathbb{R})$ is an ideal of $^\natural\mathcal{G}_\mathcal{M}(\mathbb{R})$ and moreover $\partial\{^\natural\mathcal{N}(\mathbb{R})\} \subset {^\natural\mathcal{N}(\mathbb{R})}$. Hence

$$^\natural\mathcal{G}_C(\mathbb{R}) = {^\natural\mathcal{G}_\mathcal{M}(\mathbb{R})}/{^\natural\mathcal{N}(\mathbb{R})}$$

is a differential algebra which we call the **Colombeau-Todorov algebra** of $^\natural$**generalised functions**.

Now let $[T]$ be any generalised function in the Colombeau algebra $\mathcal{G}(\mathbb{R})$, where $T : \Phi_1 \to \mathcal{C}^\infty(\mathbb{R})$ is an arbitrary representative in $\mathcal{E}_M[\Phi, \mathcal{C}^\infty]$. Then for every $\varphi \in \Phi_1$ and every $n \in \mathbb{N}$ we have $T\varphi_{(1/n)}(x) \equiv f_n(x)$ is a $\mathcal{C}^\infty(\mathbb{R})$-function, and so

$$F_\varphi = [(f_n)_{n \in \mathbb{N}}]$$

is a $^*\mathcal{C}^\infty(\mathbb{R})$ function. Moreover, since $T$ is moderate in the sense of Colombeau, $F_\varphi \in {}^\omega\mathcal{G}(\mathbb{R})$ and consequently

$$\Gamma_T = \left[ \left( {}^\Omega\text{st}(F_\varphi) \right)_{\varphi \in \Phi} \right]$$

belongs to $^\natural\mathcal{G}(\mathbb{R})$, and in fact to $^\natural\mathcal{G}_\mathcal{M}(\mathbb{R})$. It follows readily that $^\natural\mathcal{G}_\mathcal{C}(\mathbb{R})$ contains a copy of the Colombeau algebra of generalised functions.

# Bibliography

[1] Antosik, P., Mikusinski, J. & Sikorski, R. (1973) *Theory of distributions: the sequential approach*, Elsevier.

[2] Beltrami, E. J. & Wohlers, M. R. (1966) *Distributions and the boundary values of analytic functions*, Academic Press.

[3] Biagioni, H. A. (1980) *A nonlinear theory of generalised functions*, Springer-Verlag.

[4] Bremermann, H. (1965) *Distributions, complex variables, and Fourier transforms*, Addison-Wesley.

[5] Bremermann, H. (1967) 'Distributions; some remarks on analytic representations and products of distributions' *SIAM J Appl. Math.* **15**

[6] Brychkov, Y. A. & Prudnikov, A.P. (1989) *Integral transforms of generalized functions*, Gordon & Breach.

[7] Campbell, L. L. (1968) 'Sampling theorem for the Fourier transform of a distribution with bounded support' *SIAM J. Appl. Math.* **15** pp 626-636

[8] Carleman, T. (1944) *L'intégrale de Fourier et questions qui s'y ratachent*, Almqvist & Wiksells Bok.

[9] Colombeau, J-F. (1982) *Differential calculus and holomorphy; real and complex analysis in locally convex space*, North Holland.

[10] Colombeau, J-F. (1984) *New generalized functions and multiplication of distributions*, North Holland.

[11] Colombeau, J-F. (1985) *Elementary introduction to new generalized functions*, North Holland.

[12] Colombeau, J-F. & Oberguggenberger, M. 'Generalised functions and products of distributions' *Pre-print*.

[13] Corput, J. van der (1959-60) 'Introduction to the neutrix calculus' *J. Analyse Math.* **7** pp 291-398.

[14] Cutland, N. (ed.) (1988) *Nonstandard analysis and its applications*, Cambridge University Press.

[15] Dirac, P. A. M. (1930) *The principles of quantum mechanics*, Cambridge University Press.

[16] Edwards, R. E. (1965) *Functional analysis: theory and applications*, Holt, Rinehart & Winston.

[17] Egorov, Yu. V. A. (1990) 'A contribution to the theory of generalized functions' *Russian Mathematical Surveys* **45**(5) pp 1-49.

[18] Fisher, B. (1972) 'The product of distributions' *Quart. J. Math. Oxford (2)* **22** pp 291-298.

[19] Fisher, B. (1972) 'The product of the distributions $x^{-r}$ and $\delta^{(r-1)}(x)$' *Proc. Camb. Phil. Soc.* **72** pp 201-204.

[20] Fisher, B. (1973) 'Convergent and divergent products of distributions' *The Math. Student* **XLI**(4) pp 409-421.

[21] Fisher, B. (1980) 'Some notes on distributions' *The Math. Student* **38**(3) pp 269-281.

[22] Fisher, B. (1982) 'A non-commutative neutrix product of distributions' *Math. Nachr.* **108** pp 117-127.

[23] Fisher, B. (1983) 'The non-commutative neutrix product of the distributions $x_+^{-r}$ and $\delta^{(p)}(x)$' *Indian J. Pure Appl. Math.* **14**(12) pp 1439-1449.

[24] Fisher, B. (1987) 'Neutrices and the convolution of distributions' *ZBORNIK RADOVA, Pirodno - matematičkog fakulteta, Univerziteta u Novom Sadu, Serija matematiku* **17**(1) pp 119-135.

[25] Friedlander, F. G. (1982) *Introduction to the theory of distributions*, Cambridge.

[26] Gel'fand, I. M. & Shilov, G. E. (1964) *Generalized functions*, Academic Press.

[27] Güttinger, W. (1955) 'Products of improper operators and the renormalization problem of quantum field theory' *Progr. Theoret. Phys.* **13** pp 612-626.

[28] Hadamard, J. S. (1932) *Le probléme de Cauchy et les équations aux dérivées partielles linéaires hyperboliques*, Hermann.

[29] Henle, J.M. & Kleinberg, E.M. (1979) *Infinitesimal calculus*, MIT Press.

[30] Henson, C. W. & Moore, Jr, L.C. (1972) 'The nonstandard theory of topological vector spaces' *Trans. Amer. Math. Soc.* **172** pp 405-435.

[31] Hirata, Y. & Ogata, H. (1958) 'On the exchange formula for distributions' *J. Sc. Hiroshima Univ. Ser. A* **22**(3) pp 147-152.

[32] Hoskins, R. F. (1990) *Standard and nonstandard analysis*, Ellis Horwood.

[33] Hoskins, R. F. & Sousa Pinto, J. (1991) 'A nonstandard realization of the J. S. Silva axiomatic theory of distributions' *Portugaliæ Mathematica* **48**(2) pp 195-216.

[34] Hoskins, R. F. & Sousa Pinto, J. (1993) 'Nonstandard treatments of new generalised functions' *Proc. Internat. Symposium on Generalised Functions & Their Applications, Banaras Hindu Univ., 1991.*, Plenum Press.

[35] Hurd, A. E. (ed.) (1983) *Nonstandard analysis: recent developments*, Springer-Verlag.

[36] Hurd, A. E. & Loeb, P. A. (1985) *An introduction to nonstandard real analysis*, Academic Press.

[37] Itano, M. (1965) 'On the multiplicative products of distributions' *J. Sc. Hiroshima Univ. Ser. A-I* **29** pp 51-74.

[38] Itano, M. (1966) 'On the theory of multiplicative products of distributions' *J. Sc. Hiroshima Univ. Ser. A-I* **30** pp 151-181.

[39] Ivanov, V.K. (1972) Hyperdistributions and the multiplication of Schwartz distributions' *Soviet Math. Dokl.* **13**(3) pp 784-788.

[40] Jelinek, J. (1986) 'Characterization of the Colombeau product of distributions' *Comm. Math. Univ. Carolinae* **27**(2) pp 377-394.

[41] Jerri, A. J. 'The Shannon sampling theorem - its various extensions and applications: a tutorial review' *Proc. IEEE* **65**(11) pp 1565-1596.

[42] Jones, D. S. (1966) *Generalised functions*, McGraw-Hill.

[43] Jones, D. S. (1973) 'The convolution of generalized functions' *Quart. J. Math. Oxford (2)* **24** pp 145-163.

[44] Jones, D. S. (1980) 'Infinite integrals and convolution' *Proc. Roy. Soc. London* **A371** pp 479-508.

[45] Jones, D. S. (1982) 'Generalised functions and their convolution' *Proc. Roy. Soc. Edin.* **91A** pp 213-233.

[46] Jones, D. S. (1982) *The theory of generalised functions*, Cambridge.

[47] Kaminski, A. (1982) 'Convolution, product and Fourier transform of distributions' *Studia Math.* **LXXIV** pp 83-96.

[48] Kaneko, A. (1988) *Introduction to hyperfunctions*, Kluwer Acad. Publ.

[49] Katznelson, Y. (1976) *An introduction to harmonic analysis*, Dover.

[50] Keisler, H. J. (1986) *Elementary calculus; an infinitesimal approach*, Prindle, Weber & Schmidt.

[51] Keller, K. (1978) 'Analytic regularizations, finite part prescriptions and products of distributions' *Math. Ann.* **236** pp 49-84.

[52] Keller, K. (1978) 'Irregular operations in quantum field theory I: multiplication of distributions' *Reports on Mathematical Physics* **14**(3) pp 285-309.

[53] Kessler, C. (1988) 'On hyperfinite representations of distributions' *Bull. London Math. Soc.* **20** pp 139-144.

[54] Kinoshita, M. (1988) 'Nonstandard representations of distributions I' *Osaka J. Math.* **25** pp 805-824.

[55] Kinoshita, M. (1990) 'Nonstandard representations of distributions II' *Osaka J. Math.* **27** pp 843-861.

[56] Komatsu, H. (1973) 'Ultradistributions I : Structure theorems and a characterization' *J. Fac. Sci. Univ. Tokyo Sect. IA, Math.* **20** pp 25-105.

[57] Komatsu, H. (1977) 'Ultradistributions II : The kernel theorem and ultradistributions with support in a submanifold' *J. Fac. Sci. Univ. Tokyo Sect. IA, Math.* **24** pp 607-628

[58] König, H. (1955) 'Multiplikation von distributionen' *Math. Ann.* **128** pp 420-452.

[59] Korevaar, J. (1959) 'Distributions defined by fundamental sequences I - V' *Ned. Akad. Wetenschap. Proc. A* **58**

[60] Laugwitz, D. (1958) 'Eine einführung der $\delta$-funktionen' *Sitz. Ber. Bayer. Akad. Wiss., Math. - nat,* **K1** pp 41-59.

[61] Laugwitz, D. (1984) 'Infinitesimals in physics' *Papers dedicated to Professor L. Iliev's 70th Aniversary, Sofia,* pp 233-243.

[62] Lee, A. J. (1976) 'Characterization of band-limited functions and processes' *Inform. & Control* **31** pp 258-271.

[63] Li Bang-He (1978) 'Nonstandard analysis and multiplication of distributions' *Scientia Sinica* **XXI**(5) pp 561-585.

[64] Li Bang-He 'New generalized functions in nonstandard framework' *Pre-print, Int. of Systems Science Academia Sinica, Beijing.*

[65] Lighthill, M. J. (1978) *Introduction to Fourier analysis and generalised functions,* Cambridge University Press.

[66] Lightstone, A.H. & Robinson, A. (1975) *Nonarchimedean fields and asymptotic expansions,* North-Holland.

[67] Lightstone, A.H. & Wong, K. (1975) 'Dirac delta functions via nonstandard analysis' *Can. Math. Bull.* **81**(5) pp 759-762.

[68] Łojasiewicz, S. 'Sur la valeur et la limite d'une distribuition en un point' *Studia Math.* **XVI** pp 1-64.

[69] Lützen, J. (1980) *The pre-history of the theory of distributions,* Springer-Verlag.

[70] Luxemburg, W. A. J. (1964) *Nonstandard analysis. Lectures on A. Robinson's theory of infinitesimal and infinitely large numbers,* Caltech Bookstore, Pasadena.

[71] Meneses, A. S. (1967) 'As ultradistribuições de uma variavel como translatadas formais' *Exposição organizada e completada por J. P. Carvalho Dias, AEFC, Lisboa.*

[72] Mikusinski, J. (1948) 'Sur la méthode de généralization de M. Laurent Schwartz et sur la convergence faible' *Fund. Math.* **35** pp 235-239.

[73] Mikusinski, J. (1961) 'Irregular operations on distributions' *Studia Math.* **XX** pp 163-169.

[74] Mikusinski, J. (1962) 'Criteria of the existence and the associativity of the product of distributions' *Studia Math.* **XXCI** pp 253-259.

[75] Mikusinski, J. (1966) On the square of the Dirac delta distribution' *Bull. Acad. Polon.* **XIV** pp 511-513.

[76] Mikusinski, J. (1968) 'Sequential theory of the convolution of distributions' *Studia Math.* **XXIX** pp 151-160.

[77] Misra, O. P. & Lavoine, J.L. (1986) *Transform analysis of generalized functions*, North Holland.

[78] Moore, S.M. (1980) 'Nonstandard analysis and generalized functions' *Revista Colombiana de Matemáticas* **XIV** pp 73-94.

[79] Nelson, E. (1977) 'Internal set theory, a new approach to NSA' *Bull. Amer. Math. Soc.* **83** pp 1165-1198.

[80] Oberguggenberger, M. (1984) 'Products of distributions' *Journal für Mathematik* **365** pp 1-11.

[81] Oberguggenberger, M. (1986) 'Multiplications of distributions in the Colombeau algebra $\mathcal{G}(\Omega)$' *Bollettino U.M.I.* *(6)* **5-A** pp 423-429.

[82] Oberguggenberger, M. (1988) 'Products of distributions : nonstandard methods' *Zeitschrift für Analysis und ihr Anwendungen* **7**(4) pp 347-365.

[83] Oliveira, J. S. (1983) *Sobre certos espaços de ultradistribuições e uma noção generalizada de produto multiplicativo* Textos e Notas nº29, CMAF, Lisboa.

[84] Palamodov, V. P. (1987) 'From hyperfunctions to analytic functionals' *Soviet Math. Dokl.* **18**(4)

[85] Pfaffelhuber, E. (1971) 'Sampling series for band-limited generalized functions' *IEEE Trans. Inf. Th.* **IT-17** pp 650-654.

[86] Raju, C. K. (1982) 'Products and compositions with the Dirac delta function' *J. Phys. Math. Gen.* **15** pp 381-396.

[87] Raju, C. K. (1983) 'On the square of $x^{-n}$' *J. Phys. A* **16**(16) pp 3739-3753.

[88] Robert, A. (1988) *Nonstandard analysis*, Wiley.

[89] Robinson, A. (1966) *Nonstandard Analysis*, North Holland.

[90] Robinson, A. (1973) 'Function theory on some nonarchimedean fields' *Amer. Math. Monthly, Papers in the Foundations of Mathematics* **80** pp 87-109.

[91] Rosinger, E. E. (1978) *Distributions and nonlinear partial differential equations*, Springer-Verlag.

[92] Rosinger, E. E. (1987) *Generalized solutions of nonlinear partial differential equations*, North-Holland.

[93] Royden, H. L. (1968) *Real Analysis*, Macmillan.

[94] Sato, M. (1958) 'The theory of hyperfunctions' *Sûgakulo* **10** pp 1-27.

[95] Schmieden, & Laugwitz, D. (1958) 'Eine erweiterung der infinitesimalrechnung' *Math. Z.* **69** pp 1-39.

[96] Schwartz, L. (1950) 'Théorie des noyaux' *Proc. Intern. Congress of Mathematicians, Cambridge, Mass.*

[97] Schwartz, L. (1957) *Théorie des Distributions, I*, Hermann.

[98] Schwartz, L. (1959) *Théorie des Distributions, II*, Hermann.

[99] Shiraishi, R. (1959) 'On the definition of convolution for distributions' *J. Sc. Hiroshima Univ. Ser. A* **23**(1) pp 19-32.

[100] Shiraishi, R. & Itano, M. (1964) 'On the multiplicative products of distributions' *J. Sc. Hiroshima Univ. Ser. A-I* **28** pp 223-235.

[101] Silva, J. S. (1954-55) 'Sur une construction axiomatique de la théorie des distributions' *Rev. Faculdade Ciências, Lisboa, 2ª série A* **4** pp 79-186.

[102] Silva, J. S. (1961) 'Sur l'axiomatique des distributions et ses possible modèles' *Centro Internazionale Matematico Estivo. Roma. Instituto Matematico.*

[103] Silva, J. S. (1958) 'Les fonctions analytiques comme ultra-distributions dans le calcul opérationel' *Math. Ann.* **136** pp 58-96.

[104] Silva, J. S. (1962) 'Les séries de multipôlesdes physiciens et la théorie des ultradistributions' *Boletim da Academia das Ciências de Lisboa* **XXXIV**.

[105] Stroyan, K.D. & Luxemburg, W.A.J. (1976) *Introduction to the Theory of Infinitesimals*, Academic Press.

[106] Temple, G. (1953) 'Theories and applications of generalised functions' *J. London Math. Soc.* **28** pp 134-148.

[107] Temple, G. (1955) 'Generalised functions' *Proc. Roy. Soc. Ser. A* **228** pp 175-190.

[108] Temple, G. (1981) *100 years of mathematics*, Duckworth.

[109] Tillmann, H. G. (1953) 'Randverteilungen analytischer funktionen und distributionen' *Math. Zeitschr.* **59** pp 61-83.

[110] Tillmann, H. G. (1961) 'Distributionen als randverteilungen analytischer functionen II' *Math. Zeitschr.* **76** pp 5-21.

[111] Tillmann, H. G. (1961) 'Darstellung der schwartzschen distributionen durch analytischer funktionen' *Math. Zeitschr.* **77** pp 106-124.

[112] Todorov, T. D. (1985) 'Asymptotic functions as kernels of the Schwartz distributions' *Bulg. J. Phys.* **12** pp 451-464.

[113] Todorov, T. D. (1987) 'Sequential approach to Colombeau theory of generalized functions' *Pre-print, Int. Atom. Energy Ag.* **IC/87/126**.

[114] Todorov, T. D. (1987) 'Colombeau's generalized functions and nonstandard analysis' *Pre-print, Int. Atom. Energy Ag.* **IC/87/333**.

[115] Tréves, F. (1976) *Topological vector spaces, distributions and kernels*, Academic Press.

[116] Young, R. M. (1982) *An introduction to nonharmonic Fourier series*, Academic Press.

[117] Zemanian, A. H. (1965) *Distribution theory and transform analysis*, McGraw-Hill.

[118] Zemanian, A. H. (1968) *Generalized integral transformations*, Wiley.

[119] Zielezny, Z. (1968) 'On the space of convolution operators in $K'_1$.' *Studia Math.* **XXXI** pp 111-124.

# Index

Analytic representation
    of distribution 161
    of ultradistribution 194
Associated complex number 240

Banach space 26
Bandlimited functions 114
Borel measure 15
Bremermann space 172

Campbell Sampling Theorem 128
Cauchy representation 163
Cauchy-Stieltjes transform 175
Cauchy transform 167
Colombeau Algebra 232
Colombeau-Todorov algebra 291
Compatible norms 27
*continuity 268
Convolution
    integral 63
    of distributions 68
Convergence
    of distributions 70
    of ultradistributions 140
Countable union spaces 32
Countably normed spaces 28

$\mathcal{D}$-association 285
Delta sequence 75
Differential order 43
Dirac delta function 1
Dirac measure 16
Direct product 65
Distribution 9
    of exponential type 150
    of finite order 11
Distributional equivalence 272
Division problem 89
Dual space 9

Exchange formula 110, 142
External set 260

Fairly good functions 83
Filter 290
Fine function 82
Fisher classical product 218
Fisher neutrix product 222
Fourier-Carleman transform 203
Fourier transform 97
Fréchet space 30
Function
    of compact spectrum 114
    of slow growth 109
Functional 8
Fundamental sequence 80

Generalised complex numbers 240
Generalised function (Jones) 83
Good function 82

Hadamard finite part 53
Hyperfunction 184
Hypernatural numbers 256
Hyperreal number system 254

Inductive limit 31
Infinitely close 256
Integral of generalised functions 244

Internal function 261
Internal set 260
Intrinsic product 93
Iota-null 285
Irregular operation 85

Lebesgue-Stieltjes integral 16
Localisation 18

Meta-complex numbers 290
Metric space 25
Model delta sequence 79
Moderate operator
    (Colombeau) 233
    (Biagioni) 280
"moderate function 282
Moderately good function 83
Monad 256

Negligible operator
    Colombeau 234
    Biagioni 280
Neutrix calculus 59
Nonstandard analysis 253
Nonstandard extension
    of sets 256
    of functions 257
Nonstandard integral 261
Nonstandard universe 287
Norm 6, 25
Normed linear space 6, 25

Paley-Wiener theorem 117
Parseval theorem 99
Partition of unity 18
Periodic distributions 121
Periodic ultradistributions 147
Pre-distribution 265
Plancherel theorem 101
Poisson formula 105
Pseudo-function 50

Quasi-convolution operator 228

Radon measure 17
Rapid decrease 102

Regular distribution 12
Regular operation 85
Robinson field 282

Sampling operation 4
Sampling theorem (WKS) 115
Schwartz order of a distribution 11
Schwartz product 86
Seminorm 31
Silva tempered ultradistribution 152
Shadow (of a function) 268
Singular distribution
Standard part 256
    of a function 268
Standard universe 287
Strict delta sequence 78
Summability kernel 100
Superhyperreal numbers 290
Superstructure 286
Support
    of a distribution 36
    of a function 2

Tempered distribution 108
Test functions 2
Topological vector space 21
Topology 22

Ultradistribution 138
Ultrafilter 288
Unit impulse 1

Weak function 84

*series continued from front of book*

| | |
|---|---|
| Jolliffe, F.R. | SURVEY DESIGN AND ANALYSIS |
| Jollife, I.T. & Jones, B. | STATISTICAL INFERENCE |
| Jones, R.H. & Steele, N | MATHEMATICS IN COMMUNICATION THEORY |
| Jordan, D. | TOPOLOGY AND THE TORUS |
| Kapadia, R. & Andersson, G. | STATISTICS EXPLAINED: Basic Concepts and Methods |
| Kelly, J.C. | ABSTRACT ALGEBRA |
| Kim, K.H. & Roush, F.W. | APPLIED ABSTRACT ALGEBRA |
| Kim, K.H. & Roush, F.W. | TEAM THEORY |
| Krishnamurthy, V. | COMBINATORICS: Theory and Applications |
| de Lange, J., Huntley, I., Keitel, C. & Niss, M. | INNOVATION IN MATHS EDUCATION BY MODELING APPLICATIONS |
| Lindfield, G. & Penny, J.E.T. | MICROCOMPUTERS IN NUMERICAL ANALYSIS |
| Livesley, K. | MATHEMATICAL METHODS FOR ENGINEERS |
| Malik, M., Riznichenko, G.Y. & Rubin, A.B. | BIOLOGICAL ELECTRON TRANSPORT PROCESSES AND THEIR COMPUTER SIMULATION |
| Marshall, J.E., Gorecki, H., Korytowski, A. & Walton, K. | TIME-DELAY SYSEMS: Stability and Performance Criteria with Applications |
| Martin, D. | MANIFOLD THEORY: An Introduction for Mathematical Physicists |
| Massey, B.S. | MEASURES IN SCIENCE AND ENGINEERING |
| Menell, A. & Bazin, | MATHEMATICS FOR THE BIOSCIENCES |
| Moore, R. | COMPUTATIONAL FUNCTIONAL ANALYSIS |
| Moshier, S.L.B. | METHODS AND PROGRAMS FOR MATHEMATICAL FUNCTIONS |
| Murphy, J.A., Ridout, D. & McShane, B. | NUMERICAL ANALYSIS, ALGORITHMS AND COMPUTATION |
| Norcliffe, A. & Slater, G. | MATHEMATICS OF SOFTWARE CONSTRUCTION |
| Ogden, R.W. | NON-LINEAR ELASTIC DEFORMATIONS |
| Page, S.G. | MATHEMATICS: A Second Start |
| Prior, D. & Moscardini, A. | MODEL FORMULATION ANALYSIS |
| Norcliffe, A. & Slater, G. | MATHEMATICS OF SOFTWARE CONSTRUCTION |
| Norcliffe, A. & Slater, G. | MATHEMATICS OF SOFTWARE CONSTRUCTION |
| Ratschek, J. & Rokne, J. | NEW COMPUTER METHODS FOR GLOBAL OPTIMIZATION |
| Schendel, U. | INTRODUCTION TO NUMERICAL METHODS FOR PARALLEL COMPUTERS |
| Scorer, R.S. | ENVIRONMENTAL AERODYNAMICS |
| Sehmi, N.S. | LARGE ORDER STUCTURAL EIGENANALYSIS TECHNIQUES: Algorithms for Finite Element Systems |
| Smitalova, K. & Sujan, S. | DYNAMICAL MODELS IN BIOLOGICAL SCIENCE |
| Srivastava, H.M. & Owa, S. | UNIVALENT FUNCTIONS, FRACTIONAL CALCULUS AND THEIR APPLICATIONS |
| Stirling, D.S.G. | MATHEMATICAL ANALYSIS |
| Sweet, M.V. | ALGEBRA, GEOMETRY AND TRIGONOMETRY IN SCIENCE, ENGINEERING AND MATHEMATICS |
| Toth, G. | HARMONIC AND MINIMAL MAPS AND APPLICATIONS IN GEOMETRY AND PHYSICS |
| Townend, M.S. & Pountney, D.C. | COMPUTER-AIDED ENGINEERING MATHEMATICS |
| Twizell, E.H. | COMPUTATIONAL METHODS FOR PARTIAL DIFFERENTIAL EQUATIONS |
| Twizell, E.H. | NUMERICAL METHODS, WITH APPLICATIONS IN THE BIOMEDICAL SCIENCES |
| Vadja, | MATHEMATICAL GAMES AND HOW TO PLAY THEM |
| Vein, R. & Dale, P | DETERMINANTS AND THEIR APPLICATIONS IN MATHEMATICAL PHYSICS |
| Vince, A. and Morris, C. | DISCRETE MATHEMATICS FOR COMPUTING |
| Webb, J.R.L. | FUNCTIONS OF SEVERAL REAL VARIABLES |